Agrarian Distress in India

By the Same Author

— Agrarian Reforms and Institutional Changes in India (1986)
-- Sustainability of Small Holder Agriculture in India (1996)
— Land Policies for Inclusive Growth (2012)
— Empowerment of Rural Women in Developing Countries: Challenges and Pathways (2015)
— Socio-Economic Impact Assessment of BT Cotton in India (2015)

About the Editor

Dr. T. Haque, Distinguish Professor at the Council for Social Development, New Delhi and Chairman, Special Cell on Land Policies, NITI Aayog, Govt. of India is a well known Agricultural Economist. He obtained his Ph.D. degree from Patna University (India) and Post-doctoral training from Iowa State University, Ames, U.S.A. Prior to joining the Council for Social Development, Dr. Haque served as Chairman, Commission for Agricultural Costs and Prices, Government of India, ICAR National Fellow at the National Centre for Agricultural Economics and Policy Research, New Delhi; Director, Council for Social Development, New Delhi and Director, Centre for Agrarian Studies, NIRD, Hyderabad. He also served as consultant to several International organizations, including FAO, ILO and The World Bank. Dr. Haque is author of six books and about 100 research papers and reports. His areas of specialization are land reform, food security and sustainable agriculture and rural development.

Agrarian Distress in India
Causes and Remedies

Edited by
T. Haque

CONCEPT PUBLISHING COMPANY PVT. LTD.
NEW DELHI-110059

ISBN-13: 978-93-5125-257-3

First Published 2016

Published and Printed by

Concept Publishing Company Pvt. Ltd.
A/15-16, Commercial Block, Mohan Garden
New Delhi-110059 (India)
T: 25351460, 25351794, *F*: 091-11-25357109
E: publishing@conceptpub.com, W: www.conceptpub.com
Editorial Office: H-13, Bali Nagar, New Delhi-110 015, India.

Cataloging in Publication Data--Courtesy: D.K. Agencies (P) Ltd. <docinfo@dkagencies.com>

Agrarian distress in India : causes and remedies / edited by T. Haque.
pages cm
Includes bibliographical references and index.

1. Agriculture--Economic aspects--India. 2. Farmers--India--Psychology. 3. Distress (Psychology)--India. 4. Agriculture and state--India. I. Haque, T. (Tajamul), 1947- editor.

HD2072.A37 2016 DDC 338.10954 23

Contents

List of Tables

List of Figures

List of Contributors

Dr. Ankita Goyal is working as Asstt. Professor at the Council for Social Development, New Delhi. She has authored a number of research papers in the field of Agriculture Economics. Her area of specialization is agriculture, food security and sustainable farming. She is also co-author of a book entitled "Bt Cotton in India".

Dr. Archana Takte is a Research Scholar at the Centre for the Study of Regional Development, School of Social Sciences, Jawaharlal Nehru University. Her area of specialisation is natural disaster and agricultural risk management.

Dr. Brajesh Jha is Associate Professor in Institute of Economic Growth, New Delhi since 2001. He has more than two decades of research experiences of working on wide range of issues related to Agriculture and Rural Sector of India. His substantial work relates to farming system research; risk, uncertainty and sustainability of agriculture; commodity specific analysis of WTO Agreements; sources of growth in agriculture, dairy and rural non-farm sector of India. He has authored several books and around 50 research articles in professional journals, edited volumes, economic—dailies and monthlies.

Prof. C. Ramasamy B.Sc. (Ag.) and M.Sc. (Ag.Econ.) from Agricultural College, Coimbatore; did his Ph.D. from Tamil Nadu Agricultural University and Post-Doctoral Study in the University of Minnesota. Prof. Ramasamy was the Vice-Chancellor of Tamil Nadu Agricultural University. His major research work is in the area of green revolution technologies and their impact on the economy. He has published about more than sixty research papers and seven books. He is a Fellow of National Academy of Agricultural Sciences.

Prof. D. Narasimha Reddy is ICSSR National Fellow at the Council for Social Development (SRC), Hyderabad. Prior to this, he was Visiting

Professor at the Institute for Human Development and Professor and Head of the Department of Economics, Central University, Hyderabad. He has published a number of books and research papers in journals of national and international repute. He is also associated as editor and adviser of several journals.

Dr. E.V. Prakash Rao is an Associate Professor, Centre for Agrarian Studies and Disaster Mitigation, National Institute of Rural Development and Panchayati Raj, Hyderabad (India).

Dr. Gyanendra Mani is Doctorate in Agricultural Economics from G.B. Pant University of Agriculture and Technology, Pantnagar, Uttarakhand State and Certified Associate (CAIIB) of Indian Institute of Banking and Finance, Mumbai. He has contributed many articles/research papers to various national and international referred journals. He has conducted large number of studies on behalf of NABARD and published reports in form of Evaluation Studies, Commodity Specific Studies, Working Papers etc. Dr. Mani is also a National Resource Person/ 'Recognised Trainer' for 'Design of Training (DoT)' Programme being conducted by Department of Personnel and Training, Government of India. He is also looking after the editorial works of 'The Microfinance Review' in the capacity of the Managing Editor of the Journal.

Dr. K. Suman Chandra is presently working at the National Institute of Rural Development and Panchayati Raj, Hyderabad, as Professor and Head of the the Centre for Agrarian Studies and Disaster Mitigation. He is the author of several books and research papers in the field of sociology and agrarian studies.

Dr. Kingsly Immanuelraj is Scientist (Agricultural Economics) at National Centre for Agricultural Economics and Policy Research (NCAP), New Delhi. His area of research includes agricultural marketing, trade and sustainable agriculture. He teaches Production Economics at the Division of Agricultural Economics, IARI, New Delhi.

Dr. Kunwar Jalees is Research Associate at Navdayna. His field of research includes food security and sustainable agriculture.

Dr. Nivedita Sharma is a Senior Research Associate at the Council for Social Development. She was completed her Ph.D. in Agricultural Economics from IIT Roorkee in 2014. Her research interests are agriculture and rural development. She has published a number of research papers in national and international journals.

Dr. Pratap Singh Birthal is the Director at the Institute of Development Studies, Jaipur. Earlier, he served at the National Centre for Agriculture Economics and Policy Research, New Delhi and International Crops Research Institute for the Semi-Arid Tropics

(ICRISAT), Hyderabad. Dr. Birthal has made notable contributions towards understanding the nexus between agricultural diversification and farm poverty in India. He is a Fellow of the National Academy of Agricultural Sciences (NAAS), a recipient of National Fellowship and Young Scientist Award from the ICAR; D.K. Desai Award from Indian Society of Agricultural Economics; and R.T. Doshi Award from the Agricultural Economics Research Association (India) for his outstanding contributions in the field of social sciences.

Prof. Seema Bathla is Professor, at the Centre for the Study of Regional Development, Jawaharlal Nehru University, New Delhi. She has previously worked in the Institute of Economic Growth, The Energy Research Institute and Delhi College of Arts and Commerce, Delhi. She obtained her M.Phil. from Delhi School of Economics and Ph.D. from Jawaharlal Nehru University. Professor Bathla has to her credit two books (one co-authored) and more than 40 research articles in refereed national and international journals. She is on the editorial board of *Agricultural Economics Research Review and Indian Journal of Economics and Development*. She has undertaken several assignments for the World Bank, IFPRI, UNCTAD, WWF-India and IASRI. She was conferred Jawaharlal Nehru Award for Outstanding Post-Graduate Agricultural Research by the Indian Council of Agricultural Research, New Delhi.

Dr. Sharmistha Sinha is Deputy Director at the National Institute for Labour Economics Research and Development, NITI Aayog, Government of India. Previously she has worked with the Ministry of Women and Child Development, International Labour Organization, UNWOMEN and various research institutes. Her research areas include changing patterns of employment, gender based discrimination in the labour market, unpaid work, poverty. She has various national and international publications to her acclaim.

Dr. Shinoj Parappurathu is Scientist at the National Centre for Agricultural Economics and Policy Research (NCAP), New Delhi, India. He is a recipient of Norman Borlaug International Agricultural Science and Technology Fellowship awarded jointly by the Indian Council of Agricultural Research (ICAR) and United States Department of Agriculture (USDA). He works mainly in the area of public policy in India's agricultural sector with focus on agricultural marketing, price policy, trade and commodity outlook modelling. Apart from research, he is also engaged in teaching post-graduate and doctoral students at the Post-Graduate School of IARI.

Dr. Sucharita Sen is a Professor in the Centre for the Study of

Regional Development, School of Social Sciences, Jawaharlal Nehru University. She works on issues related to rural livelihoods in the neoliberal period including impact of land acquisition on the farming community, gender and work in peri-urban areas and perceptions of and adaptations to climate change.

Dr. Susmita Mitra (Ph.D. in Economics from JNU) is currently working as an Asstt. Professor in Council for Social Development. She has five international peer reviewed publications. Earlier she was awarded Junior Visiting Fellowship in Germany, Junior and Senior Research Fellowship in National Entrance Test, and Ford Foundation Scholarship during M.A.

Dr. Vaibhav Singh is a Consultant and Research Associate at Navdanya. He has co-authored three books at Navdanya. He works closely with the government for policy formulation and advice on making such policies. He is a medical graduate from the Armed Forces Medical College, Pune.

Dr. T. Haque, Distinguish, Professor at the Council for Social Development, New Delhi and Chairman, Special Cell on Land Policies, NITI Aayog, Govt. of India is a well known Agricultural Economist. He obtained his Ph.D. degree from Patna University (India) and Post-doctoral training from Iowa State University, Ames, U.S.A. Prior to joining the Council for Social Development, Dr. Haque served as Chairman, Commission for Agricultural Costs and Prices, Government of India, ICAR National Fellow at the National Centre for Agricultural Economics and Policy Research, New Delhi; Director, Council for Social Development, New Delhi and Director, Centre for Agrarian Studies, NIRD, Hyderabad. He also served as consultant to several International organizations, including FAO, ILO and The World Bank. Dr. Haque is author of six books and about 100 research papers and reports. His areas of specialization are land reform, food security and sustainable agriculture and rural development.

Introduction

T. Haque

The growth performance of India's agriculture and allied sector improved quite substantially in the past few years. The average annual growth rate of agricultural gross domestic product increased from about 2.4 per cent and 2.5 per cent during the Ninth and Tenth Five Year Plan periods to 3.3 per cent during the Eleventh Five Year Plan period, even though it fell short of the targeted growth rate of 4 per cent. However, despite improvement in agricultural growth rate in the recent years, the health of the agricultural sector in India looks quite fragile. The declining soil fertility, depleting ground water, technology fatigue, high input cost, low output prices, declining farm productivity and profitability, climate change, inadequate credit, insurance and marketing support and overall policy mismanagement have combinedly led the actual growth rate of agricultural output to be much less than the potential growth rate. Due to high pressure of population on agriculture, the per worker agricultural income is not only lower than that of non-agricultural sector, but it continues to be extremely low in several states, including Bihar, Chhattisgarh, Jharkhand, Madhya Pradesh, Manipur, Meghalaya, Mizoram and Odisha.

The low agricultural income per worker partly explains why the incidence of rural poverty is comparatively higher in most of these states, the remaining part being explained by slow growth of off-farm and non-farm income opportunities, because of inadequate infrastructure, technological and institutional innovations and policy support. There is untapped productivity potential in almost all regions. The growth rate of irrigation as well as that of technological innovations has slowed down. The productivity potential of rainfed areas has not been harnessed either. The country is still deficient in the production of pulses, edible oilseeds and coarse cereals, due to low yields, while the yield levels of major cereals like rice and wheat have started plateing due to technology fatigue as well as soil health deterioration and water depletion in the Indo-Gangetic plains region. Currently, India imports nearly 4 to 5 million tonnes of pulses and 8 to 10 million tonnes of edible oils annually. Moreover, farming as such is losing its attraction for most farmers because of declining profitability and incomes, due to yield stagnation, as well as absence of appropriate pricing, marketing and insurance mechanisms. Unfortunately, the employment and income opportunities for them outside agriculture in rural areas is also limited, forcing them to either migrate under distress to remote urban areas or stay under employed in agriculture. During 1995 to 2012, about 2.84 lakh farmers committed suicide due to distressing situation.

This volume presents both macro- and micro- pictures of economic distress and challenges faced by Indian farmers. The papers included in the volume were presented and discussed at the National Seminar on Agrarian Distress organized by the Council for Social Development, New Delhi on November 11-12, 2014 with the support from the Indian Council of Social Science Research. The seminar was organized in honour of Prof. Gilbert Etienne an outstanding social scientist who recently passed away.

There are fifteen chapters written by well-known agricultural economists and research scholars. Chapter-1 gives an overview of the agrarian distress in India, its causes and the remedies by T. Haque. According to him, majority of the farmers in India are in poverty and debt traps. Based on the latest National Sample Survey data, the average monthly income of agricultural household was only Rs. 6,426 and those having less than one hectare of land did not save anything from their incomes. Besides, net income from cultivation was only Rs. 3,081 per year per agricultural household. It is, therefore, not surprising why farmers feel depressed and distressed and tend to commit suicide due to unbearable debt burden and humiliation by money lenders. According to Narsimha Reddy, the main causes of the unfolding distress of farmers are demographic pressure on agriculture, increasing marginalization, trade liberalization, growing informal credit, declining public investment, rising costs of production and stress on subsidy and other public support system (Chapter-2).

Suman Chandra and Prakash Rao paint a gloomy picture of the state of Indian agriculture, with increasing marginalization of peasantry, declining gross capital formation and rural development expenditure, the demise of the public sector institutions and growing incidence of suicide by farmers. According to them, the main causes of farmers' distress are: (i) indebtedness to private money lenders; (ii) low returns from crops; (iii) alcoholism and personal habits; (iv) neoliberal policies; and (v) other socio-cultural factors (Chapter-3). The Vidarbha region of Maharashtra has been in the news for quite some years now, about suicide by farmers due to distress.

Sucharita Sen and Archana Takte further bring out that the liberalization of the Indian economy since the early 1990s coincide with a period of agrarian distress, one of the manifestations of which is a continuous reporting of farmer's suicides, especially in Maharashtra. The primacy of growth, sacrificing distribution, impacts the marginal farmers adversely. The withdrawal of state from investments that have public good characteristics, dilution of poverty alleviation programmes and growing corporatization of agriculture are creating conditions of agrarian distress (Chapter-4). Ankita Goyal points out that net returns from crop cultivation are highly inadequate to put the average farmers above the poverty line in most states. According to her, there is a need to provide remunerative prices to farmers, along with non-farm income opportunities in the local areas (Chapter-5).

The question is whether diversification of the rural economy will help reduce farm distress and improve the farm economy? T. Haque's paper (Chapter-6) shows that both horizontal and vertical diversification have potentials for improving farm incomes and reducing farm distress, but there are several technological, infrastructural, institutional and policy bottlenecks to such diversification. At the same time, it is seen that farmers who grow cash crops having high input costs and high fluctuation in yields and market prices are more vulnerable as they borrow money from informal sources at high rate of interest and fail to pay back in years of low prices. Accelerating the pace of horizontal diversification would require appropriate risk management mechanisms in terms of price and marketing support as well as appropriately designed insurance support to cover risks for switching from low value to high value crops and enterprises.

Parappurathu, Birthal and Immanueraj examines the unfolding process of employment and income diversification experienced by the rural farm households and contributing factors therein (Chapter-7). A number of factors such as small landholding, low agricultural productivity, surplus labour, education and access to credit motivate farmers to diversify towards non-farm sources. Sharmistha Sinha shows that non-farm employment opportunities are growing, but at a slow pace, driven mainly by the expansion of the construction sector. Chapter-8 brings out clearly the fact that female employment in agriculture has been declining since 2005 primarily because of increase in their educational participation, increase in household income, mechanization in agriculture etc. However, agro-based manufacturing, focusing on the domestic market can expand the non-agricultural sectors and can link the rural poor to growing income.

Improvement in investment and rationalization of agricultural subsidies is often considered crucial for agricultural development. Seema Bathla analyses the inter-linkages between agricultural investments, input subsidy and growth and points out that public investments along with other factors such as better prices, increase in institutional credit and productivity in agro-processing industry has helped in raising private investment and agricultural growth (Chapter-9). Even though input subsidy may negatively impinge on public investment, the impact may not be significant. It would also not be feasible to cut down subsidy in view of its positive and significant impact on input consumption especially fertilizer and power. Agrarian distress can be minimized by selecting proper strategies and prioritized investment in agriculture. Further, the suggested strategies for minimizing agrarian distress would include (i) bolstering efficiency in irrigation; (ii) improving resilience of rainfed agriculture; (iii) effective utilization of the potentials of watershed development programmes; (iv) leveraging technology; (v) availability of capital in easy terms; (vi) modernization of agricultural markets; (vii) development of alternative market channels; (viii) cushioning risks in agriculture; and (ix) improvement in governance and institutional reforms (C. Ramaswamy: Chpater-10).

Gyanendra Mani's paper brings out that institutional innovations such as SHG-Bank Linkage Programme and formation of producers' organizations in the past few years have impacted positively on the well-being of small marginal farmers and landless workers (Chapter-11). The role of technological and institutional innovations for sustained increase in farmers' income has been further anlysed by Brajesh Jha in Chapter12. Jha argues that the cost of adopting lumpy technology will drop if one form or another of co-operative arrangement of farmers exists. The integrated farming system can increase farmer's return, but the complementary relationship between various enterprises is deteriorating in the light of depleting community assets and co-operative enterprises. Any group action by farmers that helps improve the organization of the production unit and reduce the influence of middlemen is desirable.

Contract farming is often considered important for improving farmers' income through assured price and marketing support. But small and marginal farmers are not generally preferred by companies for contract farming. Besides, due to lack of appropriate dispute settlement mechanism, farmers tend to get exploited in any contractual arrangement (Chapter-13).

There have been lot of discussions on eco-friendly sustainable agriculture in the recent years. Vaibhav Singh and Kunwar Jalees has scientifically analysed the roles of seeds, soil and society in strengthening the law of returns for sustainable livelihoods (Chapter-14). The chapter by Singh and Jalees reveals how diversified organic farming is superior to conventional mono cropping. Net incomes as well as nutrition per acre of farmers practicing organic farming increase because costly external inputs such as hybrid seeds, chemical fertilizers, pesticides and irrigation are avoided. There are location specific multiple opportunities for sustainable agriculture, which should be explored and utilized by farmers for avoiding distress.

Finally, there are multiple effects of climate change on agriculture and incomes of farmers. The chapter by Sushmita Mitra (Chapter-15) shows how climate smart agriculture can help small and marginal farmers to either overcome or minimize the risks of climate change. The threats of climate change are real and small and marginal farmers need to be educated and incentivized to meet the challenges of climate change. Thus, there are multiple reasons for the growing agrarian distress and multiple strategies would be required to deal with it.

The editor is grateful to Ms. Jaya Lekshmi Nair, for her technical help in finalizing the edited volume.

1

Agrarian Distress in India

Causes and the Remedies

T. Haque

It is now widely recognized and reported that farmers and farm workers who constitute nearly 49 per cent of India's total workforce are facing increasing difficulties and a distressing situation. The most extreme manifestations of the agrarian distress are in the form of suicides by farmers, distress sale of land, distress migration and stressful living condition in the villages (Dogra, 2013, Reddy and Mishra, 2009, Suri, 2006). The National Agricultural Policy of the Ministry of Agriculture (Government of India, 2001), rightly pointed out that agriculture has become a relatively unrewarding profession due to a generally unfavourable price regime and low value addition, causing abandoning of farming and migration from rural areas. The Election Manifesto of Bharatiya Janata Party in both 1998 and 2014 Lok Sabha election, highlighted the issue of agrarian distress and promised actions to improve the economic condition of farmers, if elected (Bharatiya Janata Party, 1998 and 2014). So, there is a general recognition that farmers and other agricultural workers in India are in distress. This chapter analyses the nature and causes of the present agrarian distress and suggests some remedies.

Nature of the Crisis

Majority of the farmers in India are in poverty and debt traps. According to 70th Round of National Sample Survey for the year 2012-13, the average monthly income of agricultural households was only Rs. 6,426 and again those having less than 1 hectare of land did not save anything from their incomes. (Government of India, 2014). On average, crop farming accounted for only about 47.9 per cent of their total household incomes, followed by incomes from wages and salaries (32.2%), animal husbandry (11.9%) and non-farm business (8.0%). The farmers whose landholding was less than 0.40 hectare earned more from wages than from farming but their average monthly income ranged from Rs. 4,152 to Rs. 4,561 only. Using Tendulkar methodology, the poverty-head count ratio in rural India was still as high as 25.7 in 2011-12. It was 44.6 per cent in Chhattisgarh, 38.9 per cent in Arunachal Pradesh, 40.8 per cent in Jharkhand, 38.5 per cent in Manipur, 34.1 per cent in Bihar, 35.7 per cent in Madhya Pradesh and Odisha, 30.4 per cent in Uttar Pradesh and 33.9 per cent in Assam (Government of India, 2013a). The average daily wage rate of casual male labourers in the country was as low as Rs. 138.62 in 2011-12, and for female workers it was only Rs. 103.28. Also the working and living conditions of agricultural workers are far from decent. The incidence of poverty, malnutrition, illiteracy, ill health and indebtedness is high.

Table 1.1: Average Monthly Income and Consumption Expenditure per Agricultural Household in India by Size Class of Land Possessed During July 2012-June 2013

(In Rupees)

Size Class of land possessed (ha)	*Net Income from cultivation*	*Net Income from farming of animals*	*Income from wages/ salary*	*Net Income from net farm business*	*Total Income*	*Total Consumption expenditure*
<0.01	30	1181	2902	447	4561	5108
0.01-0.40	687	621	2386	459	4152	5401
0.41-1.00	2145	629	2011	462	5247	6020
1.01-2.00	4209	818	1728	593	7348	6457
2.01-4.00	7359	1161	1657	554	10730	7786
4.01-10.0	15243	1501	2031	861	19637	10104
10.00 and above	35685	2622	1311	1770	41388	14447
All sizes	**3081**	**763**	**2071**	**512**	**6426**	**6223**

Source: *National Sample Survey*, Government of India, 70th Round, on Key Indicators of Situation of Agricultural Households in India, December 2014.

About 52 per cent of the agricultural households in the country were estimated to be indebted. The average amount of outstanding loan per agricultural household was Rs. 47,000. The borrowing for farm business or farming is not bad as such. But majority of the farmers did not earn enough to meet their consumption needs and remained in debt trap. Above

40 per cent of the outstanding loans were from non-institutional sources. The incidence of indebtedness, i.e. the proportion of indebted agricultural households was relatively high in southern states of Andhra Pradesh (92.9%), Telengana (89.1%), Tamil Nadu (82.5%), Kerala (77.7%), and Karnataka (77.3%). Also, it was above 50 per cent in Maharashtra, Punjab, Rajasthan and Odisha.

Whatever be the immediate factors that drive the distressed farmers to desperation, it is a fact that thousands of farmers commit suicide every year, more so in the Andhra Pradesh, Karnataka, Kerala, Maharashtra and Punjab due to unbearable debt burden and humiliation by private moneylenders. During 2000-01 to 2005-06, officially 9,127 suicides by farmers were reported in Andhra Pradesh, Maharashtra, Kerala and Karnataka. The figure is much lower than the NCRB data which points out that between 2001 and 2006, there were 56,946 suicides by farmers in these four states and 1,03,982 in the country as a whole (Reddy and Mishra, 2009). Between 1995 and 2012, about 2,84,000 cases of farmer's suicide were reported (Shiva and Singh, 2014). Surprisingly, the incidence of suicides is more among farmers cultivating cash crops with high costs of cultivation and high variability in yields and output prices. It is also a paradox that India as a country is agriculturally self-reliant and the policy makers have designs and dreams about high export growth of agricultural commodities including food grains, but farmers who are the architects of these surpluses are allowed to die out of distress. Unfortunately, there is no exit strategy for their absorption outside agriculture, as off-farm and non-farm employment opportunities in rural areas are limited, forcing them to either migrate under distress to remote urban areas or remain underemployed in agriculture. Interestingly, even though there is strong attachment to land, agriculture is no longer seen as a desirable occupation. The younger generations across caste groups dislike farming (Jodhka, 2012). Although farmers are biggest sufferers of agrarian distress, there are reports that landless labourers also are committing suicide even in a developed region like Punjab, due to reduced workdays, stagnant real wages and debt trap (Bharti, 2011).

Causes of Distress

The sources of agrarian distress in India can be located in several structural, technological, institutional, social, economic and policy factors. The high pressure of population on agriculture and growing marginalization of landholding, low and stagnating crop yields, high and rising input costs, low output prices, low and fluctuating farm profitability and incomes, lack of adequate price and market support, frequent occurrence of drought, floods, cyclone, hail storm etc., climate change, lack of appropriate agricultural insurance policy, coupled with pressure on input and food subsides, slow growth or stagnation in farm investment, lack of access to quality education and skills by farm youth and consequently their poor access to off-farm and non-farm employment opportunities in the organized sector, lack of proper healthcare facilities in rural areas, inadequate institutional credit and insurance facilities for small farmers and overall agricultural policy mismanagement are the main sources of agrarian distress. The main causes of the agrarian distress can be briefly discussed as follows:

Growing Marginalization of Land Holding

Under the impact of population pressure on agricultural land, the average size of land owned per household declined from 0.725 hectare in 2002-03 to 0.592 hectare in 2012-13 (Government of India, 2014). As of 2012-13, about 75.4 per cent ownership holdings were below 1 hectare and 10 per cent holding were in the category of 1 to 2 hectare. In other words, marginal and small farms accounted for about 85.4 per cent of ownership holdings, with 53.3 per cent of total area. The large holding above 10 hectares were only 0.24 per cent, owning 5.8 per cent of total area. About 52.5 per cent ownership holdings were below 0.5 hectare.

Table 1.2: Percentage Distribution of Households and Area owned by Category of Household Ownership Holdings in 2012-13

Category of holding	*% of households*	*% of area owned*	*Average size of holding (ha)*
Landless	7.41	0.01	0.00
Marginal	75.42	29.75	0.23
Small	10.00	23.54	1.39
Semi-medium	5.01	22.07	2.61
Medium	1.93	18.83	5.78
Large	0.24	5.81	14.45
Total	**100.00**	**100.00**	**0.59**

Source: National Sample Survey Organization, Government of India, 70th Round, on Key Indicators of Land and Livestock Holdings in India, December 2014.

During the same period, the average size of operational holdings declined from 1.16 hectare in 2002-03 to 0.909 hectare in 2012-13. It would be seen from Fig. 1.1 that average size of operational holding was as low as 0.31 in West Bengal, 0.40 ha in Puducherry, 0.48 ha in Uttarakhand, 0.61 ha in Tripura, 0.63 ha in Uttar Pradesh, 0.57 ha in Odisha, 0.59 ha in Sikkim, 0.47 ha in Jammu & Kashmir, 0.54 ha in Jharkhand, 0.49 ha in Himachal Pradesh, 0.50 ha in Bihar and 0.54 ha in Goa, i.e. much below the national average size of operational holding. Although land productivity of small farms is not always lower than that of large commercial farms, the net farm incomes are lower, as they are engaged mostly in the production of low value subsistence crops on tiny holdings (Haque, 1996; Haque, 2014). Also with a meagre asset base, they are unable to withstand even minor shocks (Vyas, 2004).

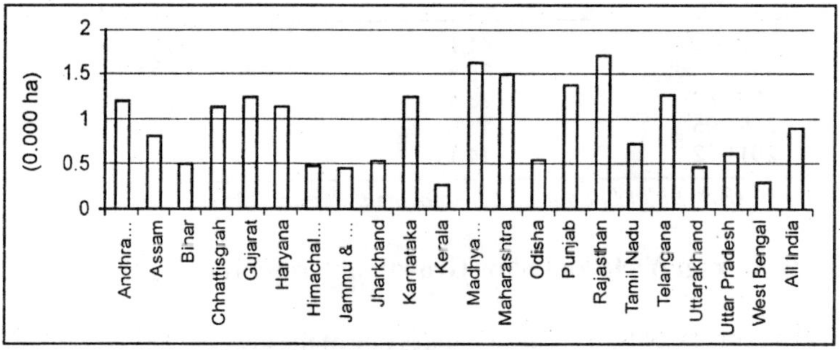

Fig. 1.1: Average area per Operational Holding

According to FAO, nearly two-thirds of the developing world's three billion rural people live in small households, many of which are poor, food insecure and malnourished with limited access to inputs and markets (FAO, 2010). About 70 per cent of the world's poor people live in rural areas and majority of them are small and marginal farmers, tenants and landless workers.

Declining Share of Agriculture in GDP and High Share in Workforce

The share of agriculture and allied sector in the country's Gross Domestic Product declined substantially overtime, while the share of agriculture in total employment dropped only marginally and consequently, the agricultural income per worker has grown not only slowly, but also it is much lower as compared to non-agricultural income per worker. The share of agriculture in GDP at 2004-05 prices dropped from 53.7 per cent in 1950-51 to 41.0 per cent in 1972-73 and 14.1 per cent in 2011-12, but its share in total workforce declined marginally from 73.9 per cent in 1972-73 to 48.9 per cent in 2011-12 (Table 1.3). The output per worker in agriculture was already lower than that in non-agriculture in 1972-73, the ratio being 1:3.6, but the difference has now significantly widened to 1:6 (IHD, 2014). In the absence of necessary education, skill and capital with farm households and slow growth of non-farm sector employment in the villages, the high dependence of workforce on agriculture, even with stressful living condition continues.

Table 1.3: Changes in Agriculture Sector's Share in GDP and Employment in India Overtime

Year	*Share in GDP*	*Share in Employment*
1972-73	41.1	73.9
1983	35.5	68.6
1993-94	28.4	64.0
2004-05	19.0	56.3
2011-12	14.1	48.9

Source: *IHD, India Labour and Employment Report*, 2014.

Low Average Yield and Slow Growth of Crop Production

The average yields of crops in India is one of the lowest in the world. The average per hectare yield of cereals in India was only 2.95 tonnes against the world average yield of 3.62 tonnes in 2012. The average yield of cereals was 5.84 tonnes in China, 7.27 tonnes in South Korea and 5.46 tonnes in Vietnam (FAO, 2013). During 2000-01 to 2011-12, the annual production growth rate of food grains in the country (calculated by the author), was about 2.4 per cent, but it was either negative or lower than the growth rate of population in Assam (1.33%), Bihar (0.23%), Himachal Pradesh (0.53%), Kerala (–2.56%), Meghalaya (1.00%), Punjab (1.36%), Mizoram (–9.39%), Uttar Pradesh (1.37%), and West Bengal (0.22%). Also the annual production growth rate of oilseeds on which the country is highly dependent on imports from outside, averaged about 2.00 per cent but it was negative in Andhra Pradesh (–0.69%), Assam (–0.60%), Himachal Pradesh (–4.83%), Karnataka (–1.07%), Kerala (–2.22%), Punjab (–2.47%), Tamil Nadu (–1.34%), Tripura (–3.46%), and Uttar Pradesh (–0.75%). During the same period, the production growth rate of fruits in the country was about 0.64 per cent, mainly due to area expansion. The annual growth rate of yield of fruits was either very low or negative in several states. The annual yield growth rate of fruits was negative in Chhattisgarh (–4.96%), Jammu & Kashmir (–3.09%), Jharkhand (–0.05%), Madhya Pradesh (–1.45%), Maharashtra (–7.79%), Puducherry (–3.00%) and Tripura (–2.98%). In the case of vegetables, the average annual growth rate of yield in the country was 2.1 per cent, but it was –0.79 per cent in Assam, 0.03 per cent in Chhattisgarh, (–0.07) per cent in Haryana, –0.15 per cent in Himachal Pradesh, –0.60 in Puducherry and 1.02 per cent in West Bengal. The average annual growth rate of milk production was about 3.28 per cent in the country, but it was much lower than the growth rate of population in Assam (0.62%), Kerala (1.12%), Mizoram (–1.20%) and Manipur (0.99%). Also the annual growth rate of fish production was 3.54 per cent in the country, but it was as low as 1.46 per cent in Gujarat, 0.46 per cent in Himachal Pradesh, 1.08 per cent in Jammu & Kashmir, 0.41 per cent in Kerala. 1.32 per cent in Madhya Pradesh, 0.84 per cent in Maharashtra, –0.29 per cent in Puducherry and –7.57 per cent in Uttarakhand. In terms of demand—supply situation, the country is in a comfortable zone with an estimated production of 265 million tonnes of food grains in 2013-14 against the projected demand of 257 million tonnes in 2016-17, but there are problems with regard to several commodities. The peak level of total production of pulses was about 19 million tonnes in 2016-17 against the projected demand of 22 million tonnes and the production of oilseeds was about 33 million tonnes in 2010-12 against the projected demand of 59 million tonnes in 2016-17. These gaps have to be filled soon. Similarly, in the case of vegetables, fruits, milk, fish and meat, there are huge gaps between what is produced at present and the projected demand in 2016-17 and beyond (Government of India, 2013b). The projected demand for fruits is 97 million tonnes against the actual production of 81 million tonnes, while the production of milk was about 82 million tonnes against the demand of 41 million tonnes. The country produced 9 million tonnes of fish in 2013-14 against the projected demand of 11 million tonnes in 2016-17.

High Input Costs and Low Return

Input costs in agriculture have increased phenomenally in the recent years, making it difficult for the small and marginal farmers to arrange money for the purchase of inputs, not to speak of making large investments. The purchased external inputs now claim larger share of farmers' operating expenses. As of 2010-11, the per hectare A_2+F_L (paid out cost plus the imputed value of family labour) for cotton ranged from Rs. 17,976 in Madhya Pradesh to Rs. 42,391 in Tamil Nadu. It was Rs. 38,359 in Maharashtra. The C_2 cost, of cotton which covers both operational and fixed costs, was as high as Rs. 59,283 in Punjab, Rs. 56,748 in Tamil Nadu and Rs. 52,584 in Maharashtra. The C_2cost per hectare in paddy was as much as Rs. 51,279 in Punjab, Rs. 50,632 in Tamil Nadu, Rs. 51,505 in Andhra Pradesh and Rs. 49,868 in Haryana. The sugarcane farmers in Maharashtra, Andhra Pradesh and Karnataka put in the range of Rs. 1,13,059 to Rs. 1,28,997 per hectare of C_2 cost. Using 1996-97 as the base, i.e. 100, the index of A_2 cost (paid out cost) in paddy cultivation per hectare increased to 356.2 in 2010-11 in Assam, 303 in Uttar Pradesh and Odisha and 278.7 in Punjab. In the case of cotton, it increased to 411.5 in Maharashtra, 354.3 in Gujarat and 371.0 in Haryana. It may be seen from Fig. 1.2 that the output-input ratio in paddy (using C_2 cost) dropped from 1.43 in 2001-02 to 1.31 in Punjab, while it remained mostly less than 1.0 in Bihar, Odisha and West Bengal. In fact, in the case of all major crops, the output-input ratio was less than one in several years in most regions, during 2001-02 to 2010-11.

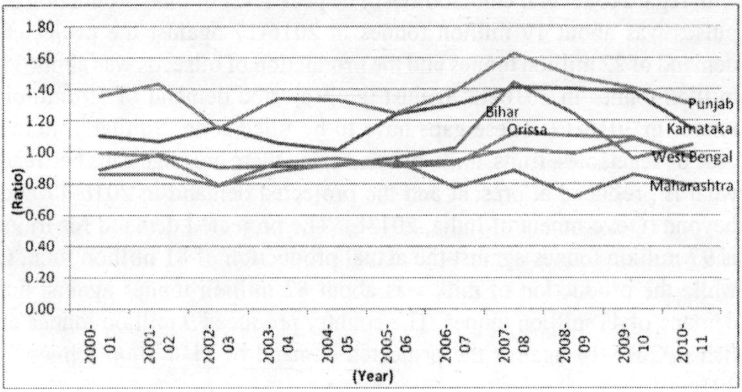

Fig. 1.2: Output-Input Ratio in Paddy Cultivation (at C_2 Cost)

During 2009-10 to 2011-12, the average net returns over cost (C_2) was negative for paddy in Assam, Bihar, Jharkhand, Maharashtra and West Bengal. Similarly, for maize, it was negative in Chhattisgarh, Gujarat, Himachal Pradesh, Madhya Pradesh, Rajasthan and Uttar Pradesh (CACP, 2014). In the case of wheat also, the net return was negative in several states including West Bengal, Himachal Pradesh and Jharkhand (CACP,

2013). Thus not only the input cost per hectare was unaffordable by small farmers, the returns from crop farming was also low and uncertain. It has been estimated that the average annual growth rate of the value of output in the agriculture and allied sector during the Eleventh Five Year (2007-2011) was 3.6 per cent, while the input costs increased at the rate of 4.3 per cent per year, showing declining trend in factor productivity (Government of India, 2013b). The high dependence on external inputs, coupled with high cost of inputs and diminishing returns due to crop loss as well as low output prices have caused havoc in the peasant economy (Rao and Suri, 2006). With the high input-low output scenario, the result has been the failure to repay the loans, attracting humiliation and harassment from the moneylenders, (Vyas, 2004).

Inadequate Price and Marketing Support

Majority of small and marginal farms produce mainly for self-consumption and sell only part of their produce to buy non-homegrown items. No doubt, this helps them to insure against food price inflation, but their lack of market orientation and in some cases, limited access to markets reduces the benefits that may arise from their participation in commercial farming. Government fixes minimum support prices for as many as 24 agricultural commodities, but barring rice and wheat in some regions, there is no effective implementation of support prices. Besides, minimum support prices (MSP) are not always remunerative, as actual costs of production in several crops in some states are higher than the MSPs fixed. According to 70th round of NSSO for 2012-13, only 20 per cent paddy growers during July-December 2012-13 and 17 per cent during January to June 2013, sold their produce in mandi and nearly 41 per cent paddy growers in the first period and 64 per cent in the second period, sold their produce to local private traders. In the case of wheat, about 44 per cent sold in *mandis* and the rest to other agencies. Majority of the farmers producing coarse cereals, pulses, cotton, jute, potato and onion sold their produce to local private traders. In terms of quantity, only 27 per cent of paddy was sold at minimum support prices (MSP) while 35 per cent of wheat was sold at MSP. In the case of other crops, negligible quantities were sold at MSP. Also majority of farmers in the country are unaware of both MSP and procurement agency. Following the recommendations of National Commission on Farmers under the chairmanship of M.S. Swaminathan, the Bharatiya Janata Party in its election manifesto for 2014 General Lok Sabha election (BJP, 2014) promised to ensure payment of remunerative prices of farm products by fixing MSP at total cost plus 50 per cent margin. But this has not yet been implemented, presumably due to pressure on subsidy cut from both within and outside. The system of MSP and PDS operations is under threat as the product specific subsidy is likely to cross the 10 per cent, the minimum level fixed by the WTO. Alternatively, direct marketing or private markets for agri-products are not adequately developed. Particularly the farmers in remote areas, fail to participate in markets due to high transportation costs. Also seasonal glut of agricultural commodities in local markets leading to their low prices and the high level of post-harvest losses incapacitate farmers and put them in distressing situation. This happens when farmers grow cash crops with large investment, using borrowed capital and fail to realize even the cost of production due to market crash. It may not be out of place to mention here that the weighted average cost (C_2) for cotton in the country was Rs. 3,480 per quintal, but it was as high as Rs. 4,103 per quintal in Tamil Nadu and Rs. 3,962 in Maharashtra, while the MSP recommended for cotton was Rs. 3,750 per quintal. The average cotton growers in Maharashtra and Tamil Nadu are, therefore, facing a distressing situation.

High Variability in Yield and Prices

The high variability in yield and prices of agricultural commodities due to erratic behaviour of weather and markets is one of the main sources of agrarian distress. Frequent occurrence of drought, floods, cyclones, hail storm, pests and diseases etc. tend to threaten the very livelihood system of farmers in some parts of the country. This is especially so because the majority of farmers in India are small and marginal who have hardly any risk bearing capacity. The problem gets compounded in the absence of an appropriate agricultural insurance or farm income insurance policy. During 1990-91 to 2011-12, the co-efficient of annual variation in the yield of food grains in the country was above 20 per cent in Andhra Pradesh, Gujarat, Kerala, Rajasthan and Tamil Nadu (the authors own calculation). Besides, the co-efficient of annual variation in the farm harvest price of rice during the same period was above 30 per cent in most of the rice growing regions, while it was in the range 40 to 60 per cent for jowar, 38 to 47 per cent for wheat, 33 to 45 per cent for maize, 35 to 42 per cent for gram, 33 to 36 per cent for rapeseed and mustard and 29 to 61 per cent for cotton. With markets becoming national and global, a low crop yield in a region does not necessarily lead to higher prices, while volatility in crop yields and low prices for several years together break havoc with farmers' lives (Suri, 2006).

Unfavourable Agricultural Trade Policy

Agricultural trade liberalization in the wake of WTO has not improved the economic condition of farmers, as was expected and at the same time it has exposed the Indian farmers to the price fluctuations at the international market. Besides, due to unfavourable tariff rates the agricultural imports increased at a faster rate than that of agricultural exports, while the pressure on price support and farm subsidies along with differential food standards put the Indian farmers at a great disadvantage. It may be seen from Fig. 1.3 that agricultural export import ratio initially declined since 1995-96 and then remained more or less stable, indicating no improvement in agriculture balance of trade.

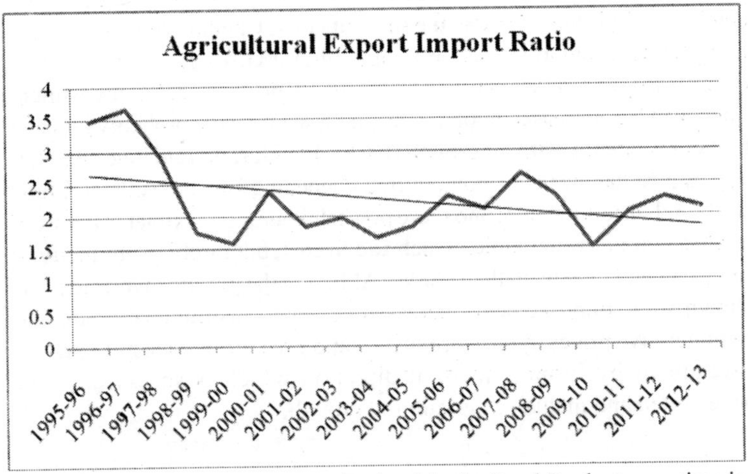

Fig. 1.3: Change in Import-Export Ratio of Agricultural Products overtime in India

In addition, high costs of production and frequent ban or restriction imposed by the Government on the export of those agricultural commodities which have the export potential based on sound economic logic, stood in the way of Indian farmers benefiting from trade liberalization. Also the world agricultural prices remained quite volatile, much more than the domestic prices in recent years, while the lack of co-ordination between domestic price policy and tariff policy, coupled with frequent ban on the exports of some agricultural commodities has hurt the Indian farmers. Also *ad hoc* policy responses have created havoc with incentive systems and agricultural growth (Alagh, 2013). The more a commodity is integrated with the world market, the higher is its adverse consequences on the farmers, especially those who are heavily dependent on export-oriented crops such as coffee, pepper etc. (Mohan Kumar S. and R.K. Sharma, 2006).

Low and Stagnating Public Investment

Public investment in agriculture has not increased much in real terms, in recent years, aside from the fact that there have been fluctuations from year to year. It would be seen from Figure 1.4 that the public investment in Agriculture and allied activities at 2004-05 prices has been more or less stagnating since 2006-07. In fact, the share of public investment in agriculture to total public investment in the economy declined from 7 per cent during 2004-05 and 2006-07 to less than 5 per cent during 2008-09 and 2011-12 (Table 1.4). Also the share of Gross Capital Formation (GCF) in agriculture and allied sector to agricultural Gross Domestic Product (GDP) dropped, marginally. The share of private sector investment improved but increased public investment would have probably induced higher private investment growth (GCF). Besides, private investment in agriculture has been for competitive deepening of borewells and tube wells and installing more and more powerful pumps to extract water from a falling groundwater table, which add to cost, but rarely to productivity improvement (Vaidyanathan, 2006). Although the private investment has shown some improvement, increased access to institutional credit and remunerative prices of farm products could give further boosts to private investment by farmers in right direction which is so essential for accelerated and inclusive agricultural growth.

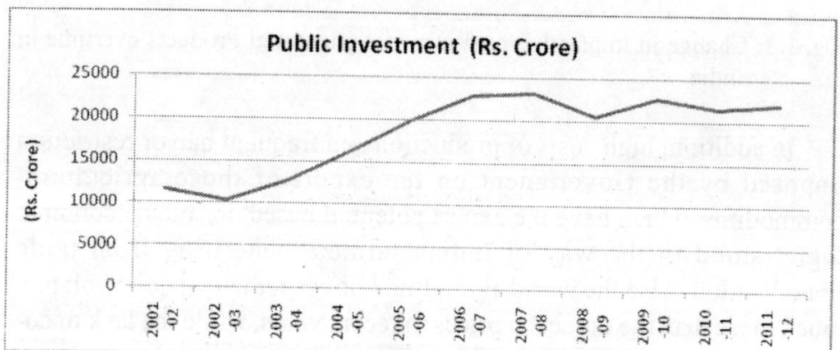

Fig. 1.4: Public Investment in Agriculture and Allied Activities (at 2004-05 prices)

Table 1.4: Gross Capital Formation (GCF) in Agriculture and Allied Sector in Recent Years (at 2004-05 prices)

Year	Per cent Share of Gross Capital Formation in agriculture in the Gross Capital Formation in the economy			GCF as a per cent of Agricultural GDP		
	Public Sector	Private Sector	Total	Public Sector	Private Sector	Total
2004-05	6.7	7.8	7.5	2.9	10.6	13.5
2005-06	7.1	7.4	7.3	3.4	11.2	14.6
2006-07	7.1	6.6	6.7	3.7	11.2	14.9
2007-08	6.1	6.7	6.6	3.5	12.6	16.1
2008-09	4.8	9.4	8.1	3.1	16.3	19.4
2009-10	5.0	8.6	7.7	3.4	16.7	20.1
2010-11	4.3	7.5	6.7	2.8	15.7	18.5
2011-12	4.6	8.3	7.5	2.8	18.0	20.8
2012-13	4.7	8.6	7.7	3.1	18.1	21.2

Source: Agricultural Statistics of a Glance 2014, Ministry of Agriculture, Government of India.

Inadequate Access to Institutional Credit

Access to institutional credit at affordable rates of interest is critical for increased input use and productivity growth. Unfortunately the small and marginal farmers' access to institutional credit is showing a declining trend in the country (RBI, 2004). The share in direct agricultural credit of loans of less than Rs. 20,000 each dropped from 92.2 per cent in 1990 to about 48 per cent in 2011, while the share of loan of more than Rs. 10 lakh each increased from only 4 per cent in 1990 to 23 per cent in 2011. Surely, large loans were advanced to finance agribusiness and not crop cultivation (Ramakumar and Chavan, 2013). Besides, the per hectare availability of institutional credit is awfully low, especially in Eastern India where there is high and slow agricultural growth due to lack of adequate institutional credit for private investment. The average amount of outstanding loan per agricultural household in 2012-13, was Rs. 3,400 in Assam, Rs. 5,700 in Jharkhand, 28,200 in Odisha, Rs. 16,300 in Bihar and Rs. 17,800 in West Bengal (Government of India, 2014). Moreover, the lending procedures of scheduled commercial banks and co-operatives are so cumbersome that farmers prefer to borrow from private moneylenders at exorbitantly high rates of interest and remain perpetually in debt trap. It would be seen from Table 1.5 that more than 53 per cent of outstanding loans by those having less than 0.40 hectare land was

Table 1.5: Per cent distribution of outstanding loans by source of loan taken for different size classes of land possessed

Size Class of land Possessed (ha)	*Per 100 distribution of outstanding loans by source of loan*								
	Government	*Co-operative society*	*Bank*	*Employer/ landlord*	*Agricultural/ professional moneylender*	*Shopkeeper/ trader*	*Relatives and friends*	*Others*	*All*
<0.01	0.4	1.6	12.9	0.6	63.7	1.4	17.5	1.8	100
0.01-0.40	1.3	14.6	31.0	0.8	32.4	2.5	14.2	3.1	100
0.41-1.00	1.7	13.9	37.6	0.8	27.4	6.6	10.6	1.4	100
1.01-2.00	2.6	14.7	47.5	0.7	23.3	1.5	7.6	2.0	100
2.01-4.00	1.9	15.6	50.0	1.4	23.8	1.2	5.8	0.3	100
4.01-10.0	3.8	17.5	50.2	0.4	18.7	1.4	6.5	1.5	100
10.00 +	1.1	14.3	63.5	0.0	16.1	0.5	3.8	0.6	100
All sizes	**2.1**	**14.8**	**42.9**	**0.8**	**25.8**	**2.9**	**9.1**	**1.6**	**100**

Source: National Sample Survey, 70th Round, NSSO, Govt. of India (2014).

from non-institutional sources and even for other marginal farmers, nearly 47 per cent loans were from non-institutional sources of which professional moneylenders dominated. Several earlier studies have identified unbearable debt burden from non-institutional sources as the main reason for farmers suicides (Deshpande, 2002; Vidya Sagar and Suman Chandra, 2003; Mishra, 2006; Vyas, 2004). This is particularly so when the level of debt and the dependence on traders and moneylenders is high and higher proportion of debt is for non-productive uses and also because the increase in net incomes from loans used for productive purposes falls far below expectations (Vaidyanathan, 2006).

Slow Pace of Technological Innovations

There is a near failure of the public research and extension system in the country to address the issue of technology fatigue and yield stagnation in the erstwhile green revolution areas and low crop and livestock yields in the remaining areas. In the past two to three decades, technological innovations has been slower than what is warranted to improve agricultural productivity. Particularly the inadequate availability of location specific high yielding, pest resistant and drought and flood tolerant varieties' of seeds is one of the major challenges. Also the nature of technological change has been neither cost effective nor environment and small farmer friendly. The total factor productivity growth indicates a declining trend in the recent years (Government of India, 2013b). Besides, the public extension system has been more or less collapsed in most regions and consequently, there are huge yield gaps which could be bridged even with the transfer of existing technologies. The inadequacy of public investment in agricultural research and extension, inefficiency of existing government research and extension system, lack of proper co-ordination between ICAR-SAUs and Central and State Governments and lack of clear policy towards bio-technology as well as participatory seed production at the village level account for the gap that exists in the field of generation and diffusion of appropriate agricultural technologies. Public investment in agricultural research including education, increased from about 0.45 per cent of agricultural GDP in the early 1990s to about 0.70 per cent during the Eleventh Five Year Plan, but it remained much lower than that in developed countries. Similarly, public investment in agricultural extension marginally dropped from about 0.15 per cent in early 1990s to 0.14 per cent of agricultural GDP in recent years. Moreover, cost-effectiveness and ecological sustainability of technology never received proper attention in the policy making.

Slow and Uncared Agricultural Diversification

The pace of agricultural diversification and value addition which could improve farm incomes and also protect agricultural environment from the loss of natural bio-diversity moved slowly in most regions, due to various technological, infrastructural, institutional and policy constraints. Still about 49 per cent of India's total workforce, is engaged in agriculture and the crop farming is the principal source of income of 63.5 per cent agricultural households in the country, followed by wages/salaried employment (22.0%), non-agricultural enterprises (4.7%), livestock (3.7%), other agricultural activity (1.1%) and others (5.1%) respectively. In fact, traditional crop farming continues to be the principal source of income for as much as 83 per cent marginal farm households and 86 per cent small farm households (Government of India, 2014). No doubt within the agricultural sector, the share of area under food grains dropped from 73.5 per cent in the early 1950s to 64.1 per cent during 2009-12 and that of area under non-food grain crops increased from 26.5 per cent to 35.9 per cent during the same period. Logically diversification in favour of non-food grain cash crops should help farmers to earn more in the years of a good crop, but farmers have no control over the markets and prices or in shaping the import-export tariffs, which are directly responsible for the distress of growers of crops such as cotton and groundnut, while the failure to get a good price even in a bad crop year makes their lives miserable (Rao and Suri, 2006).

Water Scarcity and Inefficiency of Water Use

Water scarcity and inefficient use of available water stand in the way of improvement in crop productivity. At present, the rainfed areas constitute nearly 56 per cent of India's net cultivated area. There are some common constraints to agricultural development in rainfed areas. Aside from erratic rainfall, leading to high production risks due to either drought or flood and low adoption of new technology, there is a serious problem of land degradation, due to erosion by either water or wind, including runoff in most areas. Besides, the poor drainage and water management system as well as low utilization of available ground water in high rainfall areas, especially in the eastern region and low ground water availability as well as high incidence of ground water depletion in low rainfall areas, mainly due to indiscriminate use of scarce water resources put a premium on the adoption of new technology. Furthermore, the concentration of private investment in groundwater exploitation has resulted in over-exploitation of the resource leading to negative externalities, especially in the arid and semi-arid areas where progressively larger areas are coming under the 'dark zone' (Vyas, 2004). According to an estimate, uncontrolled depletion of ground water may well put 25 per cent of India's harvest at risk (Selvarajan and Roy, 2004). In fact, the normal rainfall in North West India as well as South Peninsular region is only 615 mm and 716 mm respectively, while rapid groundwater depletion poses a real threat to rice-wheat cropping system in the entire Indo-Gangetic plain region (Ramesh Chand and Haque, 1997). During the last twenty five years, the actual rainfall was less than normal in 11 years. The country had severe drought in 2002 and 2009 which led to huge loss of crop production as well as human and animal lives in different parts of the country. It may be noted that the percentage of total cropped area irrigated was as low as 4.1 in Assam, 22.8 in Gujarat, 12.0 in Jharkhand, 20.3 in Himachal Pradesh 17.7 in Kerala, 20,.2 in Maharashtra and 28.3 in Odisha, even though the total irrigated area was about 44 per cent in the country. In fact, augmentation of water supply and their efficient and equitable utilization is the key for sustainable food security, at both household and national levels. Improving productivity per unit of water through water friendly cropping system and improvement in agronomic practices may help improve the situation, but farmers lack sufficient awareness in this regard.

Threats of Climate Change

The threats of climate change and degradation of natural resources which tend to change the production system on which the livelihoods of small farms depend are real. Adjustment to climate change requires new production patterns, adoption of new inputs with higher resilience to drought and floods and wider use of intensive agro-ecologically sustainable techniques such as conservation agriculture, SRI etc., but small farmers in India are unable to adjust to the changing environment because of the lack of adequate human and financial resources as well as proper information. Some projections indicate that more rain is likely to fall during intense monsoon periods in already rain abundant parts of the country, creating increased risk of flooding, while drought prone areas in Andhra Pradesh, Gujarat, Madhya Pradesh and Rajasthan will receive less rainfall. In Andhra Pradesh, one survey covering eight districts indicate such occurence once every 3 to 4 years, leading to losses in output value of 5 to 10 per cent, which is enough to push many farmers below the poverty line. Models for farm income in India as a whole suggest that a 2 to 3.5°C temperature increase could be associated with a net farm revenue reduction of 9-25 per cent (UNDP, 2007/2008). Also there is a distinct trend towards both drier and warmer weather in the past 15 years or so. There has been lower than mean rainfall and higher rainfall variability and also warming has increased at an accelerated pace (Government of India, 2013b).

Deterioration of Soil Health

Due to excessive and unbalanced use of chemical fertilizers, soil health has deteriorated in several places. Farmers, being unaware of nutrient status of soils in their plots, use macro- as well as micro-nutrients in an unbalanced and in some cases, excessive manner which reduces the productive capacity of soils, thereby forcing the farmers to apply more fertilizers and incur higher cost every subsequent year to obtain the same level of output. Besides, about 25 million hectare of acidic soils and 8.5 million hectare of soils affected by alkalinity and salinity in the country have very low productivity (ICAR, 2007). These soils would require to be properly treated/reclaimed for productivity enhancement. In fact, over 120 million hectares of land have been declared degraded or problem soils in the country (Government of India, 2013b).

Lack of Appropriate Organization of Small Farmers

Lack of appropriate organization of small farmers stand in the way of their participating and benefiting from market and organized supply chains. Unlike in East Asian countries, co-operatives and Farmers' Associations in India are weak and do not play a significant role in linking farmers to markets, although experiences of Joint Liability Groups (JLG) under Kudumbashree in Kerala and Farm Producer Organization (FPOs) in some states in recent years show how farmers' group action can help them to take up various farm and non-farm enterprises on viable basis.

Prevalence of Informal Tenancy

The widespread informal tenancy acts as a constraint to agricultural development. According to 70th round of National Sample Survey for the year 2012-13, the leased in area in the country constitutes about 10 per cent of the total agricultural land. This is as high as 33.8 per cent in Andhra Pradesh, 21.0 per cent in Bihar and 14.9 per cent in Haryana. These figures are probably underestimated, as many informal tenants and land owners do not report it, as leasing out agricultural land is either legally banned or highly restricted in most places, excepting for certain disabled categories of landowners. The informal tenants are insecure and do not have access to institutional credit, insurance and benefit of other government schemes and consequently the investment and productivity of leased in land remains low. The condition of informal tenants is worse during natural disasters like drought, flood, cyclone, etc. as they are not covered by any agricultural insurance for lack of any proof of crop loss as the leasing agreement is mostly oral. Besides, restrictions on agricultural tenancy reduces the rural poor's access to land through leasing and also discourages the landowners to lease out land and take up non-farm activities for upward occupational mobility, because of the fear of losing land if they lease out.

Inadequate and Insecure Land Rights of Women

The condition of women farmers is relatively worse, as majority of them do not have land recorded in their names. According to Agricultural Census, 2010-12 (Government of India) about 12.4 per cent of the total operational holding in India were operated by women, covering 9.9 per cent of the total land. Since majority of them do not have land in their names, they cannot access bank credit, insurance and other services and consequently, the productivity of land operated by them remain low (Haque, 2012). Secure rights to land and properties for women are regarded as crucial for ensuring sustainable agricultural development. According to Food and Agriculture Organization of the United Nations, if women had the access to productive resources as men, they could increase yields on their farms by 20 to 30 per cent. These gains could lift some 100 to 150 million people out of hunger (FAO, 2010). However, there are legal as well as socio-cultural barriers to land and property rights of women which need to be overcome through sustained awareness building and policy advocacy.

High Incidence of Landlessness

The high incidence of landlessness poses a threat to livelihood security of millions of rural poor households and consequently social stability in the villages. During the past few decades, the proportion of rural landless households increased from 9.6 in 1971-72 to 11.3 in 1982 and 1992 which then marginally declined to 10.0 in 2003 and 7.4 in 2012-13. In fact, the 70th round of NSSO (Government of India, 2014) shows that those with less than 0.5 hectare of land account for nearly 69.2 per cent of rural households and about 9.3 per cent rural households owned land between 0.002 to 0.005 hectare which are practically landless. About 6.7 per cent agricultural households have only homestead land.

Growing Unemployment among Educated and Semi-educated Farm Youth

The high incidence of unemployment of educated and semi-educated farm youth is perhaps the greatest potential threat to economic growth and social stability in rural areas. As of 2011-12, about 10 to 13 per cent of the rural population having graduation and post-graduation degree were unemployed (Table 1.6). The educated and semi-educated farm youth are neither interested in being employed in traditional, albeit unremunerative agriculture nor are they getting any productive jobs outside agriculture, due to lack of proper training and skills. They are a liability on the parents and the society and may result in agrarian unrest in the days to come, unless the problem is properly solved on priority basis.

Table 1.6: Rural Unemployment Rate by Education Level and Sex (UPS), 2011-12

Education Level	*Unemployment Rate*		
	Male	*Female*	*Total*
Illiterate	0.6	0.7	0.6
Below Primary	1.0	1.4	1.1
Primary	1.6	1.1	1.5
Middle	2.2	4.2	2.5
Secondary	2.6	8.8	3.5
Higher Secondary	4.3	14.2	5.5
Diploma/Certificate	10.0	25.9	12.6
Graduate	8.0	23.7	10.2
Post-graduate and above	10.0	23.2	12.6
Total	**2.1**	**2.9**	**2.3**

Source: *India Labour and Employment Report*, IHD, 2014, p. 69.

Deprivations in terms of Education and Health

Inadequate access to quality education and health care facilities in rural areas increase the cost of living of farmers and others in rural areas. In fact they have to bear extra-financial burden and stress, in case they have to send their children for schooling in remote urban centres. Also if any member of a farm household falls seriously ill, they often have to sell their assets if any, at unfavourable terms. The most bizarre manifestation of commercial health care is witnessed in the form of private clinics taking advantage of the situation when dying farmers are rushed in emergency by desperate families to access health care in good private hospitals in urban centres (Sidhu, 2002).

Remedies

The way out of agrarian distress can be long, as taking out a large number of agricultural population out of poverty and debt traps would require massive increase in investment in appropriate agricultural diversification and value addition and firm commitment and determination on the part of India's policy-makers and administrators to improve incomes of farmers through appropriate technology, infrastructure, institutional reforms and policy changes, although some short-term measures can provide immediate relief. The key remedial measures would include the following:

(i) ***Adoption of Yield Augmenting and Cost Effective Technologies:*** First, improvement in the yields of crops and livestock as well as real incomes of farmers through adoption of yield augmenting, albeit cost effective technologies, and improved agronomic practices would hold the key in ensuring food security at household as well as national levels and reducing poverty and farm distress. This would in turn require restructuring and strengthening of agricultural research and extension system to make it more participatory, and farmer friendly. Involvement of farmers groups at the village level for seed production with support from government and non-government agencies, regulation of monopolistic behaviour in the seed and other input markets and promotion of healthy competition in the supply of quality seeds and other inputs on adequate scale would be necessary components of this strategy. Besides, improvement in the yield of food crops in both rainfed and irrigated areas should receive highest priority, as it would release land for high value crops and other enterprises to achieve higher incomes for farmers without compromising on the issue of national food security. In addition raising the productivity of pulses and edible oilseeds through technological innovations would be necessary to reduce their dependence on imports and meet the growing internal demand.

(ii) ***Reducing the Cost of Production:*** Reducing the cost of production through yield improvement and overall resource use efficiency would be important for improving farm profitability. This would require the development and adoption of cost efficient technologies and agronomic practices such as balanced use of macro- and micro-nutrients, drip and sprinkler irrigation, soil conservation, bio-pesticides and small farmer friendly farm machineries, System of Root Intensification (SRI) etc.

(iii) ***Improving Market Orientation and Market Access:*** Improving market orientation and market access of farmers would enable the farmers to benefit from commercial oriented agricultural development. The key elements of market development and market access would be to (a) set-up agricultural markets in each cluster of villages through increased public and private investment; (b) provide appropriate institutional arrangements and codes of conduct for promotion of contract farming; (c) promote farm producer companies or farm producer organizations which will enable the small farmers to benefit from agro-processing and value chains; and (iv) rationalize market fees, taxes and levies.

(iv) ***Implementing Remunerative Prices Effectively:*** Effective implementation of appropriately fixed minimum support prices or price deficiency payments for all crops and in all regions, ensuring adequate margin over costs, would be necessary to incentivise the farmers to adopt new technology and improve productivity or Price Deficiency Payments. Small farmers all over the world stay in farming through Government support and hence there is no reason why that should be denied to farmers in India. The need for such support in India is even greater, as agriculture is a means of livelihood and not just business.

(v) ***Imparting Training and Skills to Farm Youth:*** There is a need to provide policy framework and support system for necessary training and skills to educated and semi-educated farm youth in high tech agriculture, agro-processing, agro-services, etc., as they are not interested to continue in traditional non-remunerative farming as an occupation. Development of off-farm and non-farm enterprises in rural areas and on the job training of farm youth with Government support would be crucial in this regard.

(vi) ***Improving Farmers' Access to Institutional Credit:*** Improving small farmers' access to institutional credit, would be necessary as above 85 per cent of operational holdings in India are small and marginal and they do not generate investible surplus of their own, and stay in farming at subsistence level or low level equilibrium. The critical interventions required in this regard would include: (a) adopting the two tier co-operative credit structure instead of the existing 3 tier, which would reduce the transaction cost and increase the flow of credit; (b) reserving 18 per cent priority sector lending only for genuine farmers, not including agro-processing or agro-services units, while there should be one specialized bank in each state to cater to the needs of commercial agriculture, including agro-processing, agro-services etc.; and (c) advancing institutional credit to all small farmers in need, at low rate of interest.

(vii) ***Derisking Agriculture:*** Derisking agriculture through implementation of properly designed agricultural insurance policy which would insure against both yield and income risks at the village and household levels and attract maximum participation of farmers on realistically fixed premium on voluntary basis, should be one of the essential elements of any strategy to improve and stabilize farm incomes. Also comprehensive insurance product should cover all crops, animals and farm equipment as well as life and health aspects of the farming population.

(viii) ***Accelerating the Pace of Agricultural Diversification:*** Accelerating the Pace of right type of agricultural diversification and value addition through appropriate technological innovations and policy support would be necessary. This would particularly require: (a) strengthening dairy co-operatives and veterinary care services for accelerated and inclusive growth of livestock; (b) rapid replacement of old senile orchards of mango, *litchi*, coconut, etc. by location specific superior varieties already developed by ICAR and SAUs for horticulture; (c) establishing strong agriculture-industry linkages through proper institutionalization of contract farming, as well as co-operative enterprises of farmers and setting-up agro-processing and agro-service centres in every cluster of villages and linking them to national and international markets. Besides, top priority should be given to development of necessary rural infrastructures, so that each village and each farm household has easy access to roads, transportation, market, banks, electricity, godowns, cold storage, etc., in the absence of which Indian farmers cannot effectively participate and benefit from diversified and market led growth.

(ix) ***Addressing the Challenges of Trade Liberalization:*** Indian farmers should be enabled to meet the challenges of trade liberalization through rationalization of subsidies, resource use efficiency and increased support to WTO permissible services such as infrastructure development, agricultural insurance, agricultural research and extension, pest and disease control, soil conservation, disaster management etc., There is also a need to find an early solution to open ended procurement and public stockholding programme for food security by way of negotiation for a better deal in WTO. Besides, Government of India must give up *adhocism* in agricultural export policy and should rationalize tariff rates and ensure that there is no frequent ban or restriction on the export of those agricultural commodities which have the export potential based on sound economic logic.

(x) ***Promoting Organic Farming and Conservation Agriculture:*** Organic farming is now recognized the world over not only for protecting human and animal health from the adverse effects of chemicals, but also as a sustainable model for farm productivity growth and alleviating poverty.

Due to excessive use of chemical fertilizers, land is turning barren in many places and yielding less every subsequent season, while wealth per acre is higher in organic mixed cropping than in conventional inorganic, albeit monocropping (Shiva and Singh, 2014). Restoring soil health through balanced use of organic, inorganic and bio-fertilizers and conservation agriculture would be necessary for sustainable productivity growth in agriculture. However, there are several challenges to development of organic farming, namely (a) development of package for organic production technologies; (b) cost and certification supports; (c) incentives for meeting organic standards; (d) extension support for promotion of organic inputs; and (e) creating adequate awareness among farmers as well as consumers about the benefits of organic products. Similarly, conservation agriculture would need to be made small farmer friendly by way of maintaining the yield level and reduction in cost through proper technology mix.

(xi) ***Implementing Innovative Land Reforms:*** In the present socio-political setting, there is hardly any scope for traditional redistribution of land, by taking away land from some farmers and giving away to the landless poor. However, it is possible as well as feasible to implement innovative land reforms such as (a) legalisation of land leasing that would increase the rural poor's access to land, improve the size and viability of small farmers as well as their access to institutional credit, insurance etc. for productivity enhancement, and also encourage the much needed occupational mobility of the farmers; (b) providing secure land right to farm women, as this would help improve their productivity and incomes; and (c) providing homestead plot of 10 to 15 cents to each landless families, *Patta* being issued either jointly in the name of wife and husband or wife only or even single woman, which would enable them not only to build a house if they do not have, but also to take up some economic activities for food and livelihood security, such as cultivation of vegetables and fruits, rearing of goats, sheep, pigs, poultry etc.

(xii) ***Improving Public Investment:*** Improving public investment in agriculture is critical for bridging the gap between actual and potential growth, and in order to achieve 6 per cent agricultural growth, in the coming years against 2 to 4 per cent achieved in the past 15 years, would require massive increase in investment. The public investment in real terms, which is showing a declining trend, must be increased adequately to meet the goal of accelerated agricultural growth. More particularly, increased investment in completing the pending irrigation projects along with modernization of irrigation systems, roads, transportation, markets, godowns, cold storage, power plants, clean energy, climate change, watershed development involving soil and water conservation, renovation of ponds/tanks, desiltation of rivers and creation of new water sources, would be necessary.

(xiii) ***Providing Support to Agricultural Practices that Promote Climate Change:*** The support to agricultural practices which promote climate change adaptation would include designing financial mechanisms to offer incentives to small farmers for safeguarding eco-system services such as watershed protection, carbon sequestration and the protection of bio-diversity. It would be essential to develop mechanisms to regulate and generate rewards for sustainable agricultural practices including payments to farmers for eco-system services or higher prices for agri-products that meet certification. The National Mission on Sustainable Agriculture launched in 2008 rightly aims at transforming Indian agriculture into a climate-resilient production system through suitable adaptation and mitigation measures in the domain of crops and animal husbandry. However, the NMSA requires adequate financial and human resources to achieve its objective. According to the Mission document, 2010, an additional budgetary support of Rs. 1,08,000 crore was required for the period 2011-12 to 2016-17. Recently due to financial constraint the Mission was dovetailed with other programmes, for efficiency and effectiveness.

(xiv) ***Providing Adequate Public Health Care in Rural Areas:*** Providing adequate, low cost, but quality health care facilities in rural areas would be an essential element of strategy to address the issue of farmers' distress. This would help reduce indebtedness and distress of those farm families which either take loans at high rate of interest from money lenders or sale assets to meet the expenses of costly private treatment and hospitalization.

Conclusion

To conclude, there are multiple causes of the present agrarian distress which should be addressed through appropriate situation specific strategies and integrated support system, as indicated in the foregoing discussion. The most important way forward would be to improve incomes of farmers through appropriate technological as well as institutional reforms and occupational mobility.

References

Alagh, Yoginder K. (2013), the Future of Indian Agriculture, National Book Trust, New Delhi.

Bharti, Vishab (2011) Indebtedness and Suicides: Field Notes on Agriculture

Labourers of Punjab, *Economic and Political Weekly*, April 2, Vol. XLVI, No. 14.

Bharatiya Janata Party (BJP, 2014) *Election Manifesto*, 2014, New Delhi.

Bharatiya Janata Party (BJP, 1998), *Election Manifesto*, 1998, New Delhi.

Chand, Ramesh and T. Haque (1997) Sustainability of Rice-Wheat Cropping System in the Indo-Gangetic Plain Region, *Economic and Political Weekly*, Mumbai, Vol. XXXII. No. 13, March 29.

Commission for Agricultural Cost and Prices (CACP, 2014), Report on Price Policy for Kharif crops for 2014-15, Government of India, New Delhi.

Commission for Agricultural Cost and Prices (CACP, 2013), *Report on Price Policy for Rabi crops for 2014-15*, Government of India, New Delhi.

Deshpande, R.S. (2002) Suicides by Farmers in Karnataka: Agrarian Distress and Poverty Alleviation; *Economic & Political Weekly*, 37 (26), June 29.

Dogra, Bharat (2013) *Farmers' Crisis in India and Urgency of Remedial Action*, Mainstream, New Delhi, July 12-18.

Food and Agriculture Organization of the UN (FAO, 2010), *Policies and Institutions to Support Small Holder Agriculture*, Committee on Agriculture, Twenty, Second Session, Rome.

Food and Agriculture Organization of the UN (FAO, 2013): *FAO Statistical Year Book, 2013*, World Food and Agriculture, Rome.

Government of India (2014), *Key Indicators of Situation of Agricultural Households in India*, National Sample Survey, 70th Round, January-December, 2013, New Delhi.

Government of India (2013a), *Press Note on Poverty Estimates*, 2013, Planning Commission, New Delhi.

Government of India (2013b), *Twelfth Five Year Plan*, Vol. II, Planning Commission, New Delhi.

Government of India (2000), *National Agricultural Policy*, Ministry of Agriculture, New Delhi.

Haque, T. (2014), *Food Security and Small Family Farming in Asia-Pacific Countries in Deep-Roots*, Food and Agriculture Organization, Rome, and The Tudor Rose, U.K.

Haque, T. (2012), *Land Policies for Social Inclusion in India in Land Policies for Inclusive Growth*, edited by T. Haque, Concept Publishing Co. (P.) Ltd., New Delhi.

Haque, T. (1996), Sustainability of Small Holder Agriculture, Concept Publishing Co., New Delhi.

ICAR (2007), State Specific Technological Interventions for Higher Agricultural Growth, New Delhi.

Institute for Human Development (IHD, 2014), *India Labour and Employment Report 2014*, New Delhi.

Jodhka, Surinder S. (2012), Agrarian Chnages in the Times of (Neo-Liberal) Crisis, *Economic & Political Weekly*, Supplement, June 30, Vol. XLVII, Nos. 26 and 27.

Mishra, Srijit (2006), Farmers' Suicides in Maharashtra, *Economic & Political Weekly*, April 22, 2006.

Rao, P. Narasimha and K.C. Suri (2006), Dimensions of Agrarian Distress in Andhra Pradesh, *Economic & Political Weekly*, Mumbai, April 22.

Reddy, D. Narasimha and Srijit Mishra (2009), *Agriculutre in the Reforms Regime in Agrarian Crisis in India*, edited by Dr. Narasimha Reddy and Srijit Mishra, Oxford University Press, New Delhi.

Reserve Bank of India (RBI, 2004), *Report of the Advisory Committee on Flow of Credit to Agriculture and Related Activities from Banking System*, under the Chairmanship of V.S. Vyas, RBI, Mumbai.

Shiva, Vandana and Vaibhav Singh (2014), *Wealth per Acre*, Natraj Publishers, New Delhi.

Sidhu, H.S. (2002), Crisis in Agrarian Economy in Punjab, *Economic & Political Weekly*, Mumbai, July 27.

Suri, K.C. (2006) Political Economy of Agrarian Distress, *Economic & Political Weekly*, Mumbai, April 22.

Selvarajan, S., B.C. Roy and N. Suresh (2004), Resource Analysis for Sustaining Water-Food Security, Proceeding of Workshop, NCAP, New Delhi.

United Nations Development Programme (UNDP, 2007/2008), Human Development Report, New York.

Vaidyanathan, A. (2006), Farmers' Suicides and the Agrarian Crisis, *Economic & Political Weekly*, Mumbai, September 23.

Vidyasagar, R. and K. Suman Chandra (2003), *Farmers' Suicides in Andhra Pradesh and Karnataka*, NIRD, Hyderabad.

Vyas, V.S. (2004) Agrarian Distress: Strategies to Protect Vulnerable Sections, *Economic & Political Weekly*, Mumbai, December 25.

2

Distress of Small Holder Agriculture

Challenges of Inclusive Agricultural Development

D. Narasimha Reddy

Introduction

For the last two decades the challenges facing the Indian agriculture have been widely debated. The debate assumes more significance because, in spite of slow but steady structural changes, India remains a predominantly rural country, and agriculture continues to be a major source of livelihood for rural population. Further, presently India, not China, is home for the largest share (833m or 24%) of world's rural population and it will remain so for decades to come (Proctor and Lucchesi, 2012). Not only that most of the rural population depends on agriculture but also most of the poor are in agriculture. While the overall share of cultivators in labour force is declining, the share of small-marginal farmers has been rising, with the average size of operational holdings reaching a very low level. Over the years, with the spread of Green Revolution, there has been shift in the agricultural technology towards more input intensity with most of the inputs, including labour, being purchased from the market where the prices have been on the continuous rise. The high cost and high risk agriculture, without appropriate institutional arrangements and declining public investment in agriculture, poor or no extension services, and declining state support systems, volatile prices, degraded land and water resources, and rising magnitude of non-institutional debt, have reduced the farming community, especially, the small-marginal farmers into a precarious condition driving many a small-marginal farmer to desperation culminating in suicide. Paradoxically, in spite of the years of reports of growing distress among small-marginal farmers, there has been no crisis in agricultural production or production of food grains. On the contrary there has been not only reasonable growth but also more than adequate buffer stocks, and even substantial exports of food grains. Thus in a sense there is no crisis of production in Indian agriculture but it is, indeed, a crisis among producers, especially small-marginal farmers. Questions have been raised as to whether a country like India, heavily dependent on Small holder agriculture, can provide food security by producing adequate output to meet the demand of the growing population? Are small-marginal farms efficient enough to bring about improvement in productivity? To what extent Small holders could earn a decent livelihood from their agriculture? A close analysis of the factors leading to distress of small-marginal farmers shows that they have not failed on efficiency but rather they were let down by the failure of supporting interventions that could have helped them out of the crisis. This chapter is, therefore, not as much concerned about food security as it is with the security of small-marginal producers of food, it is not about reducing their poverty but about as how to ensure a decent living to them. And it is not on market solutions, but on the role of state and community and collective initiatives and interventions.

The chapter is organized into sections. The introductory section is followed by a brief discussion of the changing landholding structure and the emergence of small-marginal farmers as the largest and preponderant group of farming community in India. This section also briefly refers to the growth of agriculture and its changing share in national product. It is followed by the evidence that small-farmers are not lagging in the path of improved farming practices, in spite of such initiatives, much against their resource endowments, becoming source of their distress. The third section discusses these adverse conditions with particular focus on the equity challenges that were sidelined under the neoliberal regime. The fourth and final section attempts to explore the possibilities of interventions in favour of securing sustainable livelihood for small-marginal farmers as much as sustained development of small-farmer agriculture.

Changing Landholding Structure

Past-experience of most of the developed economies has shown that growth and development would be accompanied by structural changes by way of shifts in the share of both the national product and employment from the primacy of agriculture to industry and then to services. With the growth of industry and services, spatially there would be in tandem shifts in population and workforce from rural to urban locations. But in many of the late developing countries, such clear growth-structural linkages are not evident. In India, though there has been reasonable growth for more than six decades, the share of agriculture in the national product did decline substantially from over 50 per cent to less than 15 per cent, but leaving employment in agriculture still at about 50 per cent. Spatially too, almost 70 per cent of India's population lives in rural areas, thus leaving India still as a country predominantly rural and agricultural. Within agriculture, however, there have been substantial changes in the landholding structure.

Table 2.1: Changes in the Size Distribution of Operational Holdings in India

(No. in Millions)

Farm-Size	*1980-81*	*1990-91*	*2000-01*	*2005-06*	*2010-11*
Marginal (< 1 Ha)	50.1 (56.4)	63.4 (59.5)	75.4 (62.9)	83.7 (64.8)	92.8 (67.1)
Small (1-2 Ha)	16.1 (18.1)	20.1 (18.9)	22.7 (18.9)	23.9 (18.5)	24.8 (17.9)
Semi-Medium (2-4 Ha)	12.5 (14.1)	13.9 (13.0)	14.0 (11.7)	14.1 (10.9)	13.9 (10.0)
Medium (4-10 Ha)	8.0 (9.0)	7.6 (7.1)	6.6 (5.5)	6.4 (5.0)	5.9 (4.3)
Large (>10 Ha)	2.2 (2.5)	1.7 (1.6)	1.2 (1.0)	1.1 (0.8)	1.0 (0.7)
All Sizes	88.9 (100.0)	106.6 (100.0)	119.9 (100.0)	129.2 (100.0)	138.3 (100.0)

Source: Agriculture Census 2010-11 (Government of India, 2014).

Note: Figures in parentheses indicate percentage distribution in the respective years.

Table 2.1 shows the changes in the size-wise distribution of operational holdings. Over the past three decades (1981 to 2011) there has been extensive proliferation of marginal holdings with their number and share increasing substantially. A little over two-thirds of operational holdings are marginal. The number of small farmers too increased almost by 50 per cent during the same period, though their share remained constant with a little less than one-fifth of all holdings. For understanding Indian agriculture in all its complexity, it is essential to understand the context and conditions of 85 per cent of all operational holdings which are small-marginal. Table 2.2 shows the changes in the area operated by different size of holdings and the present average size of holdings under each size category. The area operated by both marginal and small holdings together increased from about 26 per cent of the total operated area in 1980-81 to 44.6 per cent in 2010-11. Even by the turn of the twenty first century (2002-03), in 12 out of 27 states marginal and small farmers constituted an overwhelming majority of over 90 per cent of all farmers, and in 17 out of 27 states they also account for more than 50 per cent of the cultivated land (NCEUS, 2008). What is significant is that while the share of small-marginal holdings increased by about 14 per cent, the area operated by them increased by 70 per cent. In fact, there was hardly any rise of the share in the number of small holdings in the 30 year period, but there was about 56 per cent increase in the share of area operated by them. Thus, though there has been proliferation of small-marginal farmers, the average size, though already small, has not further declined in India.

Table 2.2: Changes in Size—Distribution of Operated Area in India

(in million hectares)

Farm-Size	1980-81	1990-91	2000-01	2005-06	2010-11	Average Size in 2010-11 (ha)
Marginal	19.7 (12.0)	24.9 (15.0)	29.8 (18.7)	32.0 (20.2)	35.9 (22.5)	0.39
Small	23.2 (14.2)	28.8 (17.4)	32.1 (20.2)	33.1 (20.9)	35.2 (22.1)	1.42
Semi-Medium	34.6 (21.1)	38.4 (23.2)	38.2 (24.0)	37.9 (23.9)	37.7 (23.6)	2.71
Medium	48.5 (29.6)	44.8 (27.1)	38.2 (24.0)	36.6 (23.1)	33.8 (21.2)	5.76
Large	37.7 (23.0)	28.7 (17.3)	21.1 (13.3)	18.7 (11.8)	16.9 (10.6)	17.38
All Categories	**163.7 (100.0)**	**165.5 (100.0)**	**159.4 (100.0)**	**158.3 (100.0)**	**159.6 (100.0)**	**1.15**

Source: Agriculture Census 2010-11 (Government of India, 2014).
Note: Figures in parentheses indicate percentage distribution in the respective years.

In the Indian context the increase in the number and share of small-marginal holdings, is accompanied by the increase in the area under these size-class holdings. Table 2.3 shows that over a period of 30 years from 1980-81 to 2010-11 the number of marginal farms increased by about 83 per cent, and there was almost the same extent (80%) of increase in the area under these holdings. Similarly, the number of small holdings increased by about 54 per cent during this period accompanied by an increase in the area by about 56 per cent. Thus, quite in contrast with the experience of Africa or East Asia where "small farms today are less than half the size of the small farms of the green revolution era" (Hazell, P., 2012, WDR, 2008). India occupies a unique position where the proliferation of small-marginal farms has not resulted in decrease in the average size of either marginal farms or small farms which remained at about 0.39 hectares and 1.42 hectares for marginal and small farms respectively. This, however, is not to deny the decline in the overall average size of farm from 1.84 hectares in 1980-81 to 1.15 hectares in 2010-11.

Table 2.3: Increase in the Holdings and Area under Small-Marginal (Category) and the Resulting Average Size

Years	*Increase in Number of Holdings (%)*		*Increase in Area under Holdings (%)*		*Average Size (Ha)*	
	Marginal	*Small*	*Marginal*	*Small*	*Marginal*	*Small*
1980-81	—	—	—	—	0.39	1.44
1980-81–1990-91	26.55	24.84	25.13	24.14	—	—
1990-91	—	—	—	—	0.39	1.43
1990-91–2000-01	18.93	12.94	19.68	11.46	—	—
2000-01	—	—	—	—	0.40	1.42
2000 01–2005-06	8.36	5.29	7.38	3.12	—	—
2005-06	—	—	—	—	0.38	1.38
2005-06–2010-11	10.87	3.77	12.19	9.66	0.39	1.42
1980-81–2010-11	82.68	54.04	80.40	56.03	—	—

Source: Government of India (2014) *Agricultural Census of India.*

Efficiency and Equity: Critical Dimensions of Small holder Agriculture

Efficiency of Smallholdings

In the debate on redistributive land reforms a strong case was made in favour of eliminating landlessness by reducing inequality in landholdings by fixing a ceiling on land holdings and acquisition of ceiling surplus land for redistribution among the landless and the land poor. The objection to redistributive land reforms in a relatively land scarce economies was raised on the grounds that redistributive land reforms would drastically reduce the average size into too small landholdings which would affect agricultural productivity. These objections were countered by studies which highlighted that small size of farms were justified both on efficiency and equity considerations. The efficiency of small size was sought to the proved by convincing evidence showing that crop productivity per unit of land declined with an increase in farm size, and hence the inverse relationship between the size of holding and productivity. (For a review of the debate *see* Chand, 2011, Sarthank and Mishra, 2011).

Besides the absence of economies of scale, the efficiency of small farms in the Indian agriculture arises from the fact that there prevails substantial underemployment which enables greater abundance of family labour per hectare of land, and the situation is bound to change with the expanding productive full employment opportunities, which, however, are not likely in the near future. And hence, the situation of more family workers who are motivated and likely to work beyond the considerations of hours of work not calibrated to equate returns, is likely to result in more output per hectare of land and better land productivity. It is also well documented that small farm economies not only contribute high growth but improve incomes of the lower rungs and help in reducing poverty. Small farms also contribute to greater food security, particularly in areas with poor infrastructure, they also spend higher shares of incremental income on locally produced goods and services which can stimulate employment intensive to local non-farm economy and that small farms also provide a base from which rural households can diversify their livelihoods (Hazell, 2011). These advantages of small farms may disappear as per capita incomes rise substantially, economies diversify with growing employment opportunities in higher productive sectors, and if rural wages also tend to rise with rising labour scarcity. But the conditions prevailing in rural India are nowhere near such transition in the near future, and the efficiency of the small farms still hold the prospect of improving rural development as much as small farmer improvement.

Contrary to the doubts expressed by some (Proctor and Lucchesi, 2012) on the efficiency of Small holder agriculture and the consequences on food security, the available evidence shows that the contribution of small-marginal farms to overall agricultural production has been even better than medium and large farms. Further, in spite of the rising input costs and limited access to institutional credit, small-marginal farmers have not been lagging in adopting improved agricultural practices, in spite of being exposed to high risk and vulnerability. Based on the data from the agricultural input survey of 2001-02, Chand *et al.* (2011) have analyzed the comparative performance of small-marginal with the medium-large farmers in terms of certain indicators better practices of agriculture. Table 2.4 shows that in terms extent of irrigated area, fertilizer per hectare, extent of coverage under high yielding varieties (HYVs) and cropping intensity, the performance of small-marginal farms was better than medium-large farms. Only in the case of crop diversification does there appear to be certain bias of small farms in favour of food crops, which may be considered as lag in moving towards high value crops. Apparently, these results show that Small holders are responding to improved agricultural practices as much as or even better than the medium-large farmers, and hence the small farms can ensure agricultural growth and food security. But, the critical question that remains is as to how and at what cost, both social and economics small-marginal farmers are able to participate in the growth of agricultural production. Before we turn to this question, we shall examine further evidence that small farms in India have been efficient in the making more productive use of otherwise scarce land.

Table 2.4: Efficiency Question: Farm Size and Indicators of Improved Agricultural Practices

Indicator	*Marginal*	*Small*	*Semi-Medium*	*Medium*	*Large*	*All*
1. Extent of Irrigation: (%) (2000-01)	51	39	37	37	31	39
2. Fertilizer Consumption (Kg/ha) (2001-02)	175	129	112	95	68	119
3. Area Under HYV (Irrig.) (%) (2001-02)	89	86	85	82	78	85
4. Cropping Intensity (%) (2001-02)	139	128	126	125	121	128
5. Crop Diversification (%) (2001-02)						
Food	76	71	65	66	63	68
Non-Food	24	29	35	34	37	68

Source: Chand *et al.* (2011).

The Small holders' contribution to total value of crop output was marginally higher than 50 per cent nationally although the share of land possessed was about 46 per cent. In seven out of ten high ranking states, the highest value generated per hectare of land are those dominated by marginal and small farmers in terms of land possessed for cultivation (NCEUS, 2008). The value of output per hectare on small farms continues to be higher, in spite of several disadvantages faced by the Small holders in accessing modern inputs because of rising costs due to increasing private sector domination in input markets. Still Small holders are, in general, able to keep their productivity high. Though their capacity to apply higher doses of capital is constrained, they are able to apply higher intensity of labour and other inputs, especially family labour and other owned inputs, which do indicate generally higher index of cropping intensity and diversification. The available evidence from the NSS 50th Round *Situation Assessment of Survey of Farmers (2003)* as shown in Table 2.5, clearly brings out that the value, of output per hectare is higher in marginal and smallholdings compared to medium and large holdings. It is another matter that higher yields on marginal and small farms are far from being adequate to compensate for the disadvantage of too meagre size of holdings, partly also due to high costs of production per unit (NCEUS, 2008).

Table 2.5: Farm Size and Gross Value of Output (2002-03)

Farm Size Class	*Output Value (Rs.) Per ha.*
Marginal	14,754
Small	13,001
Semi-Medium	12,903
Medium	10,655
Large	9,172
All	**12,535**

Source: NCEUS (2008) based on unit level data of *Situation Assessment Survey of Farmers*, NSSO 59th R, Report No. 497, 2005.

Equity Considerations

The overall consensus is that while the emerging agricultural technologies that have influenced farm productivity in the wake of Green Revolution are size-neutral, these are not resource neutral (Rao, 1975). It is here that equity considerations assume a decisive role in making small-farmer agriculture viable and sustainable. True, Small holders have proven capability to use resources efficiently, but their capacity to access resources, inputs, credit, technology and markets are much constrained compared to relatively large farms. It is in this context, the role of the state, on the considerations of equity as much as productivity and overall development of agriculture becomes very critical. It is also pointed out that public investment in small scale subsistence farming may be a more cost-effective alternative to other forms of income transfers and social safety nets. It is said that food aid typically costs more for each metric ton of cereals delivered in rural areas, compared with typical Small holder production costs of the same (Hazell, 2011). Policies that enabled Small holder participation in the Green Revolution included the equitable distribution of land and secure ownership and tenancy rights, alongside scale-neutral technologies input subsidies and large investments in infrastructure like roads and irrigation (Fan *et al.,* 2013). Beginning with 1960s, state in India played a critical role in creating enabling conditions for small-marginal farmers by stepping up public investment in agriculture to create irrigation and other infrastructure facilities, to promote development and distribution of improved varieties of seed, subsidized fertilizers and power supply, and also by regulating marketing. Though substantial benefits of public investment in agriculture was derived by the better-off farmers and better endowed regions, small-marginal farmers also derived certain advantages gradually, and their performance did contribute to agricultural development. Peter Hazell (2011) cites that "the best evidence on return to public investments in small-scale business oriented farming comes from the green revolution era in Asia. Fan, Gulati and Thorat (2008) have estimated the returns to different types of public investment, in agriculture in India over four decade period, beginning with 1960s". "India invested heavily in a small farm led green revolution strategy and the marginal returns to these investments in terms of growth and poverty alleviation were very favourable in the early stages of the green revolution" (Hazell, 2011). With launching of the economic reforms beginning with the early 1990s, there has been set back to the support systems to agriculture, including small-marginal farmers.

Neoliberal Regime: Growth with Distress

Reforms in Agriculture

It may sound a paradox that just as the country reached a level of self-sufficiency in the production of food grains by mid-1980s, the policy priority accorded to agriculture started declining. The strategies of structural adjustment and fiscal consolidation which became the basis of economic reforms launched in early 1990s had influenced every measure of public policy including agriculture and affected institutional support systems based on well-deliberated principles of growth with equity (Reddy and Misra, 2009). The unfinished distributive land reforms were seen as obstacles to incentives, and with it the small-marginal farms came to be considered as unviable entities that somehow had to move towards consolidation to be economical, and market incentives were expected to drive agricultural growth as much as that of the economy as a whole. A series of reforms as a part of the liberalization strategy had direct adverse impact especially on the small-marginal farming community. By the middle of 1990s foreign trade in agriculture was liberalized, which resulted in removing quantitative restrictions and bringing all transactions under tariff system.

The entire process of developing, multiplication and distribution of seeds which was in the public domain of R&D, regulation and agencies like state seed corporations, was opened for 100 per cent private investment since 1991, with the rising cost of seed and extensive penetration of spurious varieties with severe adverse consequences to the farming community. Fertilizer subsidies which were at 3.2 per cent of GDP in 1991-92 was drastically reduced to about 0.7 per cent by 2003-04. In many states, in the name of power sector reforms, tariff to agriculture was raised to recover costs. Similar measures of cost recovery were witnessed in terms of irrigation water charges. As a part of the financial reforms, 'targeted priority lending' or 'directed credit' to agriculture was put on a back burner. While there was shift in the farming systems from subsistence and self-provisioning of inputs to a system of farming mostly based on purchased-inputs, the credit needs of especially small marginal farmers become critical. With the dilution of 'social control' and the priority, the reach of the small-marginal farmers to formal institutional credit suffered a setback pushing them increasingly into the fold of high cost informal credit sources. Even as the risks of new varieties of crops increased, there was no access to crop insurance, especially to Small holders. To cap it all, public investment in agriculture was reduced, forcing farmers to make more private invest even in basic provision like irrigation, and especially growing burden of investment in precarious groundwater sources became a death knell of many a small-marginal farmer (Reddy and Misra, 2009).

The policy shift from major role of state in promoting agricultural development through substantial direct investment in irrigation, infrastructure, extension, R&D, other subsides for inputs like fertilizers and power, interest subvention, and indirect support by way of promoting institutions, regulation of markets and direction to credit institutions to increasingly towards market incentives with state tending to become a facilitator than substantive provider of resources support resulted in the change in the very nature of agricultural development in the post-reform period. What is critical to observe is the fact that the shift in agricultural policy from the primacy of state to that of market as a driven of growth has been happening at a time when there has been increase in the share of Small holder agriculture being increasingly exposed to high risk agriculture based mostly on purchased inputs. The result has been not as much as crisis in agricultural production but shocks and distress suffered by agricultural producers, especially small-marginal farmers.

Table 2.6: Growth of Agriculture and GDP: Recent Experience

Plan	*GDP Growth Rate*	*Share of Agriculture in GDP*	*Agriculture Growth Rate*
Ninth Five Year Plan	5.7	23.4	2.5
Tenth Five Year Plan	7.6	19.0	2.4
Eleventh Five Year Plan	8.0	15.2	3.7

Table 2.6 shows the growth of GDP, the share of agriculture in the national product and the growth performance during the post three-five year plans. Though growth of agriculture during the Nineth and Tenth Five Year Plans was much lower than targeted, the performance of agriculture did keep the foodgrain production and availability reasonably comfortable. In the Twelfth Five Year Plan there was a surge in the growth of agricultural production.

While the available evidence shows that small-marginal farmers have been eager participants in the adoption of improved methods and technologies that have been promoted as a part of the strategy towards faster agricultural growth, the odds were heavily loaded against them due to changing policy as well as resource conditions.

The Distress

Policy changes meant that whatever incursions were made in the past by way of specific programmes in favour small-marginal farmers did not gain the required continuity and strength (NCEUS, 2008). On the top of it the shift to neoliberal reforms further weakened support systems. The declining support systems, rising costs, volatile prices, growing risks combined with increasing pressure on the farmers in terms of meeting basic social needs such as education health needs which were increasingly ending up in private domain resulting in considerable part of domestic expenditure. There were also growing resource degradation especially in the form of depleting ground water resources which were often developed by private farmers' investment based on borrowings made from high cost private informal sources. The new crop varieties and the packages of seed-fertilizer-pesticide intensive packages also exposed farmers to high risks. A combination of these stress factors has been at the root of the unusual phenomenon of farmers' suicides in rural India specially beginning with the late 1990s. There have been extensive studies and reports on the extreme conditions of distress resulting in suicides of farmers (for example Reddy and Mishra, (2009), Mishra 2014, Vasavi 2012, Deshpande and Arora, 2010). Figure 2.1 based on the National Crime Records Bureau (NCRB) data shows that there was clear departure in the suicides mortality rates of farmer and non-farmer males since late 1990s. There are myriad cultural, social and economic factors that precipitate suicides in society. The spurt in the male farmer suicides in contrast with male non-farmers since late 1990s was a clear manifestation of the growing distress in agriculture. Figure 2.1 also shows a decline in male farmers' suicides during 2010-12. The cline is largely on account of an abrupt drop in Chhattisgarh on account of changes in reporting and non-reporting of farmers' suicides for West Bengal in 2012. However, the state of Andhra Pradesh and Maharashtra with large cotton-growing areas and with relatively higher incidence of farmers' suicides, in contrast to the all India trend, show an increasing trend in recent years (Mishra, 2014). Though the NCRB data provides separate data suicide mortality rate (SMR) of farmers and non-farmers and other categories, it is not possible to differentiate the farmers on the basis of the data. For the purpose of such differentiation, reliance has to be made on studies which are designed to investigate the type of farmers falling as victims to suicides. Table 2.7 provides data from one such study based on a sample of four districts

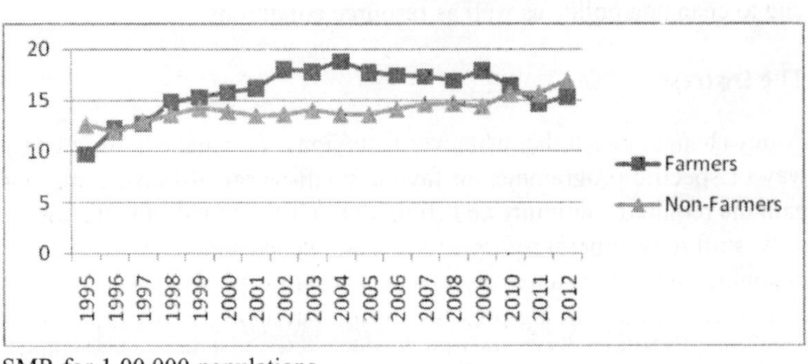

SMR for 1,00,000 populations.
Source: Mishra (2014).

Fig. 2.1: Suicide Mortality Rates (SMR) of Male Farmers and Male Non-farmers in India 1995-2010

with high incidence of suicides in the erstwhile Andhra Pradesh. It shows that most of the farmers committing suicides are small-marginal farmers. A relatively large proportion from medium holdings in Mahabubnagar and Anantapur is because of the relatively larger average size of holdings in these rainfed and drought-prone areas.

Table 2.7: Farm-Size, Social Groups and Suicide Rate (Sample of Four Districts in A.P.)

(Percentage)

	Warangal	*Mahabubnagar*	*Anantapur*	*Guntur*
I. Farm-Size				
Marginal-Small	91.2	69.1	38.2	78.1
Semi-Medium and Medium	8.5	30.1	50.1	21.8
Large	Nil	Nil	Nil	Nil
II. Social Group				
SC	15.8	19.1	9.6	13.4
ST	4.4	17.1	1.9	6.2
OBC	67.5	52.4	39.4	51.2
OC	12.3	11.4	49.0	27.2
III. Ownership Category				
Pure Tenant	7.5	Nil	3.9	8.3
Pure Owner	63.5	80.7	70.6	60.9
Owner-*cum*-Tenant	25.2	18.3	24.5	30.8

Source: Revathi (2007) in Reddy and Mishra (2009).

Main Sources of Distress

Over the years, the costs of agricultural production have been on the increase, resulting in growing capital-output ratio of agriculture. One of the main reasons for the distress in agriculture is the stagnant or declining share of public sector in the gross capital formation (GCF) in agriculture and increasing dependence of agriculture growth on private investment. Table 2.8 shows the gross capital formation (GCF) by public and private sectors in Indian agriculture over the past three decades. For the first fifteen years after the launching of economic reforms in 1991, public investment in agriculture took a big hit and remained at a low level of less than 3 per cent of agricultural GDP until 2004-05, much lower than what it was a quarter century before in 1980-81. Though there was a marginal increase in the years following 2004-05 more a response to political pressures than to bring about any substantive interventions that would make a difference to agricultural production! Public investment in agriculture never reached the pre-reform level of 4 per cent of agricultural GDP.

Table 2.8: Gross Capital Formation (GCF) in Indian Agriculture and Allied Activities

Year	GCF in Agriculture as Per cent of GDP from Agriculture (%)		
	Public Sector	Private Sector	Total
1980-81	4.0	5.2	9.2
1990-91	2.1	8.1	10.2
2000-01	1.8	8.5	10.3
Tenth Plan			
2002-03	2.0	12.2	14.2
2003-04	2.3	10.1	12.4
2004-05	2.9	10.6	13.4
2005-06	3.5	11.2	14.6
2006-07	3.7	11.2	14.9
Eleventh Plan			
2007-08	3.6	12.6	16.1
2008-09	3.1	16.3	19.4
2009-10	3.4	16.4	19.8
2010-11	3.0	17.0	20.1

Source: 1. For 1980-81 to 2000-01 (Reddy and Mishra, 2009).
2. For 10th and 11th Plans: *Twelfth Five Year Plan 2012-2017*, Vol. 2.

But during the post-reform period the share of private investment as a proportion agricultural GDP almost doubled from about 8 per cent to 17 per cent. Thus, the major driver of agricultural growth in the post-reform period is private investment. Since there is hardly any business or corporate entity in agriculture private investment in agriculture is essentially by the farming community, and since more than 50 per cent of agricultural production is from small-marginal farms, at least 50 per cent of the growing private investment could be from this latter group. How do these small-marginal farmers access such large resources? From which sources? At what cost?

Heavy reliance on private investment for agricultural growth in the context of rising costs, and continued reliance for a substantial part of the credit on private non-institutional sources placed heavy burden on the farming community, especially on the small-marginal farmers. Since many public goods like irrigation facilities, research, development and distribution costs of improved seeds, extension facilities have increasingly to be procured through private provisioning. Though agricultural credit is part of priority sector with regulated low interest, substantial part of agricultural credit continues to be from the private informal sources with high interest rates. Table 2.9 shows that the low access to institutional credit and heavy dependence on money lenders and other private sources was contained through certain radical banking reforms like nationalization and by earmarking priority lending to agriculture in 1970s. There was trend increase in the share of institutional credit to agriculture from 31.7 per cent in 1971 to 66.3 per cent in 1991, but only to be reversed since then due to certain financial sector reforms. In spite of growing political pressure in the wake of agricultural distress, even in 2012-13, the non-institutional sources accounted for as much as 41.2 per cent of outstanding debt of farmers.

Table 2.9: Source-wise Share of Borrowings of Cultivator Households

(Percentages)

Source of Credit	*1951*	*1961*	*1971*	*1981*	*1991*	*2003*	*2013*
1. Institutional	**7.3**	**18.7**	**31.7**	**63.2**	**66.3**	**57.7**	**59.8**
(a) Cooperative Banks	3.3	2.6	22.0	29.8	30.0	19.6	14.8
(b) Commercial Banks	0.9	0.6	2.4	28.8	35.2	35.6	42.9
2. Non-Institutional	**92.7**	**81.3**	**66.3**	**36.8**	**30.6**	**42.3**	**41.2**
(a) Moneylenders	69.7	49.2	36.1	16.1	17.5	25.7	25.8
Total	**100.0**	**100.0**	**100.0**	**100.0**	**100.0***	**100.0**	**100.0**

*Unspecified: 3.1

Source: S.L. Shetty (2009) in Reddy and Mishra (2009) and NSS 70th Round (2014) *Key Indicators of Situation of Agricultural Households in India.*

The situation of marginal and small farmers is much worse. The access to institutional still continues to be highly positively correlated with the size of landholdings. Table 2.10 shows that even in 2012-13, almost half of the credit was from non-institutional sources while it was more than one-third in the case of small farmers. In spite of widespread distress and wider recognition that most of the farmers' suicides are among small-marginal farmers, and often because of their involvement in borrowing at very high rates of interest from private sources. The failure to reach the small-marginal farmers with institutional credit is clearly officially recognized when the Twelfth Five Year Plan observes: "*Credit flow doubled in the Eleventh Plan but mainly by credit deepening with little increase in farmer coverage and still leaving 60 per cent of farmers without institutional credit*" (*Twelfth Five Year Plan*, Vol. 2, p. 22, 2012. Emphasis added). Beginning with 2004-05, there was a commitment by the Union Government to double institutional credit to agriculture. But the increase took the form rising share of indirect credit which meant moving from production *per se* into post-production functions. In the 2000s, banks increasingly financed sub-sectors in agriculture that were large scale, commercial and capital intensive and export oriented. The major beneficiaries of the revival of agricultural credit in the 2000s were corporate groups and others indirectly involved in agriculture and credit did not reach large number of small and marginal farmers (*Ram Kumar*, 2014).

Table 2.10: Distribution of Outstanding Loans by Source across Size Class of Land Possessed, 2003 and 2013

(Percentages)

Size class of land possessed (in hectare)	*Institutional Agencies (Banks, Co-op. Society and Government)*		*Non-Institutional Agencies (Moneylenders* and Others)*		*All*
	2003	*2013*	*2003*	*2013*	
< 0.01	22.6	14.9	77.4	85.0	100.0
0.01–040	43.3	46.9	56.7	53.0	100.0
0.40–1.00	52.8	53.2	47.2	46.8	100.0
1.01–2.00	57.6	64.8	42.4	35.1	100.0
2.01–4.00	65.1	67.5	34.9	32.5	100.0
4.01–10.00	68.8	71.5	31.2	28.5	100.0
10.00+	67.6	78.9	32.4	21.0	100.0
All sizes	**57.7**	**59.8**	**42.3**	**41.2**	**100.0**

Note: *Includes both professional and agriculturist moneylenders.
Source: Situation Assessment Survey of Farmers, NSSO, 59th R., Report No. 498 and NSS 70th Round *Key Indicators of Situation of Agricultural Households in India (2014).*

Table 2.11 presents distribution of outstanding debt by rate of interest charged by institutional and non-institutional sources. In 2002-03, 99 per cent of institutional credit carried interest rates of less than 20 per cent but as much as 343 per cent of institutional debt in the range of 15 to 20 per cent rate of interest was somewhat high side. This might be a freak phenomenon which can occur only if there were to be widespread default and fines jacking up the actual interest paid. Such a situation also raises questions about the priority sector lending to agriculture with targeted lower rates of interest. The figures for 2012-13 appear more plausible with 90 per cent of institutional credit falling in the range of interest rates of 15 per cent or below. And 26 per cent of institutional credit to agriculture with an interest rate range of 6 to 10 per cent also appears to be on the lines of policy guidelines. The problem is the persistence of non-institutional sources of credit which continues to constitute 41 per cent of total credit accessed by the farmers, and 68 per cent of it carried interest rates ranging above 20 per cent and 34 per cent was at the rates of about 30 per cent of interest. Given the fact that more than 50 per cent of small-marginal farmers depend on non-institutional sources, the cost debt for this class of borrowers becomes very heavy. Added to this their earning from agriculture are too meagre to enable them to service debt at such high rates of interest. The increasing costs of agriculture force them to seek more credit which comes from informal sources with high interest

Table 2.11: Distribution of Debt Outstanding by Rate of Interest for Institutional and Non-Institutional Sources 2003 and 2013

% Distribution of Outstanding Cash Debt				
Rate of Interest Class (%)	*Rural—All India*			
	Institutional		*Non-Institutional*	
	2003	2013	2003	2013
Nil	1.0	0.8	18	18.3
Less than 6	2.0	7.0	2	2.3
6–10	4.0	26.0	1	0.4
10–12	9.0	12.9	1	0.7
12–15	48.0	42.6	1	4.1
15–20	34.0	7.3	3	5.6
20–25	1.0	2.1	33	33.9
25–30	0	0.1	0	0.6
30 & Above	0	1.0	40	34.1
All	**100**	**100**	**100**	**100.0**

Source: NSSO, All-India Debt and Investment Survey, NSS 59th Round (January-December 2003), Report No. 501 (59/18.2/2), New Delhi, 2005, and NSS 70th Round *Key Indicators of Situation of Agricultural Households in India (2014).*

rates, rendering the debt burden beyond their meagre means of income, that triggers desperation driving some of them to end their life. But those who survive with such high burden also live a precarious life, which is clear from the facts of their income and expenditure.

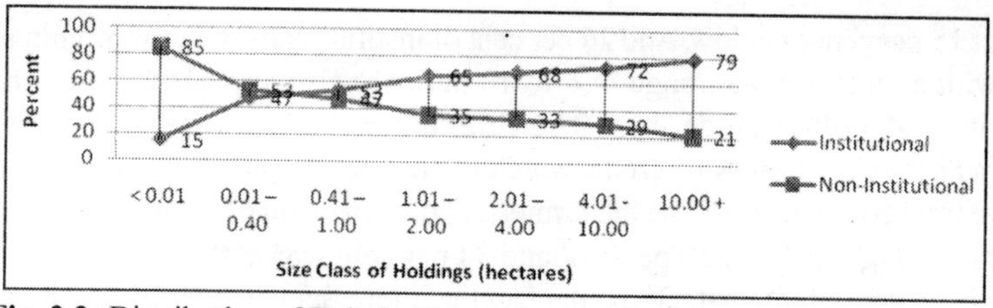

Fig. 2.2: Distribution of Debt by Source across Size Class Holdings
Source: NSS 70th Round, *Key Indicators of Situation of Agricultural Households in India*, New Delhi, MoSPI (2014).

Erosion of Support Systems

Even as the spread of new variety of crops, hybrid and even some GM crops made entry into agriculture making it highly risky, there has been comprehensive failure of extension services, a conclusion echoed by a number of commissions and committees looking into the state of agriculture (NCEUS, 2008). Even as recently as 2012-13 only about 10 per cent of the farmers had any useful access to services of public agencies like extension agent, Krishi Vigyan Kendra (KVK) or agricultural university/college. Media sources like radio or television, and progressive farmers stood better with each one accessed by about 20 per cent of farmers (NSSO-70/33, 2014).

Market and remunerative prices continues to be a major problem for small-marginal farmers. The marketing surpluses of small-marginal farmers are low, their withholding capacity is meagre and the transaction costs are high. Because of their economic vulnerability and low bargaining power, they and up selling at very low prices. The increasing liberalization of trade and the resulting volatility in market prices and poor coverage of these weaker farmers under MSP add to their problems. For instance even in 2012-13, even in the case of major crops like paddy and wheat the access to MSP appear to be confined to a few farmers. Only about 32 per cent of paddy farmers were aware of MSP, but only 10 per cent of farmers sold their produce to procurement agency. In the case of wheat farmers, about 39 per cent were aware of MSP, but only 16 per cent sold to the agency (NSSO-70/33, 2014). On the access to input markets, the weakened public extension support system has increased the dependence of the farmer on private input dealer who often act as extension adviser, output market linked provision of inputs on credit which often result not only the high cost of the tied inputs, but also inappropriate choice of crops and inputs (NCEUS, 2008). The emerging developments in the field of agricultural marketing in the form of contract farming, marketing chains linked to super markets involve new set of transaction costs like grading, meeting food standards, etc. which Small holders are not able to bear without proper regulatory and public training activities.

With the introduction of modern varieties of crops the risks of crop failure due to pest have been on the increase. The weather induced crop failures have become more frequent and widespread. Unlike in the case of traditional crops based on mostly on farmers' own seeds, manure and labour, the new varieties involve heavy doses of investment involving mostly purchased inputs and crop failure would become an episode of economic distress particularly to Small holder cultivation. Insurance becomes an essential protection. However, crop insurance has made little headway except in the case of transactions linked to cooperative or commercial bank credit. In the first instance the reach of institutional credit to small-marginal farmers, as shown earlier, is relatively limited. Second, the credit linked insurance often ends up as insurance of the credit of the institution rather than compensation realized for crop failure by the farmer. The overall reach of crop insurance continues to be dismal, with about 95 per cent of paddy and wheat crops, and about 85 to 90 per cent of high risk cotton crop not being covered under insurance even as recently as 2012-13 (NSSO-70/33, 2014).

In the past few decades of the regime of liberalization and economic reforms, two diametrically opposite trends have emerged in agriculture. On the one hand, there has been pervasive entry of new varieties of crops heavily dependent on purchased inputs resulting in high costs as much as high risks, and on the other extensive erosion of public investment and public support systems for agriculture. Under these circumstance agriculture, especially for small-marginal farmers turns out to be an activity carried on under a series of threats and uncertainties. To sum up: "In large part they [threats] arise from market failure, themselves amplified by the policy retreat from intervention that has left the private sector operating within markets as the main actor in input supply, financial services, marketing and even in technical advice and innovations. If Small holders are to survive and prosper, then they have to find ways to meet new demands in supply chains, and to obtain inputs, credit and technical knowledge from private agents at competitive prices with large scale farms" (Hazell *et al.*, 2006).

Precarious Existence

For small-marginal farmers agriculture continues to be a precarious anchor as a source of income. While it cannot be the sole source, augmenting it as a stable and substantive source becomes essential to overcome vulnerability. Table 2.12 gives details of average monthly income from different sources and consumption expenditure of agricultural households across the size classes for the year 2012-13. The income is grouped into three broad sources, viz. agriculture including animal husbandry, non-agriculture business and wage-salary from agriculture or non-agriculture. What is significant to note is that for small-marginal farmers, net income from agriculture is less than their consumption expenditure. In other words small-marginal farmers cannot meet their consumption expenditure solely from agriculture and therefore have to depend on other sources like wage-

Table 2.12: Average Monthly Income from Different Sources and Consumption Expenditure of Agricultural Households (2013)

Size Class of Land Possessed (ha)	*Net Income from Agriculture including An.Hus. (Rs.)*	*Net Income from Non-farm Business (Rs.)*	*Wages/Salary Agricultural/ Non-Agricultural (Rs.)*	*Total Income (Rs.)*	*Total Consumption Expenditure (Rs.)*
<0.01	1211 (26.6)	447 (9.8)	2902 (63.6)	4561 (100)	5108
0.01-0.40	1308 (31.5)	459 (11.1)	2386 (57.5)	4152 (100)	5401
0.41-1.00	2774 (52.9)	462 (8.8)	2011 (38.3)	5247 (100)	6020
1.01-2.00	5027 (68.4)	593 (8.1)	1728 (23.5)	7348 (100)	6457
2.01-4.00	8520 (79.4)	554 (5.2)	1657 (15.4)	10730 (100)	7786
4.01-10.00	16744 (85.3)	861 (4.4)	2031 (10.3)	19637 (100)	10104
10.00+	38307 (92.6)	1770 (4.3)	1311 (3.2)	41388 (100)	14447
All	**3081** **(47.9)**	**1275** **(19.8)**	**2071** **(32.2)**	**6426** **(100)**	**6223**

N.B: Figures in parentheses indicate source-wise% distribution of income.

Source: NSS 70th R (2014) *Key Indicators of Situation of Agricultural Households in India.*

salary or non-agricultural activities. The land-poor and the near landless marginal farmers with less than 0.40 hectares of land have their main source of income from wage-salary earnings. To meet bare consumption expenditure, even one-fourth to one-third of the upper end of small farmers has to fall back on wage-salary income in addition to the net agricultural income.

The very description of a large majority of marginal land holders as agricultural households becomes tenuous. Table 2.13 shows that only about 68 per cent of agricultural households have agriculture as the principal source of income. For a majority of marginal farmers with landholdings of 0.40 hectares or less, the main source of income is wages/salaries and to an extent other non-agricultural activities. Access to non-agricultural activities to small-marginal farmers appear to be very meagre to emerge as an important source of main income. Reading Tables 2.12 and 2.13 together shows that the average monthly earnings from non-agricultural activities are too meagre, and these activities being main source of income is also in the case of very small proportion of households. The so-called non-agricultural enterprises accessible to small-marginal farmers as of now appear to be mostly of low productive activities. The major question is how to make Small holder agriculture from precarious subsistence level to a reasonable level or stable living? (Fan *et al.*, 2013).

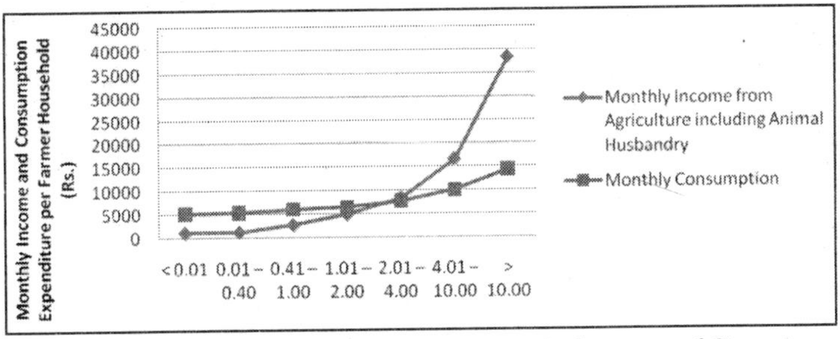

Source: NSSO (2014), NSS KI (70/33): *Key Indicators of Situation of Agricultural Households in India.*

Fig. 2.3: Average monthly Income from Agriculture and Consumption Expenditure of Agricultural Households (2012-13) All India.

Towards Equitable and Enabling Environment

As discussed earlier the efficiency of small farms may be not in doubt but the historical disadvantage of small farms in accessing credit, inputs,

Table 2.13: Size-Class and Principal Source of Income of Agricultural Households (2013)

(percentages)

Size-Class of Land	*Agriculture**	*Non-Agricultural Enterprises*	*Wages/ Salaries*	*Others*	*All*
<0.01	27.2	10.8	56.4	5.5	100
0.01-0.40	48.1	7.5	35.2	9.3	100
0.41-1.00	72.4	3.6	20.0	4.1	100
1.01-2.00	6.4	3.2	8.6	1.8	100
2.01-4.00	89.4	1.6	7.1	1.8	100
4.01-10.00	91.1	0.9	5.9	2.0	100
10.00+	96.4	1.8	1.7	1.0	100
All	**68.3**	**4.7**	**22.0**	**5.1**	**100**

*Agriculture includes cultivation, animal husbandry and other allied activities.
Source: NSS 70th Round (2014) *Key Indicators of Situation of Agricultural Households in India.*

technology and markets have aggravated with the removal or erosion of the many state agencies that served agriculture prior to market liberalization programmes under the reforms regime. he widespread conditions of distress experienced by small-marginal farmers over the recent decades has attracted not only wider public attention, but also resulted in a large number of in-depth studies as much as a number of committees and commissions at the state as well as the national level focusing on problems facing agriculture in general and small-marginal farmers in particular. There is a growing unanimity emerging from these studies and inquiries that for the sustainable and fair development of small-marginal farmers, two basic commitments, one in terms of policy and the other in terms of institution building, are needed as the premises for any intervention to succeed. The policy commitment is towards the role of state in designing and investing in creating conditions that would enable the small-marginal farmers not only to access institutional credit and insurance, fair markets, appropriate technology but also provide a certain measure of social security and quality health and education. The promotional effort should be towards legislative measures and other incentives that would facilitate formation and institutionalization of small-marginal farmers' collectives or groups that would enable them to overcome the disadvantages associated with small-size and atomistic nature of their existence. The positive outcomes of such institutions are well recognized (Rouse, 1996). In its report on special programme for marginal and small farmers, the NCEUS devoted a separate chapter for "group approaches towards strengthening marginal and small farmers" (Chapter 6, NCEUS, 2008). There are also initiatives to bring Small holders under commodity based associations to overcome their scale disadvantages on the lines of milk cooperatives (Gowda, 2015). Almost all the five Working Groups of the Twelfth Five Year Plan on agriculture not only devoted a section on the need to pay special attention to the problems of small-marginal farmers but emphasized the need for promotion of small-farmers' groups or collectives. And the Steering Committee on Agriculture of the Plan brought this institutional requirement as one of the core recommendations when it observed: "Intervention most potent is support to group farming and subsidies in government programmes by giving preference to group activity" (Planning Commission, 2012). If credit, insurance, and marketing are the two critical areas of concern for farmers, the role of small farmers organizations would be significant in accessing institutional credit through collective or joint liability, in securing storage and cash advance against the produce that awaits fair price.

There is need to redesign and enhance the reach of public investment and agricultural subsidies with a focus on small farms, addressing especially their transport and storage needs. Subsidies could be restructured towards water conserving, and soil enriching practices, using sustainable agricultural practices like the system of rice intensification (SRI) and system of crop intensification (SCI), which are bound to orient the flow of these resources more in favour of small-marginal farmers who are ideal adopters of these practices. One of the biggest drain on the resources of the small farmers is the dependence on informal sources of credit at very high cost that draws them into debt trap, one of the main causes for their distress. Besides special measures for debt relief and special priority (NCEUS, 2008), there is need to increase the flow of institutional credit by orienting it to group lending, so as to incentivize the formation of collectives or groups of Small holders and tenants. The existing initiatives in this regard requires to be scaled-up nation-wide. There are rapid changes in the marketing systems especially in the high value crops like vegetables, fruits and spices. The entry of contract farming and value chains not only need provision of training and promotion of small farmer groups but also creation of local storage and grading centres, packaging houses, small cold storages, drying and quick freezing facilities with a regulatory framework that ensures a better bargaining position for the small producers.

In recent years, there have been some favourable trends in agricultural R&D. India's share in the global publicly funded R&D in agriculture increased from 3.5 per cent in 1981 to 5 per cent in 1991 to 8.1 per cent in 2000 (Pardey *et al.*, 2006). This positive trend should be revived with orienting R&D that is relevant to Small holders and also dryland regions where most of the poor Small holders live. "There is a huge, underutilized potential to link farmers' traditional knowledge with science-based innovations through favourable institutional arrangements" (FAO, 2011). Such appropriate knowledge would not only be ideally suited for small farms but also widely sustainable. Neglect of the small-marginal farmers' problems with an assumption that these are non-viable and that with development, large hoards of small-marginal farmers that would leave agriculture that would augment the size of farming, would prove disastrous not only for Small holders but for the development of agriculture as a whole. As pointed out earlier, any such transition away from Small holder agriculture is not in the near future. What is relevant is to strengthen rural non-farm activities, since most of the marginal farmers, and substantial number of small farmers depend on multiple activities for their livelihood. There have been a number of studies which have drawn attention to the critical rôle of rural non-farm employment play in Small holder household diversification especially at the transitional stage (Onchan, 2004) and in this rural services and construction play much greater role than manufacturing (Carletto *et al.,* 2007, Haggblade *et al.,* 2002). And the recent evidence in India also proves the importance of rural non-farm activities in strengthening the viability of small-marginal farming, (NSSO, 2014; Coppard, 2001). More critical in the transition stage is to ensure better health and educational facilities to the Small holder households that would enable them not only improve agricultural productivity but also help these households to experience smooth transition to non-farm activities. Thus to ensure better livelihood for small-marginal farmers and to secure sustainable agriculture state support systems for these groups should go beyond creating institutions and facilities in agriculture, and encompass their basic needs of education, health and a certain measure of social security as well.

References

Chand, Ramesh, P.A. Lakshmi Prasanna and Aruna Singh (2011) "Farm Size and Productivity: Understanding the Strengths of Small holders and Improving Their Livelihoods", *Economic and Political Weekly*, Vol. XLVI, Nos. 26 and 27, 5-11.

Coppard, Daniel (2001) "The Rural Non-Farm Economy in India: A Review of the Literature", Natural Resource Institute, Report No. 2662, DFID and World Bank.

Eastwood, R., Lipton, M. and Newell, A. (2009) "Farm Size" in Pingali, P. and R. Evenson (eds.), *Handbook of Agricultural Economics Volume 4*, Amsterdam, Elscvier.

Fan, S., Gulati, A. and Thorat, S. (2008) "Investment, Subsidies and Pro-Poor Growth in Rural India", *Agricultural Economics*, Vol. 39, No. 2, 163-170.

Fan, Shenggen, Peter Hazell, and Sukhadeo Thorat (2000) "Government Spending, Growth and Poverty in Rural India" *American Journal of Agricultural Economics*, Vol. 82, No. 4, November, 1038-51.

Fan, Shenggen., Joanna Brzeska, Michiel Keyzer and Alex Halsema (2013) "From Subsistence to Profit: Transforming Small holder Farms", International Food Policy Research Institute, Washington.

Gowda, K. Narayana (2015) "Profitability and Sustainability of Small and Marginal Farmers through Integrated Farming System and Commodity Based Associations", in Devi Prasad *et al.* ed. *Op. cit.*, 42-52.

Gupta, H.S. (2015) "Challenges and Strategies in making Small and Marginal Farm Holdings Sustainable and Profitable" in *Agricultural Management*, Edited by Devi Prasad, J., B. Gangaiah and K. Suman Chandra, Centre for Good Governance and BSP Publications, Hyderabad, 19-30.

Hanumantha Rao, C.H. (1975) *Technological Change and Distribution of Gains in Indian Agriculture*, New Delhi, MacMillan.

Hazell, Peter (2012) "Is Small-Farm Led Development Still a Relevant Strategy for Africa and Asia?", (Accessed on 30 October, 2014).

Hazell, Peter (2011) "Five Big Questions about Five Hundred Million Small Farmers", IFAD Conference on New Directions for Small holder Agriculture, 24-25 January, 2011, IFAD, Rome (Accessed on 30 October, 2014).

Mishra, Srijit (2014) "Technology and Development through an analysis of suicides mortality rates for farmers and non-farmers across states in India 1995-2012" Discussion Paper, London School of Economics and Political Sciences, London.

National Commission on Farmers (2004), *Serving Farmers and Saving Farming*, New Delhi, Ministry of Agriculture, Government of India.

NSSO (2014) *Key Indicators of Situation of Agricultural Households in India*, NSS 70th Round, NSS, KI (70/33), New Delhi, MoSPI.

NSSO (2014) *Key Indicators of Debt and Investment in India*, NSS 70th Round, NSS, KI (70/18.2).

NCEUS (2008) *A Special Programme for Marginal and Small Farmers*, New Delhi, National Commission for Enterprises in the Unorganised Sector.

Pandey, S.M. and J.S. Sodhi (1981) *Small Farmers' Development Programme*, New Delhi, Shri Ram Centre for Industrial Relations and Human Resources.

Pardey, Philip G., Nienka Bientema, Steven Dehmer and Stanley Wood (2006) *Agricultural Research: A Growing Global Divide?* International Food Policy Research Institute, Washington, August.

Planning Commission (2013) *Twelfth Five Year Plan*, Vol. 2, *Chapter 12: Agriculture*, New Delhi, Planning Commission.

Proctor, Felicity, and Valerio Lucchesi (2012) *Small-Scale Farming and Youth in an Era of Rapid Rural Change*, London, IIED.

Ramkumar, R. and Pallavi Chavan (2014) "Bank Credit to Agriculture in India in the 2000s: Dissecting the Revival", *Review of Agrarian Studies*, Vol.

Reddy, D. Narasimha, and Srijit Mishra (eds.) (2009) *Agrarian Crisis in India*, New Delhi, Oxford University Press.

Rouse, J. (1996) "Organising for Extension: FAO Experiences in Small Farmer Group Development", Sustainable Development Department, Rome, FAO.

Sarthak, Gaurav and Srijit Mishra (2011) "Size-Class and Returns to Cultivation in India: A Cold Case Reopened", Working Paper—2011-027, Indira Gandhi Institute of Development Research, Mumbai.

Thapa, Ganesh and Raghav Ghia (2011) "Small holder Farming in Asia and the Pacific: Challenges and Opportunities" IFAD Conference on New Directions for Small holder Agriculture, 24-25 January, Rome, IFAD.

World Bank (2008) *World Development Report*, *Agriculture for Development* "Chapter 6: Supporting Small holder Competitiveness through Institutional Innovations", Washington DC, World Bank.

World Bank (2008) *World Development Report 2008*, Washington DC, World Bank.

3
Agrarian Distress and Farmers' Suicides

K. Suman Chandra and *E.V. Prakash Rao*

Background

Agriculture, the most important sector in Indian economy, is presently at the crossroads. Although the share of agriculture in the national GDP had decreased to 18 per cent during 2007 and subsequently dipped as low as to 14 per cent in 2011, yet this sector continues to provide livelihood support to more than two-thirds of the total population in India. The structural discrepancy among the sectors, i.e. agriculture, manufacturing, trade and industry depicts a large and growing difference in inter-sectoral relative productivities with productivity of workers in agriculture being one-fifth (20%) of that of those in non-agricultural occupations in 2004-05 (Table 3.1).

Table 3.1: Share of Agriculture in GDP and Employment

Year	*Share of Agriculture in GDP at 1999-2000 Prices (%)*	*Share of Agriculture in Employment—UPSS (%)*	*Ratio of Worker Productivity in Agriculture to Non-Agriculture*	*Ratio of Worker Productivity in Non-agriculture to Agriculture*
1972-73	41.0	73.9	0.20	3.92
1993-94	30.0	63.9	0.24	4.12
1999-00	25.0	60.2	0.22	4.55
2004-05	20.1	56.5	0.20	5.12

Note: GDP denotes Gross Domestic Product and UPSS denotes Usual Principal and Subsidiary Status.

Source: Central Statistical Organization (CSO), *National Accounts Statistics*, Various Years and National Sample Survey Organization (NSSO), *Employment and Unemployment Situation in India*, Various Rounds.

The slowdown in performance of the agriculture sector during the past couple of decades is a cause of concern in the overall agenda for food and income security. The aggregate productivity has been low with growth of production decelerating rapidly and, more importantly, the sector is exposed to several risks and uncertainties. More than 80 per cent of the farmers, mostly with marginal and small holdings, depend on agriculture as the primary activity and are highly vulnerable to multiple risks. The down trend in growth of total factor productivity (a measure of contribution of technology) has increased the vulnerability and induced unsustainable income. At the same time, the escalating cost of production and un-remunerative output prices have made farming increasingly unviable.

State of Indian Agriculture

The state of agriculture in India resembles a situation in which yield-increasing technology without development of adequate infrastructure (like credit, power, extension, input supplies, marketing, irrigation) is superimposed on an agrarian structure where land institutions are not rationalized. It is now an established fact that the spread of new agricultural technology has not only widened the regional imbalances but also disparities among the different class-size of farmers. Small farms still lag far behind in adopting new technology and realize only reasonable margins in productivity and profits. The problem is more intensive in semi-arid zones. Adoption of market-driven commercial and mono-cropping culture has led to severe stress on local resource base. The declining capital formation and public investment in agriculture, share of agriculture in Gross Domestic Product (GDP), the adverse terms of trade, falling return-cost ratios etc. are some of the disturbing trends which emerged on the agrarian scene in India. India that can boast about self-sufficiency in food production now is fraught with severe problems in terms of sustaining the achievements. Therefore, agriculture in India is passing through a "Severe Crisis" (Government of India, 2007).

Rallying behind the compelling rhetoric of a *Bharat*-India divide[1], the decade of 80s saw the peasants "flexing their political muscle" in the great spectacle of the "New Farmers' Movements" when tens of thousands of farmers marched to New Delhi agitating successfully, for higher agricultural prices and ever greater subsidies. However, over a period of time, there has been a definite marginalization of the agrarian interest in National Policy more specifically since the decade of 90s. The stagnation in agriculture is translating into a constellation of risks and problems for the farmers as well as suicides by distressed farmers in several regions.

Increasing Marginalization of Peasantry

Between 1960 and 2003 the number of holdings doubled from 51 million to 101 million while the area operated declined from 133 million hectares to 108 million hectares (Table 3.2). This has led to a sharp decline in the average size of the holdings leading to increasing number of small and marginal farmers. The high burden of labour force has, in addition, been falling on a slowly contracting cultivable land area.

Table 3.2: Certain Key Characteristics of Operational Holdings

	1960-61	*1970-71*	*1981-82*	*1991-92*	*2003*
	(17th)	*(26th)*	*(37th)*	*(48th)*	*(59th)*
1 Number of operational (millions)	50.77	57.07	71.04	93.45	101.27
1.1 Percentage increase	—	12.4	24.5	31.5	8.4
2 Area operated (million hectares)	133.48	125.08	118.57	125.10	107.65
3 Average area operated (hectares)	2.63	2.20	1.67	1.34	1.06

Source: *NSSO, Some Aspects of Operational Land Holdings in India,* Various Rounds cited in Report of the Expert Group on Agricultural Indebtedness, New Delhi, Ministry of Finance (2007).

The proportion of marginal landholders has increased from 39.1 per cent in 1960-61 to 71 per cent in 2003, and among them they only operate 22.6 per cent of the land (Table 3.3). This continuing skewed pattern of land ownership reflects the lack of earnest land reforms. Increasing marginalization forces the farmers into sharecropping and renting additional land. This leads to difficulties like insecurity of lease, increasing costs and inadequate returns from production, difficulties in accessing credit etc. (Assadi, 1998).

Declining Gross Capital Formation and Rural Development Expenditure

Capital formation is important for the growth of any sector. In agriculture this takes the form of irrigation, rural infrastructure etc. The share of agricultural Gross Capital Formation (GCF) in total GCF fell from 16.1

Table 3.3: Changes in the Size Distribution of Operational Holdings and Operated Area: 1960-61 to 2002-03

Category of Holdings	*Percentage of Operational Holdings*					*Percentage of Operated Area*				
	1960-61	*1970-71*	*1981-82*	*1991-92*	*2003*	*1960-61*	*1970-71*	*1981-82*	*1991-92*	*2003*
	(17th)	*(26th)*	*(37th)*	*(48th)*	*(59th)*	*(17th)*	*(26th)*	*(37th)*	*(48th)*	*(59th)*
Marginal	39.1	45.8	56.0	62.8	71.0	6.9	9.2	11.5	15.6	22.6
Small	22.6	22.4	19.3	17.8	16.6	12.3	14.8	16.6	18.7	20.9
Semi-Medium	19.8	17.7	14.2	12.0	9.2	20.7	22.6	23.6	24.1	22.5
Medium	14.0	11.1	8.6	6.1	4.3	31.2	30.5	30.1	26.4	22.2
Large	4.5	3.1	1.9	1.3	0.8	29.0	23.0	18.2	15.2	11.8
All Sizes	**100.0**	**100.0**	**100.0**	**100.0**	**100.0**	**100.0**	**100.0**	**100.0**	**100.0**	**100.0**

Source: NSSO, Some Aspects of Operational Land Holdings in India, Various Rounds cited in Report of the Expert Group on Agricultural Indebtedness, New Delhi, Ministry of Finance (2007).

per cent in 1980-81 to 7.6 per cent in 2004-2005. This was due to a decline in the share of public sector GCF from 43.2 per cent in 1980-81 to 19.2 per cent where private investment failed to compensate (Table 3.4).

Table 3.4: Gross Capital Formation (GCF) in Agriculture at Current Prices *(1999-2000 series)*

Year	*Total GCF (Rs. Crore)*	*Public GCF (Rs. Crore)*	*Private GCF (Rs.Crore)*	*Share of Public (%)*	*Share of Private (%)*	*GCF in Agr.as per cent of Total GDP*	*GCF in Agr. as per cent of Agr. GDP*	*GCF in Agr. as per cent of Aggregate GCF*
1980-81	4342	1876	2466	43.2	56.8	3.0	9.2	16.1
1990-91	15839	3586	12253	22.6	77.4	2.8	10.5	11.5
1995-96	17392	5952	11440	34.2	65.8	1.7	6.9	6.3
1999-00	50151	8670	41481	17.3	82.7	2.6	11.2	9.8
2000-01	46432	8176	38256	17.6	82.4	2.2	10.3	9.2
2001-02	60366	10353	50013	17.2	82.8	2.6	12.4	11.1
2002-03	61883	9564	52319	15.5	84.5	2.5	13.1	10.1
2003-04	61827	12218	49609	19.8	80.2	2.2	11.6	8.4
2004-05	70786	13610	57176	19.2	80.8	2.3	13.2	7.6
2005-06	83952	—	—	—	—	2.4	14.1	7.3

Note: Agr. Denotes Agriculture; GDP denotes Gross Domestic Product.

Source: CSD, National Accounts Statistics, various cited in Report of the Expert Group on Agricultural Indebtedness, New Delhi, Ministry of Finance (2007).

Simultaneously, there was a big fall in the rural share of total development expenditure from 11.7 per cent of GDP in 1991-92 to 5.9 per cent in 2000-01. This translates into less state support and hence increased expenditure by rural families in the realm of health, education, etc.

Changing Cropping Patterns

With the opening-up of the economy and the resultant increase in expectations of export opportunities as well as higher world prices for agricultural commodities led many farmers to move into cash crops, away from a mixed bag of traditional subsistence crops (Venu Menon, 2006). Devaluation of the rupee made Indian exports cheaper and hence attractive on the world markets which explain this change into cash crops (Christian Aid, 2005). On the aggregate the total area of the country's farmland growing traditional grains declined by 18 per cent in the decade after 1990-91, whereas areas growing non-food crops of cotton and sugarcane increased by 25 per cent and 10 per cent respectively (Shiva, 2005).

The Rise of the New 'Baniya'

With the collapse of public institutions the functions expected of them were taken up by the private sector. This gave rise, in particular, to a certain kind of entrepreneur who did not necessarily have the interests of the farmer at heart. This entrepreneur is not only a salesman of a particular type of seed but also of a brand of fertilizer as well as brand of pesticide. In the absence of adequate extension services, he could establish himself as the main channel of information and source of advice on cropping practices exposing the farmers to further exploitation. Unsuspecting farmers are lured into buying expensive HYV seeds through aggressive marketing techniques and were opportunistically advised to use more fertilizers and pesticides promising unrealistically high yields. The non-judicious fertilizer use is causing serious long term soil damage (Christian Aid, 2005).

Understanding Suicides in a Sociological Perspective

The understanding of suicides has come a long way from the nineteenth century view that equated suicidal behaviour with "insanity". "Two concomitant revolutions in the late nineteenth century—one in the field of sociology associated with Emile Durkheim and the other, the psychoanalytical movement led by Sigmund Freud have been synthesized in the modern view of suicide and suicidal behaviour". Social and cultural variables amplify any biological and psychological predisposition a person might have. Thus, the causes for suicides are "multi-factoral, interlinked and progressive over time pushing an individual through stages of helplessness, hopelessness and worthlessness".

Durkheim's monograph on suicides provides a good insight from a sociological perspective for understanding the phenomenon of suicides. Durkheim viewed suicide as one of the most private acts an individual could perform and, therefore, hence possible to explain that action in terms of social forces. Durkheim further stated that "collective beliefs, emotions and tendencies are not caused by certain states of consciousness of individuals but by the conditions in which the social group in its totality is placed". Suicides have therefore been identified as a *social fact* calling for an explanation in terms of other social facts. Durkheim in his holistic approach sought to explain suicides as a social fact in terms of social distribution avoiding the individual explanations. Thus, causal generalization linking social solidarity and suicides could be established. He found suicides were more likely when a person lacked social bonds or had got social relationships disrupted through a sudden change in status such as unemployment. Durkheim categorises suicides as egoistic, altruistic and anomic. Egoistic suicides occur when social disintegration takes places. Exactly opposite to this is altruistic that occurs because of too much integration into society where an individual feels loss of identity. Anomic suicides are rampant whenever there is a collapse of the rules and conventions that pose realistic limits on the desires, explanations and aspirations of the individual (Durkheim, 1966). Therefore, the sociological perspectives and social factors contributing to farmers' suicides is an important area of concern while addressing the problem.

Magnitude

Large number of suicides by farmers in various parts of the country is the most distressing phenomenon observed over the last few decades. These farmers' suicides reached almost epidemic proportions in certain pockets of the country. In fact, media social activists and enquiry commissions that were set-up by the State generated some kind of data base though they were unilateral or impressionistic in nature. Later several scholars and institutions felt that there was a need to probe the issue systematically and comprehensively so that a nation-wide picture emerged.

The main data provider on this issue is the "National Crime Record Bureau" (NCRB), Ministry of Home Affairs, Government of India. The data compiled from the police records furnished to the NCRB, "by the Chiefs of Police of all States/UTs and Mega cities" are being put out in the annual publications since 1967. Earlier reports provided basic data on the number of suicides in different States. The reports after 1967 provided details such as distribution of suicidal deaths by sex and age, causes of suicides, marital status, educational level, method/means adopted and most importantly the occupation/profession of the victims. The category of self-employed persons (farming/agriculture)—which can be taken as representing the farmers—was added for the first time in 1995 (Nagaraj, 2008). In fact, official reports initially denied the farmers suicides but as more and more information came to light the government began to accept that farmers in India were under severe stress.

At all India level more than 17,500 farmers a year killed themselves between 2002-2006, according to the experts who have analysed government statistics. Most of the farmers' suicides occurred in the States of Andhra Pradesh, Maharashtra, Karnataka, Kerala and Punjab. In 2006 the State of Maharashtra with 4,453 farm suicides accounted for over a quarter of the All-India total of 17,060 according to the National Crime Records Bureau. NCRB also stated that there were atleast 16,196 farm suicides in India in 2008 bringing the total since 1997 to 1,99,132.

It is worth mentioning that while the number of farmers' suicides increased since 2001, the number of farms has declined as thousands abandoned agriculture in distress. At least 17,368 farmers killed themselves in 2009, the most appalling figure for farm suicides according to the data revealed by the National Crime Records Bureau (NCRB).

Suicide Rates and Trends

There is a high degree of variations in terms of volume (Nos.) as well as suicide rate of farmers across different States in the country (Table 3.5). A similar variation also exists for general suicides. In fact, there seems to be a strong relationship between general suicides and farmer suicides in terms of these variations across States. Thus, it appears that those states which are suicide-prone in a general sense are also the ones which are prone to farm suicides.

Table 3.5: Number of Farmers' Suicides and All-Suicides in India, 1997-2006

Year	*Farmers' Suicides*		*All Suicides*	
	Number	*As a per cent of all suicides*	*Number*	*Suicide Rate per 100,000 poulation*
1997	13622 (100)	14.2	95829 (100)	10.0
1998	16015 (118)	15.3	104713 (109)	10.8
1999	16082 (118)	14.5	110587 (115)	11.2
2000	16603 (122)	15.3	108593 (113)	10.6
2001	16415 (121)	15.1	108506 (113)	10.6
2002	17971 (132)	16.3	110417 (115)	10.5
2003	17164 (126)	15.5	110851 (116)	10.4
2004	18241 (134)	16.0	113697 (119)	10.5
2005	17131 (126)	15.0	113914 (119)	10.3
2006	17060 (125)	14.4	118112 (123)	10.5
Total number of suicides in the period 1997-2006	166304	15.2	1095219	—
Annual Compound Growth Rate (in%) between 1997-2006	2.5	—	2.4	—

Note: Figures in brackets give indices with 1997 as the base.

Source: Various issues of Accidental Deaths and Suicides in India (ADSI), National Crime Records Bureau (NCRB), Ministry of Home Affairs, Government of India.

In terms of distribution of the number of suicides, the extent of concentration in certain States and regions—in the country seems to be higher in the case of farm suicides compared to general suicides. Thus, the top five States in terms of the number of farm suicides in 2001 viz. Maharashtra, Karnataka, Andhra Pradesh, Chhattisgarh and Madhya Pradesh—account for nearly two-thirds of the suicides in the country. The top five States in terms of the number of general suicides only partially overlap with this set, i.e. Maharashtra, West Bengal, Karnataka, Tamil Nadu and Andhra Pradesh and they account for more than 55 per cent of the total general suicides in the country. At the other extreme there are a number of States—situated largely in the Northern part of the country, the Gangetic Plain in particular like Bihar, Uttar Pradesh, Rajasthan, Punjab, Jammu & Kashmir, etc. where the general suicide rates as well as farm suicide rates are very low.

The top five States which account for a high number of farm suicides also have high rates of farm suicides: Karnataka with a farm suicide rate of 36.4 in 2001 comes next to Kerala, with Chhattisgarh (33.7) and Maharashtra (29.9) not much behind. The farm suicide rate in Andhra Pradesh (19.2) is also significantly higher than the All-India average. It is also worth recognizing that other States also have high or above average general suicide rates in comparison with the All-India rate (Table 3.6).

Table 3.6: Number and Rate of Suicides for the General Population and Farmers in the Major States of India, 2001

Sl. No.	State	No. of Suicides		Farmers' suicides as a per cent of all suicides	Suicide rate (per 100,000 members)		
		Among farmers	Among all population		In general population	Among Farmers	
						With all cultiavators considered	With only main cultivators considered
1.	Maharashtra	3536	14618	24.2	15.1	29.9	34.7
2.	Karnataka	2505	11881	21.1	22.5	36.4	40.5
3.	Andhra Pradesh	1509	10522	14.3	13.8	19.2	20.4
4.	Chhattisgarh	1452	4025	36.1	19.4	33.7	41.6
5.	Madhya Pradesh	1372	6860	20.0	11.4	12.4	15.4
6.	West Bengal	1246	13690	9.1	17.1	22.0	27.3
7.	Kerala	1035	9572	10.8	30.1	142.9	176.5
8.	Tamil Nadu	985	11290	8.7	18.1	19.3	20.8
9.	Uttar Pradesh (incl. Uttarakhand)	709	3827	18.5	2.2	3.0	3.7
10.	Gujarat	594	4791	12.4	9.5	10.2	12.6
11.	Rajasthan	505	3195	15.8	5.7	3.8	5.3
12.	Orissa	256	4052	6.3	11.0	6.0	7.5
13.	Assam	167	2647	6.3	9.9	4.5	6.2
14.	Haryana	145	2007	7.2	9.5	4.8	6.5
15.	Pondicherry	91	529	17.2	54.3	834.9	865.8
16.	Bihar (incl. Jharkhand)	88	853	10.3	0.8	0.7	0.9
17.	Punjab	45	648	6.9	2.7	2.2	2.4
18.	Tripura	41	854	4.8	26.7	13.1	16.2
19.	Himachal Pradesh	22	307	7.2	5.1	1.1	2.0
20.	Goa	18	256	7.0	19.0	35.7	60.7
21.	Jammu & Kashmir	15	153	9.8	1.5	0.9	1.6
	All India	**16415**	**108506**	**15.2**	**10.6**	**12.9**	**15.8**
	Coefficient of variation (percentage) (n=21)	116.3	93.4	—	82.3	299.0	280.6

Source: Accidental Deaths and Suicides in India (ADSI), National Crime Records Bureau (NCRB), Government of India, 2001.

Farmers' Suicides—Major Causes: A Review

With the Structural Adjustment Programme (SAP) initiated in 1991 the policy of globalization was vigorously introduced in India. Based on this policy and the directives of the World Bank, International Monetary Fund and World Trade Organization, the Indian economy was substantially overhauled. The Export-Import policy was liberalized; the import and customs duties of many products were drastically lowered or totally removed so that they could be imported without any restrictions. The government started reducing its investment in agriculture and the industrial sector allowing the private sector to take over. All such policy initiative measures had implications for the farm sector.

The most prominent manifestation of this policy is seen in the drastic decline in the growth rate of food grains. The rate of growth of agricultural output was gradually increasing during 1950-1990 and it was more than the rate of growth of the population. In the 1980s the agricultural output grew at about four per cent per annum. Thus, India became self-sufficient in food grains and started exporting wheat and rice. But during the 10-year period after the start of liberalization, the rate of growth declined to two per cent.

Unemployment in the agricultural sector increased during the reform period as agriculture was not profitable due to the fall in the prices of farm products. As a result, the number of people who are employed in the primary sector and the area under cultivation decreased which, in turn, caused a decline in rural employment. Suicides of farmers are the other fall-out of stagnation in agriculture. When agriculture was no longer yielding remunerative income, the life of the farmers became very miserable. Many of them were thus compelled to commit suicide as a last resort. Perhaps the reasons for the agrarian crisis can be attributed to (i) liberal import of agricultural products; (ii) slashing of agricultural subsidies; (iii) lack of easy and low-cost loan to agriculture; (iv) decline in government investment in the agricultural sector; (v) restructuring of the public distribution system (PDS); and (vi) special economic zones. Keeping in view the broad policy-related impact of these new economic reforms in agriculture, the reasons for farmers' suicides are presented below:

The most affected regions for suicides by farmers have been in the Vidarbha region in Maharashtra; Deccan and Hyderabad in Karnataka, i.e. Northern Karnataka, Telangana and Rayalaseema in the then united Andhra Pradesh; Wayanad, Idduki and Kannur in Kerala. It has been found by most of the studies that these regions (except Kerala region) are dry areas with poor irrigation facilities facing adverse erratic climatic conditions. The regions of Kerala are mainly hilly regions; however, during the period of crisis even these regions reported deficit rainfall. The crops cultivated in these regions were mainly commercial crops or cash crops. In Northern Karnataka regions the main crops were *jowar*, *bajra*, cotton, *ragi*, wheat, paddy and *tur*. Commercial crops also have a significant presence in the cropping pattern. The presence of at least one/two such cash crops can be seen across the size classes. In Kerala the major crops cultivated were coconut, rubber, tea, coffee, arecanut, spices etc. which are essentially commercial crops or export-oriented crops. It has also been noticed that the farmers were fascinated by the lucrative commodity prices promoting them to switch over from cultivation of multiple crops to mono crops. The major crops of Vidarbha were cotton, oilseeds, sugarcane and soybean. However, extensive cultivation of cotton has been reported. Shift towards cash crop production has also been seen in case of Andhra Pradesh where major crops cultivated were tobacco, cotton, chillies and groundnut. Many studies revealed that the victims have been mostly marginal and small farmers in most of these regions who were rendered highly indebted particularly to private money-lenders. The victims were mainly male and quite young. In Karnataka, it was observed that victims belonged to ages between 25 to 35 years and were mainly OBCs though there were also cases of farmers committing suicide hailing from dominant castes such as *Lingayats* and *Vokkaligas*. In Maharashtra (Yavatmal) it was found that the age group of victims was between 30 to 50 years and mainly Hindus. In case of Kerala, majority of the victims were male (90%) and belonged to the age group of 41-60 across all castes and religious groups. A survey in Andhra Pradesh on the socio-economic characteristics of farmers who committed suicide in 2003 and 2004 in four districts shows that the majority of the victims were male in the age group of 30-50 and majority of them belonged to backward communities.

Indebtedness

The primary cause as revealed from most studies is 'indebtedness' which again can be linked to mismatching of 'expectations'. A farmer gets indebted when his/her return from the crop is not as much as he/she expected and he/she becomes unable to pay the loans, which he/she might have taken for productive purpose, social purpose or to meet his/her immediate consumption needs. Indebtedness is quite evident in most of the affected districts. In Kerala indebtedness is higher compared to the national average due to availability of good informal and formal credit facilities, cash crop production and higher value of assets per households.

The failure of co-operative institutions has further made the large number of farmers to fall back on the moneylenders who charged exorbitant rates of interest. The interest rate varied from 36 per cent to 60 per cent. Farmers borrowed from these money lenders due to the cumbersome process followed by institutional financing as the farmers found it almost impossible to comply with their lending norms and procedures/documentation. In the case of Maharashtra and Andhra Pradesh too, indebtedness was reported to be one of the major factors.

Agriculturist moneylender unlike the professional moneylender is a native of the village and powerful in the socio-economic structure as well. In many cases he is the landowner who leased out land to the peasants. It is this combination of functions as that of a landowner, moneylender and trader that makes him a formidable force in the village enabling him to exploit the helplessness of the helpless farmers. The commission agents, dealers and sub-dealers of pesticide companies are other inputs also provide credit at astronomical rates of interest. At time, they coerce the farmers into distress-sale of their produce at the farm gate itself in discharge of the debt.

Indebtedness is only the tip of the iceberg! The secondary factors which led to indebtedness and eventually agrarian crisis in India can be attributed to climatic factors and misuse of resources, low yield, neo-liberal policies, seed sector liberalization crops, lack of alternative livelihood as well as social factors are no less important. These factors get interrelated together to form a complex process and causes agrarian distress.

Huge Debts

The ***debt trap*** is the main cause of farmers taking the extreme step of committing suicide. Debt trap is getting tightened up because of the agrarian crisis on the one hand and inaccessibility of institutional credit on the other hand. Maybe, the outstanding loan of victims is not just because of what they have invested in loss making cultivation but also because of other social issues. But it has been the general practice for ages that the farmers borrowed money for other purposes too in anticipation of the income that they would derive from agriculture.

Borrowing for social purposes like marriages, education, health care, house construction and so on are common across different segments of the society. This is more so among the middle class. There are so many institutional schemes for lending money at a low rate of interest for the middle class who have stable income. Banks are competing with each other to lend money for housing schemes, consumer durables and educational purposes for the middle class. However, the crux of the matter is that the middle class access institutional credit only on the basis of their professional income. They suffer only when their jobs are jeopardized, which is less likely to occur. These institutions will be reluctant to lend money to the farming community due to the uncertainty of their income and the resultant apprehensions in their repayment capacity thus driving the farmers to the private money-lenders.

Farmers do not have any access to Debt Relief Acts as industrialists do. Indebtedness is a matter of personal shame and not a business phenomenon in the case of peasants. We hardly hear about suicides of industrialists though even if indebted. They simply invoke insolvency laws that protect them. Farmers have no such advantage within the closed village community even as they depend more on non-institutional credit.

Alcoholism and Personal Habits

Alcoholism is so common in most part of the country—thanks to Government policies that allow free flow of branded and local brews at the cost of other needs. This is more common among poorer sections in rural areas and urban slums. Even though alcoholism is not directly related to farm distress, it adds to the problem. It is also a result of psychological depression owing to crisis in their lives. In turn alcoholism accentuated problems in the family like abuse of wife and children. Thus, alcoholism can be treated only as a symptom of the crisis and not as the cause for the crisis itself.

Neo-Liberal Policies

Neo-liberal policies were also cited to have aggravated the problem. Impact of liberalization has been felt in terms of declining output prices and increasing input prices. Due to withdrawal or decrease in Governmental support for agriculture in terms of declining subsidy, investment in irrigation and encouragement of private seed companies has led to increase in the costs of inputs for production. The overall public capital formation in agriculture is declining over time. Dry land agriculture has been long neglected. The public investments in these regions are relatively low when compared to endowed regions. It is true that the magnitude of investment both in public and private investment in money terms has slowed down. But the bulk of public investment is on surface irrigation and concentrated on projects, which have taken an unconscionably long time to complete. Costs have risen and outlays have not resulted in any significant additions to irrigated area. The other much smaller component of public investment has been in land and water conservation in rain-fed areas through watershed development.

Non-Availability of Seeds

In the present context, the multinational seed industries are seeking total control over seeds. This has made farmers felt threatened as the barriers are raised in terms of growing, distributing and exchange of seeds. Again the shift towards cash crop production, which requires monetised inputs including fertilisers and subsidies, has further increased the cost of cultivation. Coupled with these, the output prices were also falling since mid-1990s causing huge losses and indebtedness among farmers. Even when the output prices started increasing, there was no real gain because of increasing cost of cultivation and depleting yield.

Socio-cultural Factors

Socio-cultural factors also played a catalytic role in the process of agrarian crisis and farmers' suicides. Low quality of public education and health facilities drives rural poor to private institutions involving high costs. In some cases the productive loan has been used in meeting social obligations such as dowry, community functions, etc. In case of Punjab it has been noticed that maintaining status symbols (e.g. having tractor) and peer-group pressure have contributed to unproductive spending and hence indebtedness.

As a result, substantial debts have been incurred due to non-productive purposes as well. In Andhra Pradesh leasing-in land for cotton cultivation by small farmers also aggravated the plight of cultivators. It has also been found in some studies that support from family, friends, relatives, panchayat members etc. helps one to relieve the stress they faced. Not discussing one's problem with others leads to closing an avenue for letting out one's pent up feelings and frustration. Societal support, public participation and closely-knit intra-family relationship help in averting such incidents by providing necessary with confidence. However, in recent times it was found that the support-system that was all along provided by the family and the village community has been eroded to a large extent.

Land Reforms

The irony is that in spite of all the crisis the peasants are clinging on to their land and also leasing in small parcels of land for cultivation. On the one hand land is seen as the basic life line and on the other the peasants are not left with any other option except agriculture. Thus, the issue is to protect the interests of these resource-poor farmers. There is a need to bring radical changes in the existing Land Reforms Act. As the tenant farmers are leasing in land on contractual basis they are not eligible for institutional credit and crop insurance. Thus, they have to depend on private sources of money lenders at high rates of interest. Instead of making the small and marginal farms more viable with appropriate supportive measures, the suggestions for the so-called 'radical' changes in the land reforms would only push the resource-poor farmers out of their land adding to their misery. Experiences in other countries like China, South Korea etc. should be taken into account for strengthening the small farms instead of eliminating them.

Mobilization of Farmers and Farmers' Movement

During the 1980s and early 1990s India witnessed powerful mobilization of farmers in different parts of the country. They were based on the demands of remunerative prices, changes in the Government policies towards institutional credit, reduction in electricity tariffs and so on. In essence, the movements were focused against the State and State policies. However, the movement is at a low ebb since mid-1990s mainly due to the assimilation of this movement by the political parties and the farmers' organizations experienced a number of splits on non-issues.

Since, the main reason for the farmers' suicides is the debt trap, farmers are individually tied to traders-*cum*-money lenders. Assadi (1998) has suggested that on earlier occasions of agrarian distress, farmers have taken recourse to agitation mode like burning of sugarcane or dumping the surplus produce on the main roads. It was easy for the sugarcane farmers to get organised because of specific geographical concentration of farmers and single market (sugar factory) uniting all these farmers together. Whereas in the case of crops that the deceased farmers cultivated there was no such fixed market. This alone does not explain the lull in farmers' mobilization. For example ground nut farmers in Anantapur have a uniform feature that they all cultivate the same crop for which the MSP is fixed by the Government. They could have come together to protest for higher MSP. But this did not happen mainly because irrespective of the MSP (which is already lower than the cost of cultivation) the small and marginal farmers get very low prices because of their debt trap. As resource-poor farmers are individually tied to money lenders, this acts as a deterrent for them to come together. They seek solution individually including taking recourse to suicides.

Suggestions/Recommendations

To mitigate the existing agrarian crisis, the following suggestions are proposed:

1. Public investment in agriculture should be quadrupled. The Union and State Governments should enact Public Investment Responsibility Acts to ensure a minimum fixed share of budgets going into agriculture and related areas like irrigation, flood control and rural development.
2. Set-up an Agricultural Tariff Commission, on the lines of Agricultural Costs and Prices Commission (ACPC), with representation from peasant organizations, to decide on tariff levels and protection for different agricultural commodities in response to price fluctuations in the international markets.
3. Apart from fixing the MSP (minimum support price) for state procurement, the CACP should also enforce payment of MSP for cotton, sugarcane and tobacco directly procured by spinning mills, sugar mills and tobacco companies.
4. MSP should cover all agricultural commodities and ensure a comfortable margin over all costs. Procurement should be made in all regions in all states directly from the farmers and there should be no ceiling on procurement.
5. Regulate prices of other inputs like seeds and pesticides. A price control authority for these inputs should be set-up on the lines of Drug Price Control Authority. All seeds and pesticide companies whose products fail should be penalised to the extent of damage suffered by the farmers.

6. All agricultural loans—short-term as well as long-term loans—should be provided at low rates of interest not exceeding 4 per cent per annum, and loans up to Rs. 2 lakh should be provided in adequate measure without collateral even to tenants and share-croppers.
7. The Town and Village Industries (TVIs) model of China may be adapted to Indian conditions so as to promote employment-intensive rural industrialization. Unrestricted low-cost institutional credit should be provided for rural non-farm activities.
8. Set-up cooperative marketing federations of farmers to procure and market horticultural products with adequate cold-storage and transportation facilities to abolish middlemen. Subsidised kits of inputs should be provided to small and medium farmers through these cooperatives to remove sharp regional productivity differentials.
9. There is an immediate need to strengthen the network of *mandis* (market yards) across the country which provides farmers with a platform to sell their produce. Leaving it to markets will result in distress sale.
10. For a country which was able to build up an excellent marketing network for one of the most perishable commodities — milk— there is no reason why a similar approach cannot be adopted in providing a viable marketing network for fruits and vegetables. If the National Dairy Development Programme could ensure that milk is procured from each and every village, and then through a cooperative chain it is finally delivered to the consumers in the cities, there is no reason why India cannot carve out a marketing chain for fruits, vegetables and other farm commodities.
11. Interest-free bank loans should be extended to small and medium farmers to help them consolidate their holdings.
12. Crop insurance compensating for all forms of crop damage should be extended to cover all crops, all areas, and all farmers, including those who do not avail institutional loans. Right to compensation in case of natural disasters should also be incorporated.
13. Abolition of benami landholdings and absentee landlordism and legalization of all land held by poor and toiling peasants and agrarian labourers. Compulsory registration of tenancy and

share-cropping and conferring of ownership on poor and toiling tenants and share-croppers.

14. Agricultural extension services should be revived and strengthened to modernise farming methods. Agricultural research should be liberated from corporate clutches.
15. Rural infrastructure should be beefed up to strengthen agricultural growth. The RIDF corpus should be made unlimited and *panchayats*, cooperatives and even agricultural users' organizations should be entitled to borrow directly from the RIDF.
16. Cooperate farming need to be encouraged. Appropriate laws must be framed to make cooperatives more independent and effective. Drawing from the experience of the Amul Cooperative in dairy farming, a similar system needs to be adopted for vegetables/ fruit farming.
17. Agriculture, dairy and forestry should be integrated. Agricultural growth should not only be measured in terms of increase in food grain production but should be seen in the context of the village eco-system as a whole. This will also shift the focus to low external input sustainable agriculture (LEISA) practices. At the same time such an approach will limit the ecological footprint.
18. Climate change is certainly going to affect agriculture. But instead of looking at strategies only aimed at mitigating the impact on agriculture and making farmers cope with the changing weather patterns, the focus should also be to limit greenhouse gas emissions from agriculture. Considering that agriculture share in greenhouse gas emission is about 25 per cent, the thrust must shift to reducing the application of chemical fertiliser/ pesticides in farming. In the dryland regions of the country, for instance, at present hybrid crops which required almost twice the amount of water than normal crop varieties, are grown. Common sense tells us that in rainfed regions, which occupy 65 per cent of the cultivable area, crops requiring less water should be grown. But it is just the opposite in reality thereby accentuating the water crisis at times of rainfall delay.

Conclusion

It is quite true that the agrarian crisis cannot be solved through economic packages alone. What is needed is social interventions so that the farmers realize that suicide is not the way out... They should understand that they need to develop self-confidence. The future generation should have the mental strength to face life's challenges. What can be done is: Social support should be provided as a buffering effect for stress. The greater the support from friends, family, local communities, national policies and society as a whole, the less impact stress will have on individuals and families. It is crucial for Governments and NGOs to work together, on a local, national and global level, to address and solve this critical issue.

Note

1. Bharat is the Hindi name for India, nationally subsuming the oppressed rural many and urban few; and India, the English name representing the dominant.

References

1. Aerthyil, Mathew (2008), Agrarian Crisis in India is a Creation of the Policy of Globalization, *Mainstream*, Vol. XLVI, No. 13.
2. Assadi, Muzaffar (1998), Farmers Suicides—Signs of Distress in Rural Economy, *EPW*, 33 (14).
3. Balamuralidhar, P. (2009), *Crisis in the country side: Farmer's Suicides and the Political Economy of Agrarian Distress in India*, Working Paper Series, Development Studies Institute, London School of Economics and Political Science, London, U.K.
4. Christian Aid, 2005, *The Damage Done: Aid, Death and Dogma*, London: Christian Aid.
5. Durkheim, Emile (1966), *Suicide: A Study in Sociology,* The Free Press, New York.
6. Government of India (GoI), 2007, *Report of the Expert Group on Agricultural Indebtedness*, New Delhi: Ministry of Fianance
7. Gupta, H.S. (2014), "Challenges and Strategies in Making Small and Marginal Farm Holdings Sustainable and Profitable", IARI, New Delhi (unpublished paper)
8. Nagaraj, K. (2008), *Farmer's Suicides in India: Magnitude, Trends and Spatial Patterns*, MIDS, Chennai.
9. Sharma, Devindra (2014), The Loud Cries of Farmers' Widows have been lost in the din of the Narendra Modi Coronation, 27th May, 2014: http://www.dnaindia.com/analysis.

10. Shiva, V., (2005), *Earth Democracy: Justice Sustainability and Peace*, Cambridge, MA: South End Press.
11. Suman Chandra, K. *et al.* (2014), *Agrarian Crisis in India: The Way Out*, Academic Publishers, New Delhi.
12. The Sanhati Collective (2012), Farmers Suicides in India: A Policy-induced Disaster of Epic Proportions—Internet based article.
13. Venu, Menon S. (2006) *'Globalization, State and Disempowerment: A Study of Cotton Farmers' Suicides in Warangal'*, Munich Personal RePEc Archive Paper No. 1633: *http://mpra.ub.uni-muenchen.de/1633/*(Last accessed August 21, 2008).
14. Vidyasagar, R. and Suman Chandra, K. (2004), *Farmer's Suicides in Andhra Pradesh and Karnataka*, NIRD, Hyderabad.

4

Spatialities of Agricultural Risk and Farmers' Suicides in Maharashtra

Sucharita Sen and *Archana Takte*

Introduction

The liberalization of the Indian economy during the early nineties coincided with a period of agrarian distress in India, one of the manifestations of which was a continuous reporting of farmer's suicides. This chapter, through analyzing the spatial patterns of risk and farmers' suicides in Maharashtra argues that this coincidence is not an event related only to climatic uncertainties and poverty, or even reflective of socio-psychological state of farmer's minds, but is a direct and inevitable fallout of the process of developmental growth path that the country has followed for nearly the last two and half decades.

The reform process did not directly have much for the agricultural sector, except in two ways that it was expected to benefit from. The first was a favourable terms of trade *vis-à-vis* the non-agricultural sectors due to greater efficiency in the manufacturing and the services sector. The second was through trade liberalization, by availing opportunities for exportable crops that had a competitive advantage in the world market. The post-reform developments in agriculture, instead, had a very different story to tell. The country instead witnessed violence rooted in rural India, termed as Maoist terrorism, large scale outmigration from rural India, particularly the more vulnerable areas, and a continuous stream of farmers' suicides (Sahay, 2014). To name a few more conditions of the crisis, the growth rate has decelerated for some crops and some regions due to commercialization have reduced diversification and are hence less capable of coping with conditions of both natural and man-made stress conditions. After initial agricultural export spurt, not much could be achieved on long term because of continuing subsidization by the developed countries and putting Indian product under the scanner of different phyto-sanitary sanctions. Also, despite a high growth in GDP led by the services sector, the shifting away of labour force from agriculture to non-agriculture has not gathered any momentum after 1987-88. The gap between the share of contribution of agriculture to income and employment is increasing, implying higher levels of relative distress on agriculture-dependent population. The reduction of public capital formation in agriculture has been reducing steadily, and though the increase in private capital formation has more than made up for the decline, these two sources of capital formation are of very different forms (Patnaik, 2003). Open access infrastructures having public good characteristics like rural roads, rural electrification, minor and major surface irrigation are results of public investment, while tube wells, bore wells, that skew access to resources like ground water and cold storage facilities that is selective in its reach are examples of private sector investments. Finally, in the new 'open' and global policy regime, there seems to be backtracking on issues like land reforms, ceiling on holdings, security of land tenure and lastly, the new Land Acquisition Policy. Increasing corporatization of agriculture through promotion of contract farming that is gradually replacing co-operative initiatives of credit and farming has distributed risks unevenly between farmers and the corporate sector, biasing the odds against the former.

Much of the processes and outcomes mentioned above are rooted in the state's capitalist strategies that is deepening with the years (Sen and Mallick, 2012). The state has withdrawn not only from the seed market in such cases, but also from the supply and quality control of inputs, which has been taken over by the private players, both small and large, very often working in tandem.

This chapter attempts to measure and analyse the spatial pattern of agricultural risk. It validates this spatial pattern by superimposing it with farmer's suicides in Maharashtra to understand whether the risk, hence measured, can serve as an indicator of agrarian distress. More importantly, it enquires whether the spatial patterns of distress can be explained by corporatization of agriculture. Finally, it attempts to understand the nature of institutional support that is there in areas with high incidence of risk and distress.

Agricultural Characteristics of Maharashtra

Agriculture and allied sector contributes to only 8 per cent of

Maharashtra's state domestic product, against India's 15 per cent. Compared to this, the dependence on agriculture for employment is still substantial. Though the land allocation in agriculture in Maharashtra has not changed much, the cropping pattern has shown a lot of dynamism in the recent past. Maharashtra had a relatively diversified cropping pattern, though in the recent years there is a change in favour of oilseeds and non-food grain crops like cotton and sugarcane, while the area under *jowar* and *bajra* is on the decline. Bt cotton has nearly taken over the entire cotton growing area in Maharashtra, like in most of the other states in India. Currently Maharashtra allocates about 30 per cent of the area to cotton and contributes to about 20 per cent of the country's production. Other than cotton, sugarcane is another important commercial crop of the state, but occupies the western part of the state, other than the Konkan region. Agro-climatically called the rain-shadow zone, though deficient rainfall, this region is one of the most developed agriculturally and has above 50 per cent area under irrigation, against the state average of 17 per cent (Fig. 4.1). The Eastern Vidarbha, on the other hand, has slightly higher rainfall but very poor irrigation covergage.

Mapping Risk Proneness in Maharashtra

Mapping spatial patterns of risk in Maharashtra is relevant on more than one count. Firstly, it is a state from which farmers' suicides due to agrarian distress has been reported in a significant way. Secondly, an exercise of mapping risk can be used as a useful tool for policy-makers to focus efforts and investments in the distress prone areas. Thirdly, comparing the risk patterns with conditions of natural and economic distress enables us to have a clearer understanding of the factors promoting risks and hence provides important policy directions.

Framework for Measurement of Risk-Proneness

Fig 4.2 demonstrates the way in which Risk Proneness has been conceptualized for this study. Three elements that have been considered contributing to risk are risk-causing factors, risk limiting factors and vulnerability factors. The risk causing factors have two sources, natural or in particular monsoon instability, and the nature of cultivation practices adopted by farmers. The farmers are exposed to greater risks if the stakes are higher, and hence higher the cost of cultivation, higher is the element of risk. Natural resource use patterns are often dictated

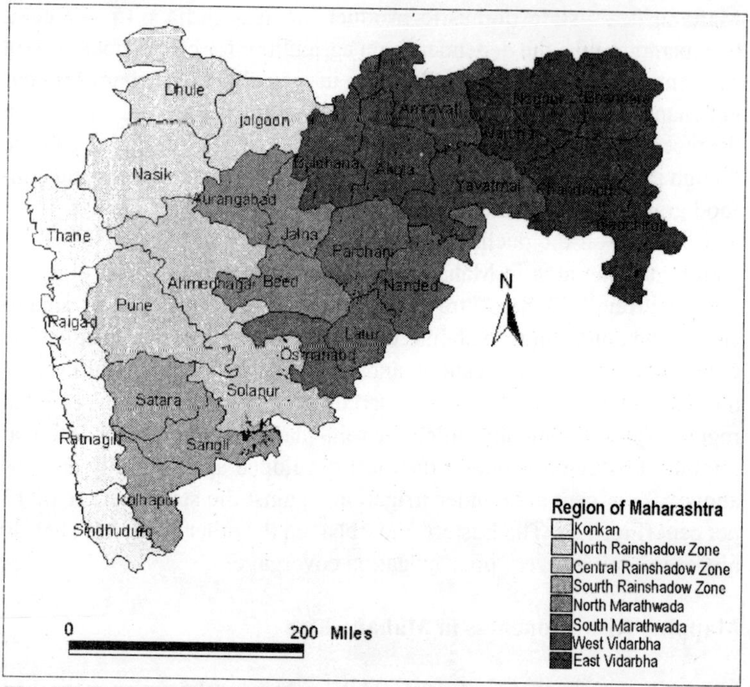

Fig. 4.1: Regions of Maharashtra

by the quantum of investments at stake. India has been increasingly depending on ground water sources for irrigation and harvesting more than the recharge of groundwater. Finally, yield instability is a function of a number of factors, and is shaped by natural, technological and socio-economic factor. The aggregation of risk causing factors has thus been done on the basis of the four indicators of rainfall instability[1], share of area under high value crops[2], ratio of groundwater utilization to recharge and yield instability.[3]

The risk limiting factors includes irrigation coverage, both the extent and intensity[4], degree of crop diversification[5] and access to regulated markets. The latter is taken as a proxy for farmer's access to remunerative prices. These factors together would ameliorate risk faced by the farmers. Access to irrigation not only enables higher yield but also enables the farmer to overcome periods of lower than expected rainfall, depending on the assuredness of the irrigation source. The more diverse the cropping pattern, the less the dependence on an individual crop. The access to regulated prices lowers the price risk faced by the farmer. The reciprocal of the aggregation of these factors was taken as a second measure of risk-proneness.

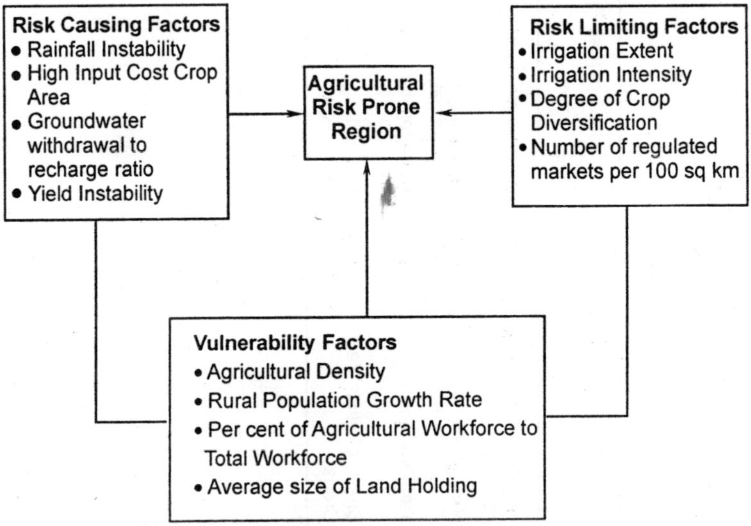

Fig. 4.2: Framework for Calculating Risk-Proneness across Districts in Maharashtra

The third set of factors called the vulnerability factors include the broader socio-economic context in which the farmer operates in. Pressure on agricultural land, high growth of rural population, low opportunities in the non-farm sectors as well as small farm sizes are characteristic features of regions that would render the farmers vulnerable. Their ability to deal with shocks or disturbances would be limited under such circumstances.

The three above-mentioned sets of factors have been aggregated to arrive at the risk proneness. We have used a multi-stage Principal Component Analysis to construct the Index of Risk-Proneness across the districts of Maharashtra.[6] The risk causing factors, the risk limiting factors and the vulnerability factors were aggregated at the first stage. The first principal component was considered as the composite index for each of these factors, to derive three indices. In the next stage these three composite indices were aggregated again to derive the risk-proneness index.

Table 4.1: Factor Loadings of Variables for Measurement of Indices of Risk Proneness

Variable	*Factor Loading*	*Variable*	*Factor Loading*	*Variable*	*Factor Loading*	*Variable*	*Factor Loading*
Risk Limiting Factors (Reciprocal)		*Vulnerability Index*		*Risk Causing Factors*		*Index of Risk Proneness*	
Irrigation Extent	0.579**	Rural Population Growth Rate	0.407*	Groundwater use to recharge	0.835**	Risk Causing Factors	0.84**
Irrigation Intensity	0.814**	Land Holding Size	0.12	Rainfall Variability	0.451*	Risk Limiting Factors (Reciprocal)	0.32
Diversification Index	0.583**	Share of Agricultural Work Force	0.911**	Yield Variability	0.420*	Vulnerability Index	0.874**
Regulated Markets	0.734**	Aricultural Density	0.447*	Share of High Value Crops	0.712**		

* Significant at 5 per cent; ** Significant at 1 per cent.

Source: Estimated by the authors.

Agricultural Risk Causing Factors

The first principal component of the four variables taken for estimate the composite index of risk causing factors explains 47.44 per cent of variation of the data-set. The factor loadings of the four variables are significant and positive indicating that they have a significant correlation with the composite risk causing factors, and have a direct association with this index (Table 4.1). The groundwater use to recharge ratio has the maximum weight, followed by the area under high value crops. These weights are dependent on the degree of association with the other variables.

The spatial pattern of the risk causing factors reveal that Western Vidarbha, Wardha and Nagpur district of the Eastern Vidarbha and Nanded of South Marathwada and Dhule of the Northern Rain-shadow zone fall

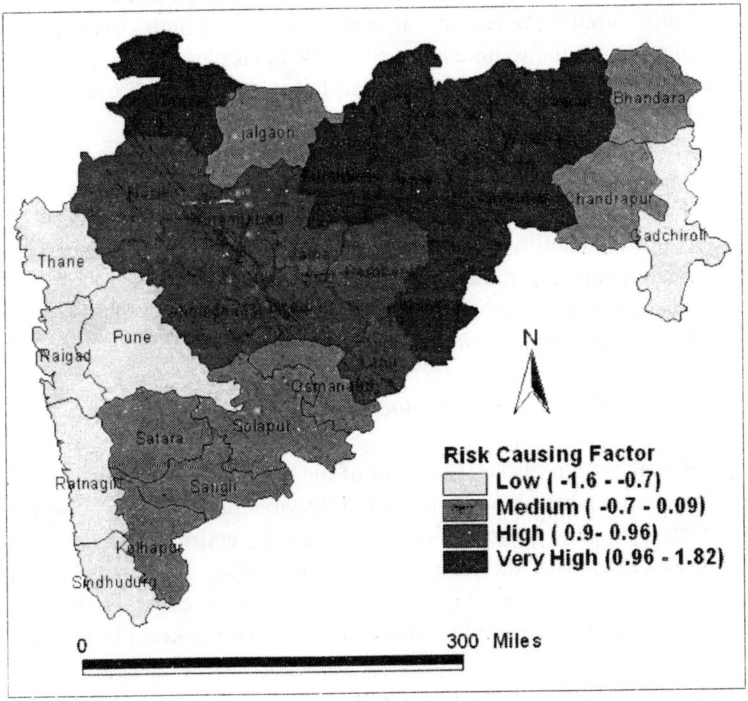

Fig. 4.3: Risk Causing Factors in Maharashtra.

in the high range (Fig 4.3). This region forms a contiguous belt with the exception of Jalagaon district of the Northern Rain-shadow zone. From the Western Vidarbha index of risk causing factor reduces towards the Konkan and Eastern Vidarbha. Pune district and South Rainshadow zone falls in medium risk causing region with Eastern Vidarbha's Chandrapur and Bhandara district. Low values of the index of risk causing factors are observed in the Konkan, Pune and Gadchiroli districts of the Eastern Vidarbha.

Socio-Economic Vulnerability Index

Unlike the clear regions that emerged in case of risk causing factors, the socio-ecenomic vulnerability index has a more homogenous pattern across Maharashtra (Fig. 4.3). Except some patches of low (Thane, Raigad) and medium (Pune, Kolhapur, Nagpur) vulnerable regions, the socio-economic context is vulnerable for most of the states. Osamanbad and Parbhani of the south Marathwada region fall in the category of high vulnerability. The important point to note here is that the areas characterized by high and very high risk causing factors also fall within the high category of socio-economic vulnerability.

The result of the Principal Component of 4 selected variables of socio-economic vulnerability factors shows that first component, namely rural population growth rate explains 49.21 per cent variation in socio-economic vulnerability. Based on the factor loadings, growth rate of rural population has the maximum weightage in the socio-economic vulnerability index (Table 4.1). The factor loading for average size of land holding is insignificant.

Agricultural Risk Limiting (Reducing) Factor

The spatial pattern of the reciprocal of the risk limiting factors closely resembles the patterns of socio-economic vulnerability. Excepting the Konkan region and Eastern Vidarbha, nearly the entire state falls under the high category of the reciprocal of risk limiting factors. Irrigation intensity, which indicates high dry season irrigation coverage, has the highest weightage followed by access to regulated markets (Table 4.1).

Index of Agricultural Risk Proneness

The result of the Principal Component of the 3 sub-indices shows that

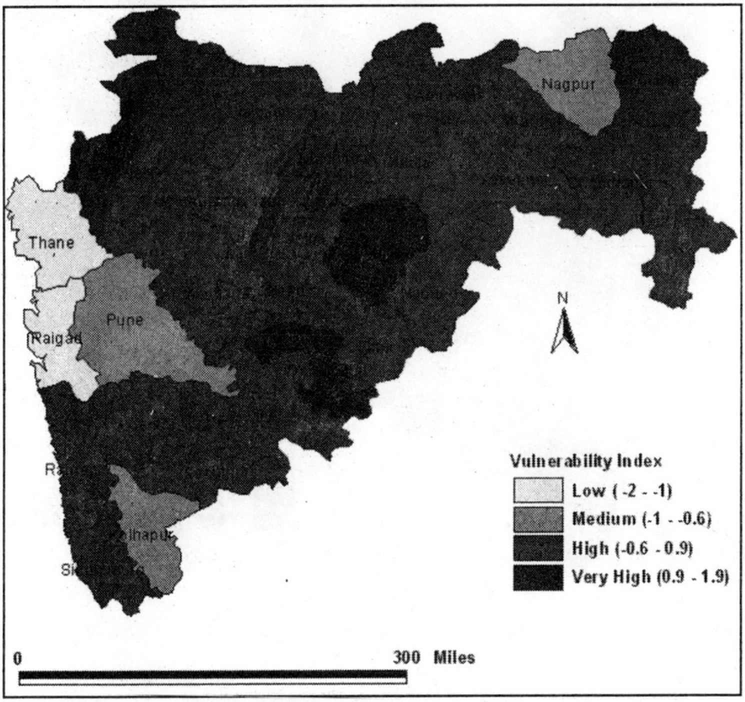

Fig. 4.4: Socio-Economic Vulnerability in Maharashtra

first component explains 72 per cent variation in Risk Prone Factor. The Konkan and Gadchiroli districts of Eastern Vidarbha are less prone to agricultural risk. Except Pune, Solapur and Kolhapur, rest of the Rainshadow zone is under the high category of risk proneness (Fig 4.6). Parbhani and Wardha are two district of very highly risk prone, which are surrounded by high risk prone region in North and South Marathwada and East and West Vidarbha.

Fig. 4.6 also brings together the spatial patterns of risk-proneness in Maharashtra and the incidence of farmers' suicides. There is a strong association between the two, though while the suicides are concentrated in Western Vidarbha, the high and very high risk prone regions include a larger area around this, as mentioned before.

The data for the farmers' suicide include only those to whom the State Government has disbursed help packages, recognizing them as suicides that have been committed as a result of agrarian distress in the period between 1991 and 2006. Farmers' suicides are concentrated in the

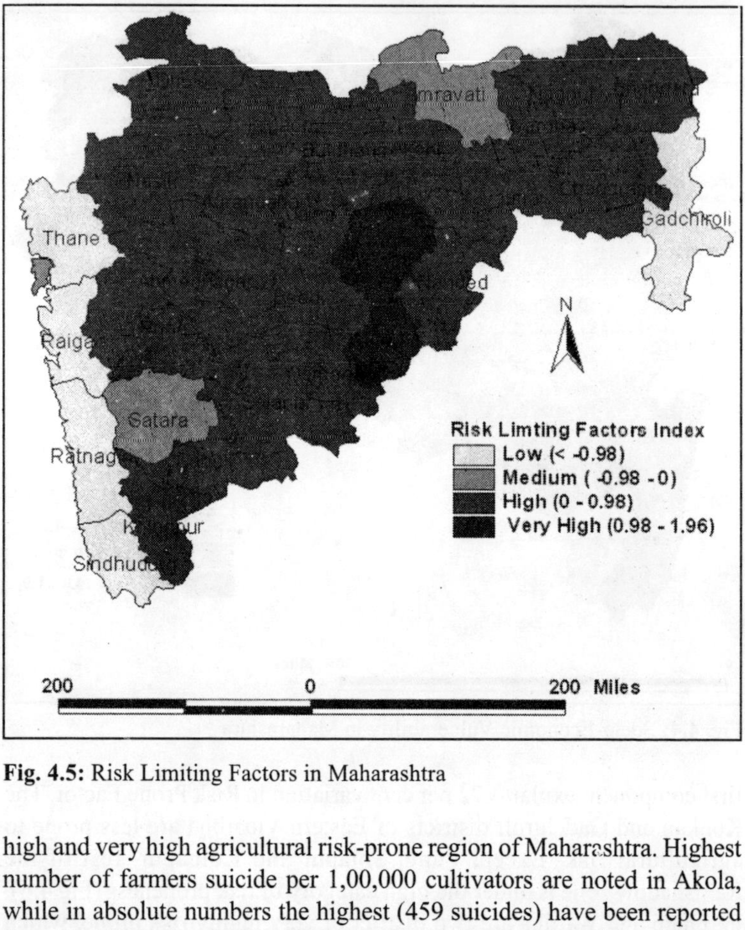

Fig. 4.5: Risk Limiting Factors in Maharashtra

high and very high agricultural risk-prone region of Maharashtra. Highest number of farmers suicide per 1,00,000 cultivators are noted in Akola, while in absolute numbers the highest (459 suicides) have been reported from Yavatmal. It may be noted that the incidence of suicides shown in Fig. 4.6 could substantially underestimate the real incidence of suicide.

Table 4.2: Correlation of Farmers' Suicides and Components of Risk Proneness

	Socio-Economic Vulnerability	*Risk Causing Factor*	*Risk Limiting Factor*	*Agricultural Risk Prone Region*
Incidence of Farmers suicides	.263	.502(**)	.101	.369(*)

* Significant at 5 per cent, ** Significant at 1 per cent.

Source: Accidental Deaths and Suicides in India (ADSI), National Crime Records Bureau (NCRB), Government of India, 2001.

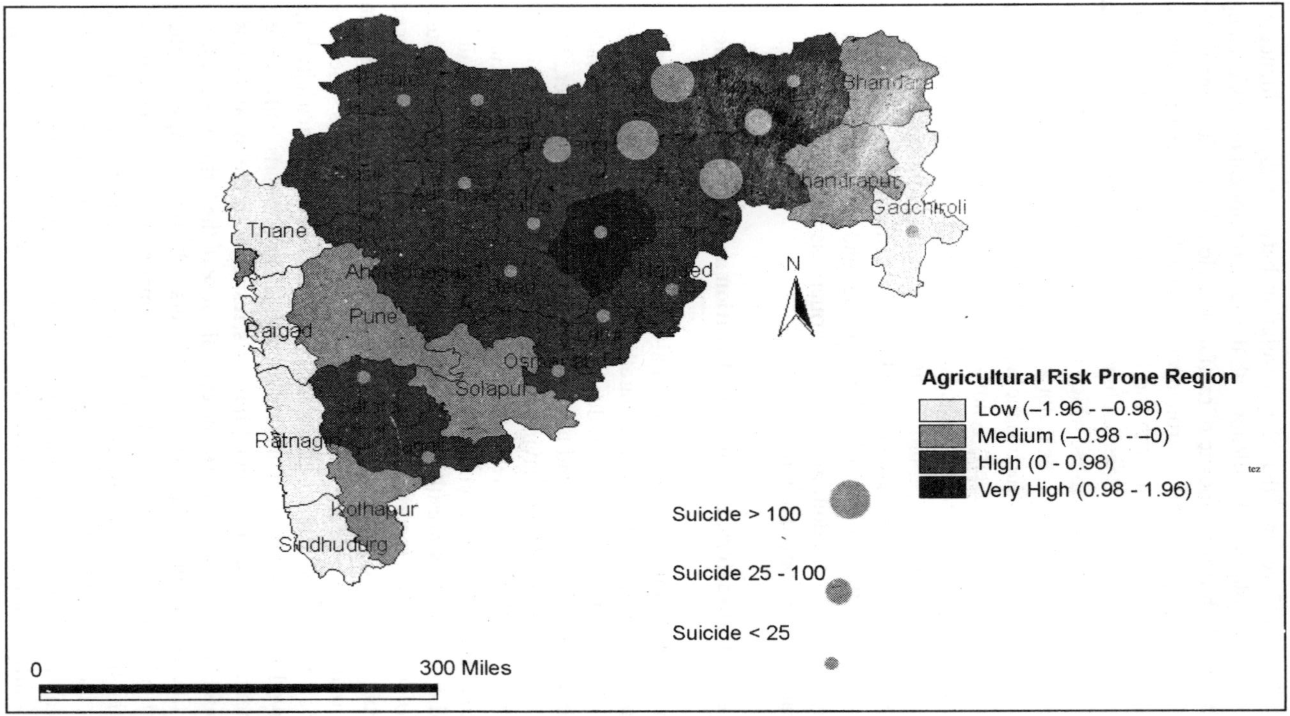

Fig. 4.6: Spatial Pattern of Risk-Proneness in Maharashtra and Incidence of Farmers' Suicides

Table 4.2 reveals that the correlation between the incidence of farmers' suicides are significant with the sub-component of risk causing factors, which is a composite of rainfall and yield instability, area under high value crops as well as groundwater utilization to recharge ratio. It has a weaker correlation with the final index of risk proneness. It is important to note that it does not have significant correlation either with the poverty related variables, which is indirectly subsumed in the socio-economic vulnerability index, nor with the developmental variables of irrigation and market, which is included in the risk limiting factors. This does not indicate that poverty is not a component part of agrarian distress, nor does it imply that irrigation and access to regulated markets does not reduce agricultural risk. It only appears that in case of Maharashtra, the conditions that pushes farmers to take the extreme step is more closely associated with high cost of cultivation, ground water scarcity, which may mean further addition of cost to farmers, as well as instabilities in weather conditions and in turn yield.

'Natural' Disasters, Bt Cotton Cultivation and Farmers' Suicides

On 9th November, 2014, DNA Analysis reported a suicide that had happened a few months back in Marathwada, by a farmer called Gurunath in Khandapur in Marathwada, due to what was termed as a crop failure due to untimely rainfall and hailstorms. Again on 4th December, 2014, NDTV reported a farmer having committed suicide electrocuting himself in Nagpur district in Western Vidarbha, a district from which the current Chief Minister comes from. The reason cited was again was erratic rainfall and crop failure as a result of that. On 12th December, 2014, a news item in *Indian Express* reported that the Chief Minister of the State announced a Rs. 7000 crore package for the drought prone areas and a long term drought package of Rs. 34,500 to provide relief to the farmers. The same report mentioned smaller grants for crop loans, waiving electricity bills as well as repaying loans of the farmers' debts. These reports, and countless others, as well as the Chief Minister's response to the agrarian crisis would have a reader interpret the central cause of the farmers' suicides to be fluctuations in rainfall. The loan waivers, which is essentially a short term relief measure, reiterates the sense of importance of the 'naturalness' of the distress. In other words, it is obviously presumed that firstly, such suicides cannot be dealt with easily since they are linked to the weather instabilities, other than partially through drought support measures, and secondly, short-term loan waivers are required only when natural phenomenon does not behave as expected. Though in a majority of the cases, farmers' suicides are triggered by drought conditions, or irregularities of rainfall, it is fallacious to assume that such fluctuations are central in explaining farmers' suicides. Assuming so does not explain the reasons due to which the suicides have started occurring in the last

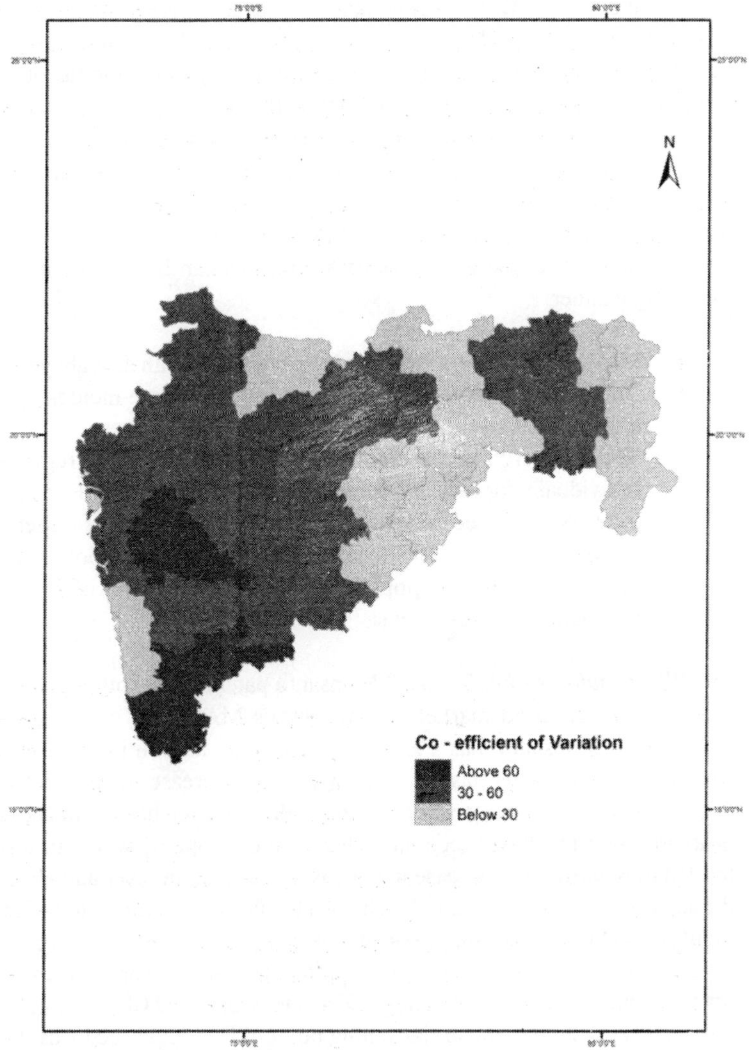

Fig. 4.7: Rainfall Variation in Maharashtra, 1980-2006

two decades with distinctly higher incidence, other than the danger of perceiving that since it is caused by 'natural' phenomenon, it is irreversible. That these relief packages do not address the root cause of farmers' suicides has been agreed upon by scholars before (Narayanamurthy, 2006).

The basic cause of farmer's suicides is not associated in weather fluctuations and this can be comprehensively understood by comparing Figs 4.6 and 4.7. While the Rain-shadow zones in Western Maharashtra experience both the lowest and most variable rainfall, the suicides are reported from Vidarbha and Marathwada for the most part. On the other hand, there is nearly a complete overlap with the reported incidence of suicides and extent of cotton growing area (Figs 4.6 and 4.8). Even in other parts of India, one of the common threads that tie many farmers' suicides is that many of them are cotton growing farmers, most of them having adopted Bt cotton varieties (Mishra, 2007).

The story that emerges in Maharashtra thus can be tied up in the following manner:

- Many of the farmers' suicides have coincided with drought years.
- But they are not necessarily in the regions where incidence of droughts is more.
- They converge almost completely with cotton growing regions. Individually farmers that committed suicides are mostly Bt cotton growers in the recent years. Post legalization of the transgenic Bt cotton seeds in 2002, most of the cotton growers took to this variety, which was popularized by expectations of higher production and lower costs.

Why should this be the case? Monsanto patent of Bt cotton gives it monopoly in the seed market. Maharashtra's MAHICO, which has a merger with Monsanto, has a significant share of distribution of cotton seed in the state. This has not only led to an increase in the cost of production due to higher cost of the seeds, the adoption has led to use of pesticides and fertilizers recommended by the company, which in turn has led to a substantial increase in the cost of cultivation, over and above the seed cost (Chindarkar, 2007). This led the farmers getting into credit-input ties with the input suppliers, at very high rates of interest. Reports of spurious supplies by these input suppliers leading to crop failures are not uncommon (Kennedy and King, 2014, Vidyasagar and Chandra, 2004, Satyanarayan, 2006). The initial selling point of Bt Cotton seed was that the farmers would save on pesticide cost since the transgenic seed itself

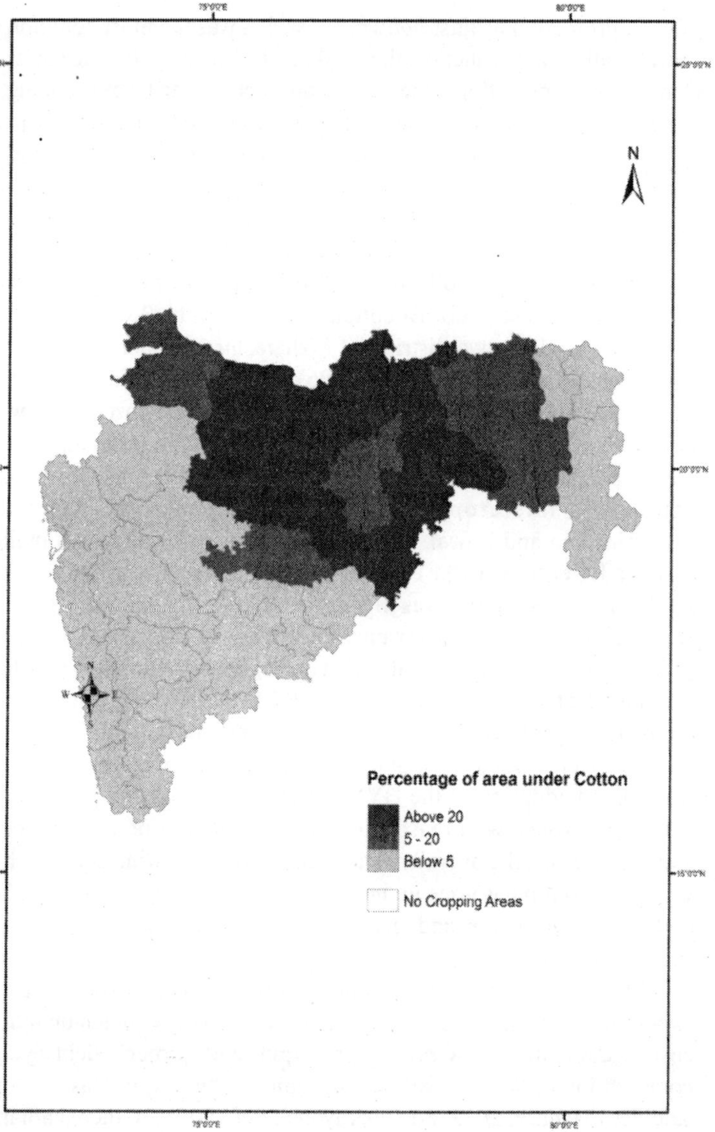

Fig. 4.8: Area under Cotton in Maharashtra, 2005-06

offers a protection against *bollworms,* which was the most common pest causing failure in production, thus making it both cost effective and stable. Contrary to expectations, the cost of production, far from reducing, has gone up over a period of time, partly due to the privatization of inputs, and partly due to secondary pests appearing in cotton production as a result of a new ecological foodchain that had removed bollworms. Also, the bollworm came back in the production system after it developed resistance to the 'Bollgard' the branded seed of Monsanto, which led the company to bring out 'Bollgard 2', at a higher price, as would be expected. The cost of cultivation of Bt cotton has found to be 35 per cent higher, based on a study in two districts of Maharashtra, Buldana and Yavatmal (Narayanmurthy and Kalamkar, 2006). The seed cost in this study has been found to be higher by 232 per cent and pesticide costs by 5 per cent. Another study based on a Vidarbha village found that though there are yield advantages for Bt cotton, these advantages declined when non-Bt cotton was intercropped with Sorghum and some other crops (Swaminathan and Rawal, 2011). The cost of cultivation was higher in case of Bt cotton, even cost of pesticides. Also, the variability of production of Bt cotton was higher than that of the hybrid cotton. This had coincided with the Government withdrawing from the seed market, a situation that is an antithetical to that during the time of green revolution. Not only did the Government invest in R&D during the green revolution, the marketing of seeds was completely done under public supervision Farmers were aided with subsidies given for fertilizers that led to widespread adoption of the HYV varieties of wheat and paddy even in the rainfed zones with reasonable success within a period of a decade and a half. Thus, the overall risk associated with this transgenic high-cost variety of cotton has gone up in a number of ways, the major ones being high cost of cultivation and increased instabilities in the agro-ecosystems (Sarthak and Mishra, 2012).

For more than a decade and a half, academic works based on experiences on a number of states have come to the conclusion that high cost of cultivation, low prices for output, and farmer's debt cycles in combination with crop failures are among the major reasons for the suicides (Madare, 2012, Satyanarayana *et al.,* 2006). Other scholars put it as a combination of low yield, low price, high input cost, high cost technology and lack of access to credit (Mishra, 2007). The corporatization of agriculture leading to stress has been articulated in terms of privatization of seeds, ineffectiveness of agricultural credit, institutions, mono-cropping limited to cash crops, followed by an increased use of pesticides, fertilizers and subsequently low land productivity (Chindarkar, 2007). The effect of this corporatization is differentiated among farmers and while the better off farmers can cope up or even reap benefits from these risks, the smaller farmers, both men and women, face acute distress (Gaurav and Mishra, 2012, Chindarkar, 2007). The plights of the left behind family, particularly the wives of the farmers that have committed suicides have been documented at length (Padhi, 2009).

From all the above, what is discounted effectively is the 'naturalness' of the cause of farmers' suicides. In the case of Maharashtra, its high convergence with share of area under cotton show that there is systematic pattern in these distress stories that cannot be explained by occurrences of droughts or isolated socio-psychological state of minds of the farmers that took this extreme step. The suicides need to be seen as the 'tip of the iceberg' of a deep-rooted agrarian distress that has a direct link with the process of development that we are situated it.

Agrarian Distress and Role of Pubic Institutional Mechanisms

The role of the state following liberalization has not only reduced to make way for the private sector, it has changed substantively to that of a facilitator for promoting an 'efficient' market. In addition, it has itself been prompted to function much like the private sector, as both the Central and the State Governments need to increasingly compete with the private sector to mobilize resources. Hence, the space to invest in areas that does not seek the inherent characteristics of 'efficiency' is reducing drastically over time.

Crop insurance is one of the measures that have been attempted by the Government to reduce distress. The Comprehensive Crop Insurance Scheme, Experimental Crop Insurance and Farm Income Insurance Scheme have been withdrawn, not too far after it was implemented. Currently, the National Agriculture Insurance Scheme is operative in various States of the country and the State of Maharashtra. The Government of Maharashtra data for 2004 reveal that the higher coverage of insurance is enjoined by the sugarcane growing areas, and not the cotton growing regions. The high risk prone regions and districts with high incidence of suicides like Akola, Buldana, Amravati, Yavatmal and Wardha, all had less than 10 per cent area under insurance coverage. The relationship between risk proneness and insurance coverage hence is weak, as observed from Fig. 4.9.

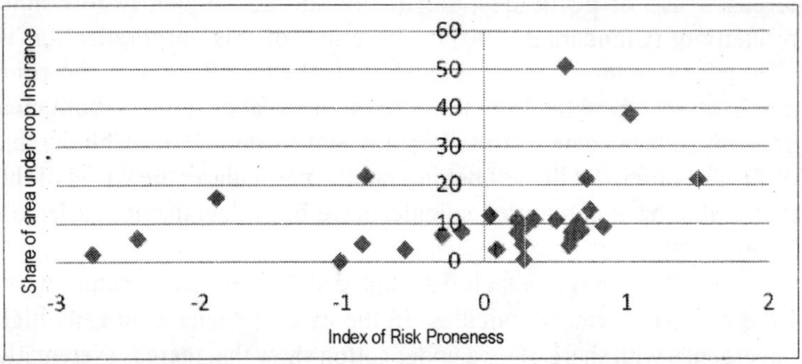

Fig. 4.9: Relationship between Risk Proneness and Insurance Coverage

Employment support programmes can provide the farmers under distress a route to diversify their income profile. MGNREGA, in spite of a lot of criticisms has promoted an enabling environment for greater economic security, increase in agricultural wages, slowing down of migration, creation of productive assets in vulnerable districts and higher levels of economic independence to women etc. (Dreze and Oldiges, 2007). Though Figs. 4.10 and 4.11 show a weak relationship between the risk-proneness and MGNREGA performance, many of the districts having high incidence of suicides have above average performance particularly in terms of share of households that worked for 100 days. In terms of jobs provided to job demanded, the vulnerable districts are lagging behind except Akola. But most importantly, the data for job demanded show that the share of this to total rural households in the high risk prone districts is higher than the state average except in Buldhana, in spite of a fair amount of commercialization in agriculture (Table 4.3). These figures are much higher compared, for example, to that in the low risk developed districts in Konkan. Notwithstanding the criticisms levelled against the scheme, the analysis indicates that the relief provided through employment could make a difference among the poor, and could enable them to cope with distress better.

The recent initiative taken by the government of changing the labour-material ratio from 60:40 to 51:49, rationing funds and restricting coverage to only one-third of the blocks instead of strengthening its implementation in the country would seriously undermine the potential the scheme has to provide relief to farmers facing acute distress (*Indian Express*, October 10, 2014).

Table 4.3: Proportion of Households that Demanded Job under MGNREGA to Total Rural Households

Low Risk Konkan District		*High Risk Vidarbha Districts With high incidence of Farmers' Suicides*	
Raigard	1.0	Akola	9.3
Tatnagiri	1.7	Amravati	12.4
Sindhudurg	5.3	Buldhana	3.1
Thane	5.5	Wardha	16.8
		Yavatmal	8.1
Maharashtra		**7.5**	

Source: Calculated from nrega.nic.in

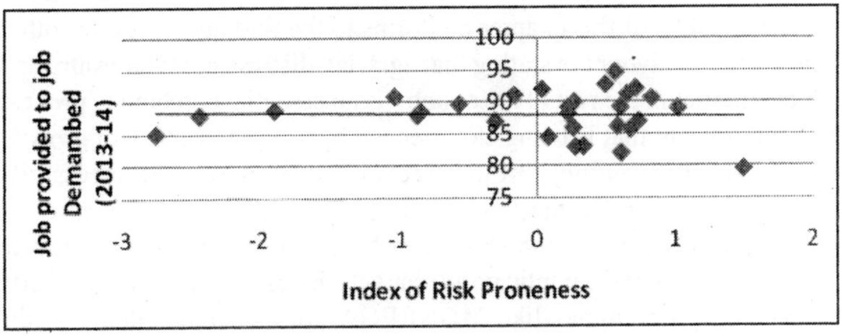

Fig. 4.10: Performance of MGNREGA (1) *vis-à-vis* Risk Proneness

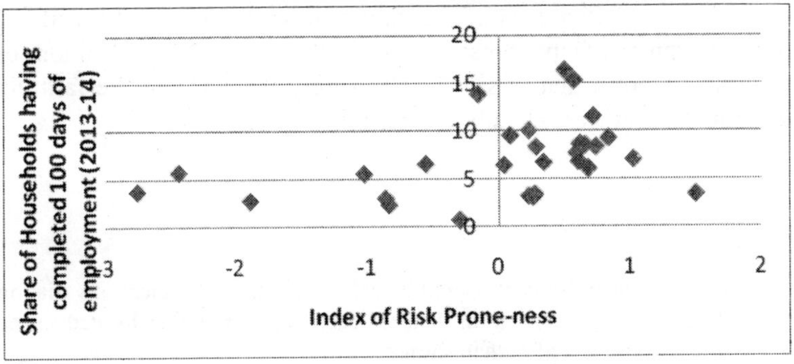

Fig. 4.11: Performance of MGNREGA (2) *vis-à-vis* Risk Proneness

Conclusion

The analysis in this chapter shows that the spatial patterns of risk proneness in Maharashtra close conforms to the incidence of farmers' suicides. Though the general reporting in the media and the government response to these suicides assume centrality of weather conditions in explaining such incidences, the study shows weak relationship with rainfall variability and farmers' suicides. To the contrary, both risk proneness and incidence of farmers' suicides almost completely coincide with the spatial distribution of area under cotton cultivation to total cropped area. The study reveals that the existing public institutional support as crop insurance cover can be far better targeted towards the districts prone to acute distress. Going by the demand for jobs under MGNREGA, it is observed that the risk prone districts mostly have higher than average job demand indicating the need for such schemes to enable the farmers to cope with acute distress.

The results of this chapter conforms to the findings of several other academic endeavours in stating that agrarian distress in Maharashtra and for that matter in the rest of the country is an essential outcome of process of development that is being experienced by us. The primacy of growth, sacrificing distribution, through seeking 'efficiency' will necessarily impact the farmers on the margin adversely, as the continuing spree of suicides post-liberalization show clearly. The withdrawal of state from investments that have public good characteristics and dilution of poverty alleviation programmes like MGNREGA are essential fallouts of the changing role of the state. Corporatization of agriculture through monopolistic private seed markets, contract farming and the like on the one hand and withdrawal of the public sector from the seed markets, credit systems and input subsidies on the other, are creating conditions of agrarian distress that is slowly and surely pauperizing the farming community that were already vulnerable.

Notes

1. Rainfall instability is measured by calculating the coefficient of variation of 25 years annual rainfall data for each weather station located in the different districts of Maharashtra.
2. From the Commission of Agricultural Costs and Prices, Ministry of Agriculture, Government of India, the cost of cultivation data was used to include cotton, sugarcane, groundnut, sunflower and safflower to compute the area under high value crops.
3. A price weighted average of 10 most important crops was taken for every district to compute yield instability over 10 years.

4. While the irrigation extent is measured by net irrigated area/net sown area, the irrigation intensity is a ratio of gross irrigated area and net irrigated area.
5. Crop diversification has been measured by 1 minus Herfindahl's index $\left(H = \sum_{i=1}^{N} s_i^2\right)$, where s_i is the share of crop i in the gross cropped area, and N is the number of crops.
6. In case of multi-stage PCA, the composite index formed at a lower level is used as a variable in the next step for computing the composite index and so on.

References

1. Drèze, J. and C. Oldige (2007). "Commendable act." *Frontline*, 24.14: 14-27.
2. Kennedy, J., and L. King (2014). The political economy of farmers' suicides in India: indebted cash-crop farmers with marginal landholdings explain state-level variation in suicide rates. *Globalization and Health,* 10(1): 16-16.
3. Madare, S.N., (2012) Global perspective on Farmer Suicides in India, *Indian Streams Research Journal*, 2 (II): 1-4.
4. Mishra, Srijit (2007). *Risks, Farmers' Suicides and Agrarian Crisis in India: Is There A Way Out?* Working paper-2007-014, Indira Gandhi Institute of Development Research, Mumbai, September: 1-20.
5. Narayanamurthy A. and S.S. Kalamkar (2006). Is Bt Cotton Cultivation Economically Viable for Indian Farmers? An Empirical Analysis, *Economic and Political Weekly,* 41(26): 2716-2724.
6. Narayanamurthy, A. (2006) Relief Package for Farmers: Can It Stop Suicides? *Economic & Political Weekly*, 41(31): 3353-3355.
7. Padhi, R. (2009). On Women Surviving Farmer Suicides in Punjab, *Economic & Political Weekly*, May 9, 2009, 44(19): 53-59.
8. Patnaik, Utsa (2003). "Global capitalism, deflation and agrarian crisis in developing countries." *Journal of Agrarian Change*, 3(1-2): 33-66.
9. Revathi, E. (1998). Farmers' Suicide: missing issues. *Economic & Political Weekly*, 33:1207.
10. Sarthak G. and S. Mishra (2012). *To Bt or Not to Bt? Risk and Uncertainty Considerations in Technology Assessment,* Indira Gandhi Institute of Development Research, Mumbai, January: 1-36.
11. Sahay, G.R. (2014). Globalization, Liberalization and Agrarian Distress: A Study of Suicides among Farmers in India, Presented in *Global Labour University Conference,* Berlin: 1-41.
12. Sen, S. and C. Mallick (2012). Understanding Gujarat's Agricultural Growth

in a Liberalizing Environment: Signs of a Redefined Margin? In Atul Sood (ed.) *Poverty amidst Prosperity, Essays on the Trajectory of Development in Gujarat,* Aakar Publication, Delhi: 88-127.

13. Swaminathan, M., and V. Rawal. (2011). Are there benefits from the cultivation of Bt cotton? A comment based on data from a Vidarbha village, *Review of Agrarian Studies,* 1.1 (2011): 101-124.
14. Vidyasagar, R. and K. Suman Chandra (2004). Farmers' Suicides in Andhra Pradesh and Karnataka, *Research Report Series-64, NIRD,* Hyderabad, April.

5

Trends and Variability in Costs and Returns in Crop Farming in India

Ankita Goyal, CSD

Agriculture is a way of life in India, with more than 64 per cent of the rural population still dependent on it for its source of income and livelihood. Thus, it has been and will continue to be central to all the policies to bring about socio-economic equality and development of the country. Over the last few years there has been an understanding among scholars that agriculture is becoming less remunerative as a primary occupation and thus farmers, especially the new generation are looking for opportunities to come out of agriculture.

This chapter seeks to look at the trends and variabilities in the costs and returns in the cultivation of major crops in India in the post-reform period. The statewise time series data of Cost of Cultivation for various crops were collected for the year 1996-97 to 2010-11 from the Ministry of Agriculture, Government of India and used for analyzing the trends and variability in costs and returns in the post-liberalization period, while the unit-wise data of the same source was used to find out the pattern of costs and returns in different size group of farmers at two points of time, i.e. 2003-04 and 2010-11, which were both normal agricultural years. The unit-wise for earlier period was not available.

Trends and Variabilities

Two major food crops of India, Paddy and Wheat, and two major cash crops, namely Cotton and Sugarcane have been studied using two commonly used cost concepts, A2+FL and C2. While A2 + FL represents the input costs, namely labour costs (including hired and family labour), fertilizer and pesticide costs, seed costs, irrigation charges etc., C2 cost represents variable cost plus fixed cost including imputed value of owned land, rental value of leased in or leased out land, depreciation cost etc. Value of total product of each of the crops studied has been calculated as value of main product plus the value of by-product. Table 5.1 gives us the trend growth rates and coefficient of variation of the costs of cultivation and the value of total product (main product + by-product) for the major Cotton growing states across India. It is seen from the table that for the period 1996-97 to 2001-02, Andhra Pradesh, Gujarat and Madhya Pradesh

Table 5.1: Trend Growth Rates and Coefficient of Variation of Costs of Cultivation and Value of Total Product of Cotton from 1996-97 to 2010-11

States	*Items*	*Trend Growth Rate (%)*				*Coeff. of Variation*			
		1996-97 to 2000-01	*2001-02 to 2005-06*	*2006-07 to 2010-11*	*2002-03 to 2010-11*	*1996-97 to 2000-01*	*2001-02 to 2005-06*	*2006-07 to 2010-11*	*2002-03 to 2010-11*
Andhra Pradesh	A2 + FL	–5.7	6.7	4.5	7.0	13.2	13.0	8.5	19.3
	C2	–5.7	4.8	7.9	6.4	13.7	16.6	12.7	19.7
	VOP	–0.7	–0.9	7.6	6.9	12.5	28.9	13.5	25.9
Gujarat	A2 + FL	–2.1	15.9	16.2	12.7	10.7	26.6	24.1	33.2
	C2	–4.9	18.3	19.0	14.0	14.7	28.9	28.0	37.1
	VOP	–16.5	30.1	27.0	17.9	34.9	41.8	43.1	51.8
Haryana	A2 + FL	8.9	11.6	11.7	12.2	15.0	17.7	19.2	31.9
	C2	8.8	16.8	13.2	12.3	15.8	24.1	20.1	32.1
	VOP	14.3	38.0	23.9	16.9	27.4	46.6	32.3	46.5
Karnataka	A2 + FL	7.6	3.3	18.4	15.6	15.9	13.1	29.0	42.9
	C2	3.8	2.3	20.9	15.5	13.2	10.0	33.9	46.1
	VOP	–3.2	1.4	27.4	16.5	16.8	24.1	51.0	61.8
Madhya Pradesh	A2 + FL	–16.9	22.0	–2.7	2.1	37.3	31.5	7.0	13.1
	C2	–11.2	25.6	3.3	6.5	23.6	39.4	10.2	21.7
	VOP	3.6	36.8	18.0	19.3	16.9	71.5	31.1	49.8
Maharashtra	A2 + FL	6.5	3.7	22.1	10.9	11.9	6.5	34.3	37.1
	C2	5.9	4.0	24.1	11.2	11.8	8.3	36.3	38.9
	VOP	0.2	6.3	30.8	14.8	9.8	18.1	44.5	50.8
Punjab	A2 + FL	9.4	2.1	9.3	5.4	19.4	8.9	14.3	17.1
	C2	8.1	7.8	13.0	9.5	20.9	12.6	19.2	26.3
	VOP	11.3	23.6	17.8	14.4	37.7	32.6	26.5	37.9
Rajasthan	A2 + FL	0.7	5.7	14.9	11.6	10.8	9.9	23.8	33.7
	C2	3.5	8.0	20.4	13.5	10.6	12.8	31.7	40.4
	VOP	2.4	2.7	30.2	16.0	10.3	9.1	49.1	57.1
Tamil Nadu	A2 + FL	3.5	–1.6	20.8	8.3	5.7	6.6	30.5	30.9
	C2	6.4	–3.1	20.8	7.2	10.7	8.7	30.9	29.2
	VOP	0.6	–5.6	28.8	15.1	3.2	15.4	42.0	51.1

Source: Calculated from Various CACP Reports, DACNET, Ministry of Agriculture, Government of India.

experienced negative rate of growth of costs A2+FL and C2 as well as for the value of output in these states except Madhya Pradesh which although had a low growth rate but was positive. The highest growth rate of value of output was in Haryana and Punjab which also had costs growing at a comparatively higher rate. In the period 2001-02 to 2005-06 the growth rate of value of output has been much higher for a number of states like Gujarat, Haryana, Madhya Pradesh and Punjab at around 30 per cent. The costs have grown at highest rate in Madhya Pradesh while they have grown at a negative rate in Tamil Nadu where the value of output also fell overtime. For the most recent period that is from 2006-07 to 2010-11, the growth rates for all items studied here have been positive for all states. The costs have increased sharply in Karnataka, Maharashtra, Rajasthan and Tamil Nadu but so have the value of total output except in Madhya Pradesh and Punjab where it increased at a lower rate than the previous period. To see the fluctuations in the costs and value of output of cotton in these three periods, we have calculated the coefficient of variation for each of these periods for all states. The fluctuations in the costs seem to be fairly low and uniform in the period 1996-97 to 2000-01 in all states except in MP where it fluctuated at 37 per cent and 24 per cent for the costs A2+FL and C2 respectively. In the next period studied, Madhya Pradesh continued with high degree of variation in costs while in the recent period, 2006-07 to 2010-11, the fluctuations declined to only 7 per cent in Madhya Pradesh but were high in Karnataka, Rajasthan and Tamil Nadu. As for the value of output, while the variations were quite modest for most states in the first period except Punjab, Haryana and Gujarat, in the next two periods, there was much higher fluctuation in most states except in Rajasthan, Tamil Nadu and Andhra Pradesh.

The period since 2002-03 has been very important for cotton production in India as this marked the introduction of Bt seed in most cotton producing states which is said to have changed the input costs as well as returns drastically. To see the effect of Bt cotton, we have calculated the growth rates of costs and value of output and the variability in the period 2002-03 to 2010-11. While the growth rates of the three items is positive for all states, the costs have grown at the highest rate in Karnataka at 16 per cent which is the same as the rate of growth of value of output in that state. The value of output grew at 15 per cent per annum or more for all states except Andhra Pradesh. Thus it can be safely said that the value of total product grew at a higher rate than the costs of cultivation since the introduction of Bt cotton. Coming to fluctuations, it is seen that

Table 5.2: Trend Growth Rates and Coefficient of Variation of Costs of Cultivation and Value of Total Product of Paddy from 1996-97 to 2010-11

States	*Items*	*Trend Growth Rate (%)*			*Coeff. of Variation*		
		1996-97 to 2000-01	*2001-02 to 2005-06*	*2006-07 to 2010-11*	*1996-97 to 2000-01*	*2001-02 to 2005-06*	*2006-07 to 2010-11*
Andhra Pradesh	A2 + FL	6.5	0.5	16.8	10.5	1.6	24.2
	C2	6.5	1.5	15.2	10.4	3.3	22.5
	VOP	6.6	3.8	14.1	11.6	8.4	23.3
Assam	A2 + FL	11.8	3.7	13.7	18.7	6.5	20.5
	C2	11.2	3.3	14.8	17.2	5.6	21.6
	VOP	9.8	0.7	20.4	16.1	6.0	28.0
Bihar	A2 + FL	9.3	7.9	11.7	14.5	12.3	18.5
	C2	7.1	6.6	9.4	11.0	10.1	14.6
	VOP	2.8	6.3	5.4	9.7	10.0	14.5
Haryana	A2 + FL	6.8	5.2	10.7	10.7	12.9	16.6
	C2	4.9	6.4	13.2	8.0	11.5	19.7
	VOP	9.0	6.7	13.3	15.3	10.6	21.2
Karnataka	A2 + FL	18.0	0.1	15.0	25.4	7.0	22.8
	C2	10.1	0.4	13.5	15.1	5.8	19.8
	VOP	–5.2	2.1	10.5	8.8	8.3	16.9
Kerala	A2 + FL	3.0	3.1	9.2	4.9	5.1	14.6
	C2	4.1	2.9	10.5	6.7	4.8	16.0
	VOP	8.4	0.6	16.1	13.5	5.2	23.1
Madhya Pradesh	A2 + FL	4.6	0.0	16.3	8.2	6.6	23.5
	C2	4.4	0.6	19.6	7.2	9.0	28.3
	VOP	–2.1	3.6	30.2	13.0	15.8	44.4
Odisha	A2 + FL	11.1	2.4	13.0	16.4	3.9	19.5
	C2	10.4	3.3	12.3	15.5	5.5	18.1
	VOP	7.3	3.8	12.9	11.1	11.1	19.9
Punjab	A2 + FL	7.2	5.0	17.6	11.0	14.8	25.1
	C2	5.5	5.8	15.3	8.9	10.7	22.3
	VOP	9.2	4.4	12.4	15.9	9.8	19.4
Tamil Nadu	A2 + FL	4.8	2.0	11.2	7.9	3.6	17.4
	C2	5.7	2.2	9.9	10.1	5.3	15.4
	VOP	0.6	–2.7	13.7	1.7	9.8	20.6
Uttar Pradesh	A2 + FL	8.9	5.8	11.9	14.0	9.2	18.4
	C2	7.9	6.5	13.8	12.9	10.1	20.6
	VOP	2.9	7.0	17.3	8.3	12.1	24.5
West Bengal	A2 + FL	5.5	2.0	15.0	9.8	4.2	22.7
	C2	3.3	3.3	13.4	8.2	5.7	20.2
	VOP	–2.7	6.7	14.7	13.8	10.5	21.9

Source: Calculated from Various CACP Reports, DACNET, Ministry of Agriculture, Government of India.

the coefficient of variation has been much higher for the entire Bt cotton cultivation period taken together.

Table 5.2 gives the trend growth rates and coefficient of variation of the costs of cultivation and the value of total product (main product + by-product) for the major Paddy growing States across India. The growth rates have been positive throughout the period but there has been a jump in the rate of growth in the recent period, over the previous two periods, for both the costs of cultivation, in all the twelve states studied here. Similarly for the value of total output from Paddy cultivation, the period 2006-07 to 2010-11 saw an accelerated rate of growth as compared to the other two periods studied for all states and more specifically for Madhya Pradesh where value of output of Paddy registered a rate of growth of more than 30 per cent per annum. While Punjab and Haryana showed the highest growth rates in the earlier period; the rate of growth was not exceptional in the period from 2006-07 to 2010-11 when compared to other States like Madhya Pradesh where rate of growth of

Table 5.3: Trend Growth Rates and Coefficient of Variation of Costs of Cultivation and Value of Total Product of Sugarcane from 1996-97 to 2010-11

State	*Items*	*Trend Growth Rate (%)*			*Coeff. of Variation*		
		1996-97 to 2000-01	*2001-02 to 2005-06*	*2006-07 to 2010-11*	*1996-97 to 2000-01*	*2001-02 to 2005-06*	*2006-07 to 2010-11*
Andhra Pradesh	A2 + FL	8.0	3.4	12.6	12.8	16.5	19.9
	C2	4.9	5.8	13.6	7.8	16.8	21.2
	VOP	1.2	10.6	18.4	3.7	20.5	30.5
Haryana	A2 + FL	15.8	1.1	8.3	30.6	15.0	24.3
	C2	19.2	2.7	12.5	32.9	5.5	29.0
	VOP	18.1	4.0	21.8	28.8	13.1	42.7
Karnataka	A2 + FL	16.4	14.4	12.2	28.1	20.8	26.4
	C2	11.1	13.5	16.4	19.6	20.1	28.8
	VOP	2.2	15.9	25.2	7.7	23.8	40.2
Maharashtra	A2 + FL	15.2	9.5	12.8	21.9	18.3	21.6
	C2	14.3	10.1	17.8	20.7	17.9	28.5
	VOP	8.0	15.5	31.1	17.3	30.7	50.7
Tamil Nadu	A2 + FL	6.1	6.0	10.9	10.6	13.6	17.8
	C2	7.4	4.3	8.9	13.4	11.3	14.4
	VOP	1.6	4.8	16.9	3.5	17.5	26.2
Uttar Pradesh	A2 + FL	10.1	2.7	13.5	15.4	14.1	23.8
	C2	7.0	7.2	13.7	10.8	14.5	22.5
	VOP	4.2	13.7	17.6	8.1	24.7	36.9

Source: Calculated from Various CACP Reports, DACNET, Ministry of Agriculture, Government of India.

value of output went from a negative 2.1 per cent to a positive 30 per cent per annum. Similar improvement was seen in Karnataka, Tamil Nadu, West Bengal and Uttar Pradesh. Bihar has experienced the lowest increase in the rate of growth of value of output in the entire period. It is also noticed that some States like Bihar, Karnataka and Punjab continued to have costs increasing at a higher rate as compared to the rate of growth of value of output.

Also the fluctuation in the costs and output value, during 2001-02 to 2005-06 was quite modest in terms of the value of coefficient of variation as compared to the other two periods and especially the most recent period where the average fluctuations in the values is more than 20 per cent in

Table 5.4: Trend Growth Rates and Coefficient of Variation of Costs of Cultivation and Value of Total Product of Wheat from 1996-97 to 2010-11

States	*Items*	*Trend Growth Rate (%)*			*Coeff. Of Variation*		
		1996-97 to 2000-01	*2001-02 to 2005-06*	*2006-07 to 2010-11*	*1996-97 to 2000-01*	*2001-02 to 2005-06*	*2006-07 to 2010-11*
Bihar	A2 + FL	8.3	1.2	12.8	12.7	1.9	19.8
	C2	7.6	0.9	12.1	12.3	2.7	18.1
	VOP	5.8	3.7	9.6	13.4	6.0	14.9
Gujarat	A2 + FL	10.8	1.1	7.9	16.5	4.2	12.2
	C2	9.7	1.7	8.8	15.3	3.8	13.5
	VOP	4.9	3.8	11.5	9.1	7.5	18.8
Haryana	A2 + FL	6.7	4.9	9.2	12.1	7.8	13.9
	C2	7.7	4.7	11.1	12.5	8.1	16.5
	VOP	7.5	2.4	10.6	12.7	5.3	16.2
Himachal Pradesh	A2 + FL	6.5	8.1	–3.9	13.4	12.5	16.1
	C2	5.8	11.0	–6.9	12.7	17.1	20.2
	VOP	–0.6	15.8	–11.0	14.5	24.5	32.7
Madhya Pradesh	A2 + FL	5.5	4.4	7.8	9.8	7.4	12.7
	C2	5.4	6.2	8.6	11.2	10.4	13.4
	VOP	3.0	10.6	10.0	13.6	17.3	16.5
Punjab	A2 + FL	6.9	2.7	9.0	11.0	4.9	14.5
	C2	6.8	3.6	9.2	11.1	7.2	14.0
	VOP	10.2	2.1	7.3	16.9	7.0	12.3
Rajasthan	A2 + FL	3.9	0.6	6.3	7.8	5.3	10.4
	C2	6.5	1.9	7.4	11.7	7.5	11.6
	VOP	5.8	7.3	8.5	12.4	14.2	13.3
Uttar Pradesh	A2 + FL	5.9	8.7	8.1	9.9	13.2	12.4
	C2	5.7	8.2	10.8	10.0	12.5	16.5
	VOP	2.3	5.1	7.8	6.5	9.3	12.6

Source: Calculated from Various CACP Reports, DACNET, Ministry of Agriculture, Government of India.

all states taken together. Here also Madhya Pradesh stands out with coefficient of variation of the value of output being 44 per cent in the period 2006-07 to 2010-11.

As regarding sugarcane, trend growth rates of value of output during 1996-97 to 2000-01 was very low as compared to the rate at which costs of cultivation grew in all the sugarcane growing states, except in Haryana and to some extent in Maharashtra. The trend was reversed in the next period where value of output showed decent growth, except in Haryana and Tamil Nadu, and the trend continued in the latest period with all the states showing rate of growth of value of output greater than that of cost. Amongst all the six States, Maharashtra and Karnataka have shown the highest growth in the value of output while the increase in costs remained lower.

During 2006-07 to 2010-11, all the States have shown consistently higher growth rate in the value of output than in the costs of production, but fluctuations in the annual growth rate of value of output was much higher than in the costs of cultivation (Table 5.4).

Table 5.4 gives the trend growth rates and coefficient of variation of the costs of cultivation and the value of total product for wheat in the major wheat growing States across India. The growth rate of both the costs and the value of output has been positive throughout the periods studied, except in Himachal Pradesh in 2006-07 to 2010-11 where the value of all three have shown a deceleration. The value of output declined at about 11 per cent per annum. The rate of growth was quite modest in all the three periods, with costs growing at the highest rate in Bihar at around 12 per cent in 2006-07 to 2010-11 and value of output increasing at the highest rate in Gujarat at 11.5 per cent per annum in the same period. Punjab, Haryana and Uttar Pradesh have invariably experienced higher rate of growth of costs as compared to the growth rate of value of output in all the three periods. Fluctuations have been moderate in all the three periods and quite similar in all the States for all the three variables considered here.

Having looked at the rates of growth and variations in the costs of cultivation and the value of output of the four crops overtime, it is now important to see how the returns from cultivation fared in these periods.

In the tables that follow three years moving average of the net returns for Paddy, Wheat, Cotton and Sugarcane in the major producing States for the period 1996-97 to 2010-11, was calculated. In case of Cotton it is seen that while all other States faced negative returns over C2 in the earlier years, Rajasthan fared much better all through the period. Since

Table 5.5: Three-Year Moving Average of Gross and Net Returns per Hectare in Major Cotton Producing States

States	*AP*	*Gujarat*	*Haryana*	*Karnataka*	*MP*	*Maharashtra*	*Punjab*	*Rajasthan*	*TN*
Net Returns over (A2+FL) (Rs/Ha)									
1996-98	7875	11051	4823	3862	3278	4520	1077	12440	5785
1998-2000	8816	5500	8622	1744	2892	3624	2260	13834	4976
2000-02	12970	2374	3587	5202	1546	1502	3145	13702	4183
2002-04	19519	13463	11939	7632	–370	4652	12200	15496	3174
2004-06	12451	16099	10723	5385	9858	3071	19535	15418	3898
2006-08	21359	22299	20531	10506	15699	7804	29413	20965	10648
2008-10	25705	42312	34739	20056	29658	17027	44287	44784	23695
Net Returns over C2 (Rs/Ha)									
1996-98	–241	6692	–307	708	42	1585	–4037	8731	–947
1998-2000	1838	2093	2253	–934	–322	300	–3746	8814	–3879
2000-02	3807	–1573	–1614	813	–2235	–2651	–4567	9136	–5662
2002-04	6319	7692	3658	2376	–5300	–940	1695	9804	–5427
2004-06	1583	8664	1237	505	405	–1949	6634	8932	–3718
2006-08	7788	12767	8500	4242	4920	1133	11798	13188	3429
2008-10	8414	26711	18541	10309	15752	5957	19987	31147	12534

Source: Calculated from Various CACP Reports, DACNET, Ministry of Agriculture, Government of India.

2006-07 it is seen that all States have had positive returns over C2 cost which is also the period since when Bt cotton came into Indian agriculture in a big way. While it is noticed that Punjab has come a long way from earning the highest negative returns over C2 till 2002-03 to positioning itself among the top three net gainers in the recent period, Maharashtra which fared pretty well in 1996-1998 has the lowest returns among all other States since 2006-08 to present. Also, on comparing with other States, Andhra Pradesh has not experienced high net returns from cotton production. It is also worth mentioning that Rajasthan, which has been the leader throughout has almost 40-50 per cent of the cotton area in Rajasthan under desi variety of seeds rather than Bt seeds.

Table 5.6 illustrated the net returns to Paddy farmers in major States producing Paddy across India. Apart from Haryana, Punjab, Andhra Pradesh and Karnataka, all other States experienced negative returns over C2 in earlier years. While Punjab and Haryana are undoubtedly major Paddy producing States, Andhra Pradesh, Karnataka and Kerala have also fared well in recent years. The eastern States of Assam, Bihar, Odisha and West Bengal still lag behind in terms of returns to Paddy farmers.

Table 5.7 shows the net returns per hectare in major Sugarcane producing States. Sugarcane is a major annual cash crop for many farmers and the high returns justify the farmers' decision to cultivate it. While Tamil Nadu and Karnataka led the other States in terms of the net returns in the earlier period, Karnataka moved much ahead by 2008-2010. Maharashtra was the lowest gainer and was way behind other states until 2006-2008 but has picked up in the latest period to come at par with others. Amongst all other States, Andhra Pradesh seems to have gained the least in terms of returns over C2 earned per hectare. It would also be seen from Table 5.7 that the net returns even at current prices fluctuate widely from year to year in all the States.

Table 5.8 gives the average net returns per hectare in major wheat producing states. All the six States where most of the wheat of the country is produced have similar positive net returns with Rajasthan gaining the maximum returns since 2002 onwards. The lowest net returns over C2 were earned in Bihar throughout the period. Also net returns varied from year to year.

Variations in Land Size Classes

It is understood that growth is not the sole objective of economic policy. It is necessary to ensure that the benefits of growth accrue to all sections

Table 5.6: Three-Year Moving Average of Gross and Net Returns per Hectare in Major Paddy Producing States

States	*AP*	*Assam*	*Bihar*	*Haryana*	*Karnataka*	*Kerala*	*MP*	*Odisha*	*Punjab*	*TN*	*HP*	*WB*
Net Returns over (A2+FL) (Rs./Ha)												
1996-98	8722	4155	4798	8962	22210	3184	3112	4383	11397	11350	6073	7937
1998-2000	10281	4628	4361	11061	14705	6088	2595	4328	14336	9834	5754	5920
2000-02	10724	2878	2891	12723	10523	6690	774	3117	17066	10142	4542	1659
2002-04	14609	2010	3086	12983	10397	4853	1335	3005	18242	9348	5614	2831
2004-06	14435	1138	3940	17629	13216	6265	2556	4132	22250	6283	6160	5252
2006-08	20880	3677	8906	34829	22114	14625	8790	8411	34907	12650	12385	7450
2008-10	25575	6229	6965	41957	24643	23383	15927	9990	40233	17984	17918	9939
Net Returns over C2 (Rs./Ha)												
1996-98	1199	1473	1286	1925	14713	–1015	–393	816	3545	5339	3809	2071
1998-2000	1599	1412	204	3757	7573	1025	1641	8	5973	2619	3172	243
2000-02	1232	–559	–1405	4313	3170	1372	3175	1405	7505	1854	–611	4076
2002-04	4105	1534	–1305	2886	2423	–186	2079	2039	8284	575	–585	3613
2004-06	379	2554	–894	6240	5088	265	–1191	–1371	9932	2973	–317	1824
2006-08	679	–1096	3148	13160	11807	6459	3135	758	18442	2144	3752	528
2008-10	7858	–252	527	20725	12397	12528	7046	881	39486	5952	5931	448

Source: Calculated from Various CACP Reports, DACNET, Ministry of Agriculture, Government of India.

Table 5.7: Three-Year Moving Average of Gross and Net Returns per Hectare in Major Sugarcane Producing States

States	*AP*	*Haryana*	*Karnataka*	*Maharashtra*	*TN*	*UP*
			Net Returns over (A2+FL) (Rs/Ha)			
1996-98	28940	26113	41599	20400	50120	24475
1998-2000	25208	38277	38133	19129	48374	24105
2000-02	22731	37344	37457	12649	40636	23134
2002-04	27031	39316	50301	24355	31551	29131
2004-06	42721	50042	62648	35783	44032	44913
2006-08	41639	51077	61526	28271	53778	41245
2008-10	75011	99838	121674	94972	84906	70805
			Net Returns over C2 (Rs/Ha)			
1996-98	10911	12681	26816	10278	33360	13168
1998-2000	7276	17109	22072	6469	27429	12697
2000-02	4053	14199	20341	–2166	19974	11227
2002-04	7992	16882	29007	6531	12608	14186
2004-06	15344	22258	39457	14498	22067	26545
2006-08	10677	23946	37799	5582	30707	21379
2008-10	32483	58885	84176	53450	59547	43790

Source: Calculated from Various CACP Reports, DACNET, Ministry of Agriculture, Government of India.

Table 5.8: Three-Year Moving Average of Gross and Net Returns per Hectare in Major Wheat Producing States

States	*Bihar*	*Gujarat*	*Haryana*	*MP*	*Punjab*	*Rajasthan*	*UP*
			Net Returns over (A2+FL) (Rs/Ha)				
1996-98	6009	9557	14185	5673	13725	9898	10556
1998-2000	6905	9563	17146	6287	18308	12183	10704
2000-02	4997	10452	15823	5450	17496	12899	9355
2002-04	5275	10648	14928	7234	15397	15420	9283
2004-06	7548	13703	18487	11524	20043	20216	11769
2006-08	12523	18726	30448	18099	29570	25888	20567
2008-10	13541	24360	36588	21925	32188	31811	22913
			Net Returns over C2 (Rs/Ha)				
1996-98	1900	5934	6605	1755	5481	5929	5398
1998-2000	1868	5385	8213	1684	8644	6928	4924
2000-02	114	4977	6666	666	7614	6700	3066
2002-04	206	5774	5738	1803	5973	8932	2353
2004-06	2509	8371	7500	4622	8036	12563	3865
2006-08	7018	12234	15248	8846	13556	16288	10390
2008-10	6786	16130	16880	10910	13355	19749	9249

Source: Calculated from Various CACP Reports, DACNET, Ministry of Agriculture, Government of India.

of the society. So in this section, we see how the farmers belonging to different land size classes cultivating these crops have performed. Since Paddy and Wheat are cultivated in rotation and constitute the major crops grown in a year, we take them together for States like Bihar, Haryana, Madhya Pradesh, Punjab and Uttar Pradesh. The other crops studied here are Sugarcane, which is an annual cash crop, and cotton which is another major cash crop grown.

For this analysis unit level data of the cost of cultivation studies for the two crop years 2003-04 and 2010-11 which were normal agricultural years without any floods or droughts and thus comparable, were used.

Table 5.9 illustrates the net returns for different categories of farmers' cultivating Paddy and Wheat crop in rotation. In 2003-04, the net returns from paddy and wheat cultivation taken together were negative for all states for all size classes except small and semi-medium farmers in Madhya Pradesh. In 2010-11, also the net returns continued to be negative for all states in all size categories except Haryana and Madhya Pradesh which showed decent positive net returns. While in Madhya Pradesh the net returns increased with size class, it was highest for the medium category of farmers in Haryana. Since the average is only indicative of the mean of all size classes taken together, it is imperative to look at the percentage of population belonging to each category that has value of product below the average of the state. Also, assuming that farming is the primary occupation, the net returns from major crop cultivated must be looked at *vis-à-vis* the poverty line of that state.

Table 5.9 also gives the percentage of households with net returns below the poverty line of that state for the two years 2003-04 and 2010-11. The results show that in 2003-04 Bihar and Madhya Pradesh had nearly 100 per cent of all groups of farmers gaining net returns below poverty line. Uttar Pradesh is no better. The only states doing slightly better were Haryana and Punjab which also deteriorated by 2010. Thus, the crop rotation known to be most remunerative to farmers especially in Punjab and Haryana is also no longer yielding results.

The percentage of households falling below average value of output, it is observed that the proportion has somewhat increased in 2010-11 from what was observed for 2003-04 for all States. More than 75 per cent of the small farmers in Haryana and Madhya Pradesh fall below the average VOP of the state. This gives us an idea about the inequality in the state and shows us that the average does not represent all farm categories.

Moving over to Sugarcane cultivation by different categories of farmers, it is seen from the table 5.10 that in 2003-04, among the four

Table 5.9: Size Class-wise Net Returns from Paddy and Wheat Together and Proportion of Farming Households with Net Returns Below Poverty Line and Total VOP of Paddy and Wheat Below Average for 2003-04 and 2010-11

Size Group	Net Returns Rs./Ha				% of HH with			
	VOP-C2		VOP-(A2+FL)		Net Returns Below Poverty Line		Total VOP Below Avg VOP	
	2003-04	2010-11	2003-04	2010-11	2003-04	2010-11	2003-04	2010-11
				Bihar				
Marginal	−11376.8	−8079.1	4241.6	19643.0	100.0	100.0	53.6	43.8
Small	−9329.1	−1363.9	218.5	20517.7	99.5	100.0	52.9	50.7
Semi-Medium	−8741.0	−2263.8	−575.9	21288.0	100.0	100.0	57.4	60.5
Medium	−3904.2	−5414.6	1065.7	24899.2	100.0	100.0	68.3	66.0
Large	−10172.2	−2573.9	3654.9	19956.9	98.2	100.0	61.5	73.3
				Haryana				
Marginal	−2550.3	15155.2	28745.5	78792.8	58.3	89.5	41.7	78.9
Small	−14784.7	27636.4	23117.2	85312.2	67.6	80.6	58.8	48.4
Semi-Medium	−4979.4	20803.2	24907.9	88092.5	57.5	81.3	35.0	56.3
Medium	−21325.1	31917.1	27124.9	90270.1	45.0	65.9	42.5	58.5
Large	−10776.2	20579.2	27170.5	87107.7	37.1	100.0	32.3	65.9
				MP				
Marginal	−2426.8	9668.2	4510.3	37359.1	100.0	80.0	52.2	75.0
Small	2586.4	12127.9	2905.8	40494.8	100.0	76.0	51.9	72.0
Semi-Medium	1647.1	14379.8	2645.1	45003.0	93.3	71.9	56.7	65.6
Medium	−583.4	13545.7	5716.2	50739.1	90.3	58.3	51.6	60.4
Large	−813.1	16712.1	4781.2	52437.5	87.8	60.9	43.9	54.3
				Punjab				
Marginal	6159.5	−8417.4	32909.9	78847.4	42.0	86.4	38.0	54.2
Small	−5845.3	−34172.6	27856.8	78512.1	41.9	93.3	47.7	43.3
Semi-Medium	−20222.3	−33072.0	31837.7	81958.8	25.9	82.4	43.1	49.1
Medium	−27550.2	−44956.5	30821.5	87703.7	29.7	76.0	59.4	49.6
Large	−29147.9	−33397.3	32119.1	89747.1	15.5	79.0	50.7	44.6
				UP				
Marginal	−11671.3	−12543.6	16318.3	43524.9	76.0	81.1	54.5	66.2
Small	−33758.0	−42889.2	12439.4	45799.2	68.7	84.2	45.6	52.2
Semi-Medium	−38071.9	−31409.4	16183.2	52573.2	58.9	67.1	44.5	69.6
Medium	−31697.7	−37042.0	14557.5	48860.0	71.1	71.1	60.1	58.7
Large	−33055.4	−42068.4	12948.5	53461.0	66.7	63.9	59.1	63.9

Source: Calculated from Various CACP Reports, DACNET, Ministry of Agriculture, Government of India and Poverty Line Estimates of the Planning Commission.

states studied, Maharashtra and Uttar Pradesh showed negative net returns over C2 for all farm classes. Of the two states, the magnitude was much higher in Maharashtra. In 2010-11 it is observed that in Tamil Nadu the returns over C2 were almost similar for all size classes of farmers having somewhat declined with size of farm, while in Andhra Pradesh the returns

Table 5.10: Size Class-wise Net Returns from Sugarcane and Proportion of Farming Households with Net Returns below Poverty Line and Total VOP of Sugarcane below Average for 2003-04 and 2010-11

Size Group	*Net Returns (Rs./Ha)*				*% of HH with*			
	VOP-C2		*VOP-(A2+FL)*		*Net Returns Below Poverty Line*		*Total VOP Below Avg VOP*	
	2003-04	*2010-11*	*2003-04*	*2010-11*	*2003-04*	*2010-11*	*2003-04*	*2010-11*
				Andhra Pradesh				
Marginal	3086.3	7356.0	24273.8	56370.1	80.0	36.4	90.0	54.5
Small	14041.7	32259.8	39574.8	80411.9	40.0	26.7	73.3	46.7
Semi-Medium	14011.7	22664.7	36663.0	72078.9	20.0	22.7	53.3	59.1
Medium	13916.5	24206.7	36099.7	79430.8	28.6	20.0	57.1	55.0
Large	5221.8	32107.9	34858.1	88864.9	24.0	12.5	60.0	50.0
				Maharashtra				
Marginal	–36513.9	39332.2	–172.0	109905.4	95.2	12.0	52.4	56.0
Small	–21354.1	–5347.0	2886.4	87142.1	94.3	28.0	57.1	56.0
Semi-Medium	–34172.3	28614.3	9299.2	107725.4	87.8	20.9	53.7	45.3
Medium	–14629.5	1717.5	16819.6	95632.0	76.9	18.0	44.2	51.7
Large	–14182.8	12582.3	16370.6	117789.8	81.6	11.1	63.2	48.5
				Tamil Nadu				
Marginal	6071.1	72107.7	27109.9	112687.1	57.1	0.0	71.4	25.0
Small	5426.6	70956.1	30218.1	115502.8	29.4	3.1	47.1	34.4
Semi-Medium	–182.3	64505.5	31727.9	104108.1	36.6	8.5	56.1	47.9
Medium	6686.6	63283.7	36342.6	106696.3	28.2	2.2	56.4	50.5
Large	10492.5	54507.6	34197.0	102860.4	41.4	0.0	55.2	56.0
				Uttar Pradesh				
Marginal	–6499.1	21267.5	24343.4	69601.1	57.5	25.0	70.0	52.5
Small	–6547.0	12876.9	26753.2	70472.3	52.6	13.4	60.8	54.6
Semi-Medium	–11060.1	4952.2	26891.9	75759.5	52.4	17.9	56.0	47.9
Medium	–4828.3	2130.6	29859.4	73267.7	41.2	12.5	36.8	42.1
Large	–3017.4	22257.4	34866.0	67068.5	23.7	16.2	18.6	64.9

Source: Calculated from Various CACP Reports, DACNET, Ministry of Agriculture, Government of India and Poverty Line Estimates of the Planning Commission.

were lowest for the marginal farmers and for the small farmers in Maharashtra. In Uttar Pradesh, semi medium and medium farmers were the least gainers. Looking at the percentage of households with net returns below poverty line, it was found that Andhra Pradesh and Maharashtra were much worse off than Tamil Nadu and Uttar Pradesh in 2003-04 but the proportion has declined over time.

As for the percentage of farming households with total output value below average, it was much higher overall in 2003-04 as compared to 2010-11. In Andhra Pradesh, almost 90 per cent of the households under marginal category had their VOP below average. Similarly, more than 70 per cent of the marginal households in Tamil Nadu and Uttar Pradesh had the value of sugarcane output below the average VOP of Tamil Nadu.

Table 5.11 illustrates the returns from cotton and proportion of farming households with VOP less than poverty line and VOP of various size classes *vis-à-vis* average VOP. In 2003-04, except Andhra Pradesh and Gujarat, in all others states cultivating cotton, the net returns are seen to be negative for all farm classes. In Karnataka, apart from small and medium farmers all others received positive net returns.

Table 5.11: Size Class-wise Net Returns from Cotton and Proportion of Farming Households with Net Returns below Poverty Line and Total VOP of Cotton below Average for 2003-04 and 2010-11

Size Group	*Net Returns Rs./Ha*				*% of HH with*			
	VOP-C2		*VOP-(A2+FL)*		*Net Returns Below Poverty Line*		*Total VOP Below Average VOP*	
	2003-04	*2010-11*	*2003-04*	*2010-11*	*2003-04*	*2010-11*	*2003-04*	*2010-11*
				AP				
Marginal	586.3	5390.1	16177.1	25651.5	100.0	82.1	100.0	57.1
Small	6936.5	4508.9	24831.9	25316.0	66.7	83.3	44.4	73.3
Semi-Medium	5012.7	4116.6	22502.6	28831.0	50.0	75.9	41.7	55.2
Medium	6111.7	10998.2	22022.4	36039.6	66.7	72.2	55.6	41.7
Large	10285.6	5729.3	28510.3	26594.1	75.0	77.1	62.5	48.6
				Gujarat				
Marginal	10295.7	44716.3	20479.7	71382.3	74.4	29.0	66.7	56.5
Small	9544.9	43927.8	22192.4	70470.9	73.2	26.0	55.4	54.5
Semi-Medium	8684.0	44577.3	22474.5	73027.8	65.5	25.4	56.0	55.3
Medium	3575.5	41650.7	22262.5	72147.7	67.3	32.1	57.1	54.3
Large	6008.5	39659.7	24540.5	72429.5	69.3	25.9	58.6	53.5

(Contd.)

(*Contd.*)

Size Group	Net Returns Rs./Ha				% of HH with			
	VOP-C2		VOP-(A2+FL)		Net Returns Below Poverty Line		Total VOP Below Avg VOP	
	2003-04	2010-11	2003-04	2010-11	2003-04	2010-11	2003-04	2010-11
				Haryana				
Marginal	–780.7	7879.4	15052.4	28948.9	87.0	66.7	47.8	50.0
Small	–2969.0	22207.4	12849.6	43889.0	95.0	60.0	60.0	46.7
Semi-Medium	–4976.3	14546.1	12911.8	38328.2	90.6	65.5	62.5	65.5
Medium	–11101.9	19410.8	14310.9	42144.2	95.1	64.0	46.3	52.0
Large	–8875.9	19619.6	12605.5	44867.7	92.9	64.0	66.7	52.0
				Karnataka				
Marginal	661.8	14763.7	4159.7	30351.9	100.0	60.0	100.0	70.0
Small	–4277.2	29406.0	–2156.6	45545.9	100.0	41.7	100.0	50.0
Semi-Medium	3281.7	21525.7	12423.3	38570.7	80.0	57.9	60.0	63.2
Medium	–337.9	24006.5	6122.9	49080.4	85.7	42.9	71.4	42.9
Large	2205.6	25910.4	11366.3	44428.0	71.4	47.6	50.0	66.7
				Maharashtra				
Marginal	–1397.1	–11439.3	8038.3	30561.3	97.8	75.0	48.9	42.9
Small	–2126.6	–3025.9	8833.9	27740.7	98.1	75.3	51.9	56.2
Semi-Medium	–2707.2	–2741.4	10683.5	25686.0	92.9	75.6	52.9	61.4
Medium	–4245.1	–8340.7	10622.4	30740.6	94.7	72.7	62.8	55.9
Large	–2403.4	–3191.9	9233.7	26516.1	98.3	79.2	68.4	60.1
				Punjab				
Marginal	–3060.1	195.0	13528.3	34562.3	89.5	69.2	42.1	61.5
Small	–29.7	1624.8	18908.6	40084.8	82.4	54.5	41.2	54.5
Semi-Medium	–9567.2	–2656.9	13464.9	52126.5	95.8	52.9	50.0	52.9
Medium	–25535.2	–3772.0	13065.6	46313.0	93.3	50.0	53.3	50.0
Large	–18975.8	–8443.5	17319.1	58598.2	90.9	40.0	36.4	35.0
				Rajasthan				
Marginal	–4208.9	41969.3	4427.1	64484.0	100.0	20.0	57.1	60.0
Small	–6770.5	48599.8	6482.4	73165.6	88.9	0.0	77.8	50.0
Semi-Medium	–97.0	40982.6	12549.5	68796.5	93.3	9.1	53.3	36.4
Medium	860.3	36996.0	16709.9	78734.1	76.2	7.1	42.9	35.7
Large	–332.8	29985.4	15128.1	71871.8	90.0	21.4	45.0	50.0
				Tamil Nadu				
Marginal	–10422.0	25627.6	11103.4	41297.3	81.8	25.0	54.5	50.0
Small	–10646.6	14399.8	4117.8	34324.2	100.0	40.0	62.5	80.0
Semi Medium	–10762.9	23456.0	9748.9	36659.3	90.0	75.0	60.0	75.0
Medium	–7028.2	49985.4	11530.1	63144.1	100.0	37.5	61.5	50.0
Large	–8365.8	–6418.8	18233.0	33301.7	85.7	60.0	28.6	50.0

Source: Calculated from Various CACP Reports, DACNET, Ministry of Agriculture, Government of India and Poverty Line Estimates of the Planning Commission.

In 2010-11, all farm classes in Maharashtra continued to earn negative returns from cotton cultivation along with medium and large farmers of Punjab. The rest of the states had all classes earning positive net returns with Gujarat and Rajasthan being the highest gainers. The positive net returns earned by all classes of farmers in Andhra Pradesh were much less as compared to their counterparts in other States.

Looking at the percentage of households with net returns below poverty line, we find that returns from cotton are no better. It seems to have slightly improved over time but still majority of farmers are not able to earn even above the minimum poverty line. Only Rajasthan and Tamil Nadu and to some extent Gujarat seems to be doing better in recent years. Thus cotton growers are much better off in Rajasthan as compared to Maharashtra and Andhra Pradesh, where the maximum farmer suicides have been reported.

The percentage of total farming households with value of cotton produced below the poverty line, was quite high in all the States in 2003-04. Andhra Pradesh seems to be a little better amongst all other states with only 25 per cent of the semi medium and 11 per cent of the medium farmers having value of output below the poverty line. By 2010-11 the situation improved in all other states while it worsened in Andhra Pradesh. Rajasthan is observed to have no household with value of output of cotton falling below the poverty line of that state while it continued to be high in Maharashtra and worsened in Andhra Pradesh. In 2010-11, the worst off farmers are in Andhra Pradesh and Maharashtra and the situation was similar across size classes.

Conclusion

Thus, it may be concluded by saying that the situation has improved somewhat over the years in terms of the returns from crop cultivation for certain sections in some states but it is still not at par with the profits and standards of living offered by other occupations when compared with the amount of efforts and risks involving in agriculture. It is for this reason that the existing farmers are looking for opportunities to move out of agriculture, if not completely then partially by no longer keeping it as their primary occupation. Apart from providing remunerative prices to farmers for their output produced so that it covers the input cost as well as the risk undertaken and gives them an incentive to continue in agriculture, the need of the hour is to build agro-industries and other non-farm business opportunities in local areas. Employment of locals in these industries would give them an alternative income opportunity without having to leave agriculture.

6

Will Economic Diversification Help Reduce Farm Distress?

T. Haque

It is now widely recognized that the size of land holding of majority of Indian farmers is so low that it cannot generate adequate employment and income for their sustenance based on crop cultivation only (Haque, 1996). This is particularly true if small farmers cultivate only low value subsistence crops and do not undertake other off-farm and non-farm activities such as horticulture, animal husbandry, fishing, agro-prcessing, agro-services, trade, etc. At the same time, it is also seen that farmers who grow cash crops like cotton, groundnut, sugarcane, etc. having high input costs and high fluctuations in crop yields and market prices, suffer most as they have to borrow money from formal and informal sources and fail to pay back in years of low prices (Dogra, 2013; Rao and Suri, 2006). This chapter analyses the pace and pattern of horizontal and vertical diversification in India's rural economy and examine whether such diversification will help improve farmers income and reduce distress and if so what constrains them.

Rationale for Diversification

Horizontal diversification in the context of agriculture, refers to diverse agricultural activities undertaken by farmers for either increasing or stabilising their income levels or both. In a macro-sense, it refers to structural changes in the production pattern in the agriculture and allied sector, including crop production, horticulture, livestock, fisheries, etc. Such diversification enables the farmers to take avantages of production complimentaries and to reap the benefits of synergism through appropriate choice of crop and non-crop enterprises (Taylor, 1994). *Ceteris paribus*, diversified farms generally have higher growth and stability of income, as the risks due to fluctuating production and market prices would be minimum, as all crops/enterprises on the same farm would not face adverse weather or price situation at the same time. Besides, vertical diversification which increases farmers access to farm employment helps improve household income and reduce poverty (Sheila Bhalla, 1998).

Trends and Impact of Horizontal Diversification

There has been a significant change in the cropping pattern as well as in the relaive share of various crops in the aggregate value of crop output in the past few decades.

The proportion of area under rice and wheat increased significantly in the States like Punjab, Haryana, Uttar Pradesh presumambly under the impact of irrigation, new seed varieties and assured price and marketing support. The proportion of area under coarse cereals declined over time in almost all the States, while that of pulses and oil seeds increased in most States. Also the proportion of area under sugarcane and cotton, increased over time in most states (authors own calculation, based on data compiled from the Directorate of Economics, Ministry of Agriculture, Government of India).

No doubt, dominant rice-wheat cropping system in Punjab and Haryana along with cotton and sugarcane have improved the farm productivity and income quite significantly, but their adverse impacts in terms of depletion of ground water and erosion of soil health have been widely reported as these are all high water consuming crops. Despite high costs of production, these crops do not yield well in unirrigated areas and therefore, the farmers in rainfed areas have not benefitted much from either rice-wheat cropping system or Bt cotton and sugarcane. Besides, the relative profitability of rice and wheat crops vis a vis pulses and coarse cereals was the main factor behind the increase in rice and wheat area in Punjab and Haryana.

Diversification towards Horticulture

India has an area of about 6.6 million hectares under fruits with a production of 75.3 million tonnes, while area under vegetables constitutes about 8.8 million hectares with a production of 150.6 million tonnes (as of 2011-12).

The share of fruits and vegetables in the total value of crop output at constant prices increased from about 23.6 per cent during 1999-02 to 27.3 per cent in 2008-11. Considering the country as a whole less than 5 per cent of the net sown area is under fruits and vegetable crops, but it contributes about 27 per cent to the total value of crop output, indicating the importance of horticulture in the agricultural economy of India. It would not be out of place to mention here that fruits and vegetables have untapped potentials for raising farm incomes and reducing distress.

Currently there is a huge gap between potential yield and actual yield of almost all fruit and vegetable crops. The results of ICAR demonstration plots show that there is scope for raising the yield levels quite substantially through appropriate technologies and managerial interventions (ICAR, 2007).

First, the present level of technological development for fruit and vegetable crops is not only inadequate, but also the available technology does not equally suit various agro-climatic situations (Haque, 1996).

Second, infrastructure variables such as roads, markets, cold storage, cool chains, transportation, are significant determinants of horticultural diversification (Joshi *et al.*, 2007).

Third, most of the fruit trees yield only after a time lag and therefore, small and marginal farmers on their tiny holdings and having low investible surplus, may find it difficult to diversify in favour of fruit crops, especially *vis-à-vis* large farms.

Fourth, from the demand side, the roles of growth of income and urbanization in influencing horticultural diversification are well recognized because of increased marketability of high value commodities in urban centres and high income groups.

Fifth, *Ceteris paribus*, it is the economics of fruit and vegetable cultivation which determines the pace of horticultural diversification. The relative returns from fruits and vegetables are comparatively higher in most places (Singh and Sahoo, 2007). Also the recent trends in prices show that the whole sale price index of fruits and vegetables (base 2004-05=100) increased from 108.0 in 2005-06 to 257.33 in 2014-15, while that of rice rose from 105.2 in 2005-06 to 241.4 in 2014-15 and wheat from 105.0 in 2005-06 to 211.7 in 2014-15. Also the per hectare value of output from fruits and vegetables is several times higher than that of cereals and pulses. However, farmers may not always have the ability to invest in crops which require high investment of capital and labour inputs. Besides, proper market and price information do not exist for some of the high value, perishable commodities which constrain the process of diversification (Haque, 1996). Moreover, lack of local processing and value addition stand in the way of horticultural diversification.

Diversification towards Livestock

According to Livestock Census, 2007, India had about 530 million livestock population. Besides, it had 469 million poultry birds and 19 million dogs. Livestock sector employs about 5.5 per cent of the total workforce (Planning Commission, Government of India, 2007). In the arid and semi-arid areas, it also acts as an insurance against the vagaries of nature, such as drought, famine and other calamities.

The livestock's share in the total value of output of agriculture and allied sector increased from about 16.7 per cent in the 1950s to 27.2 per cent during 2000-01 to 2005-06 and then marginally dropped to 25.6 per cent during 2008-11.

Considering the role of some principal factors (Table 6.1), it was found that access to fodder, veterinary care services, road connectivity and membership of milk co-operatives explained nearly 44 per cent of the total variations in the share of livestock output in different regions, of which the first two variables turned out to be statistically significant. In fact, the role of dairy co-opeatives in promoting milk production in several states is well known. Even in a relatively under developed state of Bihar, the Sudha Dairy Co-operative has brought about revolution in milk production in recent years. Other factors contribute as much as 56 per cent to the total variations, of which technology, price and management practices would probably be more important. Productivity improvement in livestock and poultry requires adoption of improved germplasma, crystoscope assisted artificial insemination for improving conception rate, complete feedblock and area specific mineral supplement, silvipasture based farming system for increased fodder production, hormonal

Table 6.1: Principal factors influencing Inter-regional variation in the share of Livestock Output to Total value of Agriculture and Allied Sector output

Explanatory Factors	*B Co-efficient*
% share of fodder crops	0.299*** Significant at 10 per cent level
% villages with all weather road	0.260
Distance of veterinary care services	0.706*** Significant at 10 per cent level
Membership of milk Co-operatives	0.67
R2 = 0.442	Constant: –0.900

Source: Estimated by the author.

modulation of poultry for increased production of eggs and value addition of milk, poultry products and hairs of sheep and yak. Besides, development of veterinary care services and milk co-operatives would hold the key (ICAR, 2007). Once these constraints are removed, farmers will be able to increase their incomes by way of more intensive and extensive livestock rearing.

Development of Fisheries

The fisheries sub-sector shared nearly 4.15 per cent of total value of output of agriculture and allied sector in 1999-02 and 4.52 per cent in 2008-11. Considering the country as a whole the contribution of fisheries to total GDP of agriculture and allied sector was 5.2 per cent, during 2010-13. The total fish production in India increased from about 28.80 million tons in 1985-86 to 49.5 million tons in 1995-96 and 84.2 million tons in 2010-11.

The available data show that there are untapped yield potentials in fisheries, the utilization of which can increase fish production substantially.

First, the average fish production in about 3.15 million hectares of water reservoirs is only about 29.7 kg/ha which can be increased to 77.7 kg/ha, with appropriate cultural practices (ICAR, 2007). Similarly, shrimp culture can be increased from the current level of 0.14 million hectare to 1.19 million hectare. The key challenges include: (i) adequate stocking of quality seeds; (ii) improved carp and catfish culture in Andhra Pradesh, Assam, Chhattisgarh, Haryana, Orissa, West Bengal and Bihar; (iii) shrimp and scampi farming, where possible such as Goa and Gujarat; (iv) fish-duck-pig culture in Assam; (v) fresh water prawn culture in Bihar, Haryana, Karnataka, Odisha, Punjab, Uttar Pradesh and West Bengal; (vi) rice-fish farming in Odisha, Bihar and West Bengal; (vii) carp culture in Madhya Pradesh, Maharashtra, Odisha, Rajasthan and Uttar Pradesh; (viii) setting up of adequate aquafeed units and disease diagnostic centres; and (ix) development of technical skills and human resources for sustainable fisheries.

Second, various infrastructural constraints, are—(i) disorganized marketing structure, along with dominance of middlemen; (ii) lack of appropriate, albeit adequate transportation; (iii) unhygienic condition of the market where fish is generally sold; (iv) poor access of the fishers to institutional credit for upgradation of fishing vessels; and (v) inadequate infrastructure for the production of fish seed.

The other constraints to fisheries development include: (i) Lack of appropriate sanitary and phyto sanitary measures, in the absence of which Indian fishes often lose market in developed countries; (ii) Pollution of rivers, streams and other water bodies with industrial and community wastes which affect fish production; and (iii) the pre-occupation of the fish industry with export that results in the neglect of domestic market development.

Non-Farm Vertical Diversification

The existing data base for analyzing the pace and pattern of non-farm rural diversification is either inadequate or dated. Nevertheless various rounds of National Sample Survey by National Sample Survey Organisation (NSSO) generate data which throw some light on non-farm sources of employment and income of rural households. It would be seen from Table 6.1 that the share of non-agricultural employment in rural India increased from about 19.8 per cent in 1993-94 to 35.9 per cent in 2011-12. The per cent share of manufacturing sector in total employment increased from 6.8 per cent in 1983 to 7.0 per cent in 1993-94 and 8.7 per cent in 2011-12. While the share of trade sub-sector increased from 3.5 per cent in 1983 to 4.3 per cent in 1993-94 and 6.8 per cent in 2011-12. The contribution of other services increased from 4.9 per cent in 1983 to 5.7 per cent in 2011-12. The share of construction sector increased from 1.7 per cent in 1983 to 2.3 per cent in 1993-94, 4.9 per cent in 2004-05 and 11.1 per cent in 2011-12. However, the present pace of rural non-agricultural diversification can at best be said modest. Also there were regional variations in the pattern of non-agricultural diversification (Appendix-1). From the point of view of rural non-agricultural employment, various Indian States/Union territories could be grouped into three distinct zones. First, States like Arunachal Pradesh, Gujarat, Karnataka, Madhya Pradesh, Maharashtra, Nagaland, Sikkim and Mizoram had less than 30 per cent employment in the rural non-farm sector. Second, Assam, Bihar, Jammu & Kashmir, Himachal Pradesh, Meghalaya, Punjab, Rajasthan, Tamil Nadu, Uttar Pradesh and West Bengal had employment in the non-farm sector in the range of 31 to 50 per cent. Third, Delhi, Goa, Kerala, Tripura, A&N Islands, Jharkhand, Manipur, Odisha, Punjab, Chandigarh, D&N Haveli, Lakshadweep, Puducherry and Tamil Nadu also had relatively higher percentage share of rural non-farm employment above 50 per cent.

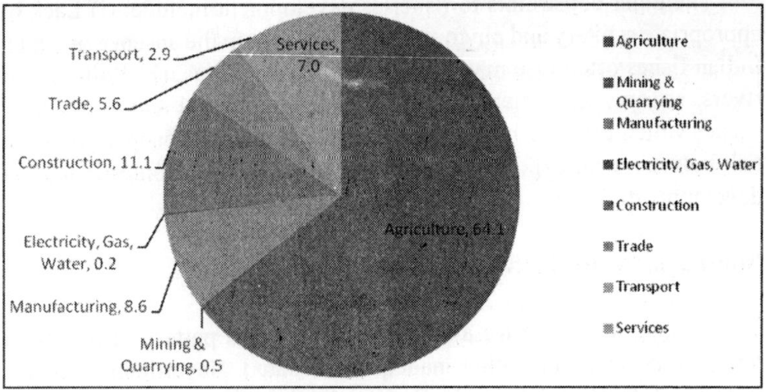

Fig. 6.1: Shares of Various Non-Agricultural Employment in Total Rural Employment (as of 2011-12)

During 1993-94 to 2011-12, the share of non-farm employment declined in Sikkim and Nagaland, while it increased in slow speed in several states including Arunachal Pradesh, Andhra Pradesh, Gujarat, Karnataka, Maharashtra, Uttar Pradesh. The recent NSS data (Government of India, 2013) further show that 55.9 per cent of rural workers are self-employed, while 8.7 per cent are regular employees and 35.4 per cent are casual laboures. The average daily earnings of regular employees was Rs. 298 and that of casual labourer was Rs. 139 only. Also most of the rural enterprises were small. On average, there were about 2 workers per enterprise. Hardly 1 per cent of enterprises engaged more than 10 workers.

It may be further seen from Appendix - 2 that the income from crop cultivation accounted for only 47.9 per cent of the toal income of an average agricultural household. About 32.2 per cent of the total income came from wage employment. In fact income from wages was as high as 41.5 per cent in Andhra Pradesh, 45.9 per cent in Himachal Pradesh, 57.8 per cent in Jammu & Kashmir, 44.2 per cent in Kerala, 43.1 per cent in Manipur, 40.2 per cent in Mizoram, 53.7 per cent in Nagaland, 45.8 per cent in Sikkim, 41.6 per cent in Tamil Nadu, 40.2 per cent in Tripura and 53.4 per cent in West Bengal. The proportion of non-farm business income was relatively high in States like West Bengal (16.3%), Tamil Nadu (15.2%), Odisha (10.8%), Maharashtra (11.3%), Kerala (21.3%) and Jammu & Kashmir (11.7%). But the economic condition of farmers in all these states is not that good, because even rural non-farm wages and business incomes are not high as these are low paid informal sector employment.

Determinants of Non-Farm Diversification

According to Vaidyanathan (1986), the non-farm diversification phenomenon was primarily distress induced as population pressure aggravated the land constraint and there emerged low wage rural residual sector. This is further corroborated from the recent NSS data which shows that landless and semi-landless households earn more through wages (about 63.6%). Similar was the case with those with less than 0.40 hectare land who earned only 16.5 per cent of their total income from farming and 57.5 per cent through wages (Government of India, 2014). For large farmers in the size category above 10 hectares, crop cultivation (86.2%), followed by livestock (6.3%), farm business (4.3%) the main sources of income. Wage earnings accounted for only 3.2 per cent of their total incomes. It seems that farmers having high land size do not get so much attracted to non-farm business. A study by Sheila Bhalla (2005a) shows the high farm income regions depicting increasing rural non-farm work specialization, along with rapidly diversifying corridor districts spread along road and rail lines between distant urban industrial conglomerations. According to Sheila Bhalla (2005b), in less prosperous regions and drought afflicted villages, multiple activities are generally a symptom of distress, while in more prosperous villages, this may be a method of getting one step ahead of other households. The inter-regional variations in non-farm rural employment and income could be largely explained by rural literacy rate, road connectivity, land productivity and proportion of marginal farms (Table 6.2). These factors together explained nearly 73 per cent of the total variations in employment and 69 per cent in non-farm business income of which rural literacy rate and proportion of marginal farmers were statistically

Table 6.2: Co-efficient of Principal factors Determining Inter-regional variations in Rural Non-farm Employment and Income

	Elasticity Co-efficient	
	Non-Farm Employment	*Non-Farm Business Income*
Rural Literacy Rate	0.544*	0.353* significant at 1%
Road connectivity	0.66	0.088
Land Productivity	–0.149	0.613* significant at 1%
% Marginal Farmers	0.721*	0.025
Constant	–2.636	–0.260
R2	0.728	0.693

Source: Author's own calculation.

significant in the case of non-farm employment and literacy rate and land productivity in the case of non-farm business income.

In addition to the above mentioned Principal factors that determine rural diversification, there are other factors which play significant roles, but not properly captured in any quantitative analysis. For example, contract farming has opened up new opportunities for strong farm-firm linkages for poultry, vegetables and agro-processing sector in recent years. In fact in several cases, pricing and institutional factors seem to have played very important roles in off-farm and non-farm rural diversification. Likewise, when we talk of access to market for processed products, it is not only the regional or domestic market which matters, but also the export market. Along with rising domestic market, due to growth of income and urbanization, the growing export market is equally important. Besides, in order to create new potentials for non-farm diversification, some policy gaps still exist. The recent initiatives for establishment of food parks, agri-export zones and human resource development provide necessary, but not the sufficient conditions for agro-processed based sustainable diversifications. For instance, the Central Government may have to move towards zero taxation for both processed food as well as inputs used in the manufacture of these products for quite some time. Moreover, marketing, financing and technological needs of the rural non-farm sector, especially in the unorganized sector have been largely neglected by the government which needs to be relooked. In addition, the level of the infrastructure for agro-processing has to be improved both qualitatively and in size. Furthermore, lack of adequate training centres stand in the way of development of rural entrepreneurship.

Impact of Diversification

It is generally recognized that diversification towards non-farm activities increase the rural poor's income and therefore, help in poverty reduction. However, it is not yet fully established whether both horizontal and vertical diversification would equally reduce income inequality. Diversification towards livestock may reduce inequality as small farmers have relatively higher shares, especially in goat, sheep etc. and tend to gain more, while non-farm business may increase inequality. This could be so if large farmers have relatively more access to non-farm business than marginal and small farmers. In fact, the ownership of rural non-farm business is in the hands of relatively better off section, whereas marginal farmers and landless labourers normally seek non-farm wage employment in the informal sector, where the wages also are relatively low. Even in the so-called organized agro-processing sector, the real wages of the workers have remained more or less stagnant (Rao, 2009). However, the effect of rural non-farm diversification on poverty is obvious even from the state-wise aggregate data. It would be seen from Table 6.3 that states like Kerala, West Bengal, Tamil Nadu and Haryana, where the percentage of rural non-farm workers is high, the rates of decline in rural poverty also was higher. However, in absolute term, the incidence of rural poverty was still higher than national average in West Bengal, but low in Kerala and Haryana and medium in Tamil Nadu.

In the context of North West India, where rice, wheat cropping system is not ecologically sustainable, the farmers will shift to non-rice-wheat cropping system only when their relative prices and marketability improve.

Similarly non-farm diversification would need strong farm-firm linkages through contract farming, development of agro-processing and

Table 6.3: Rural Poverty Levels by Major States during 2004-05 and 2011-12

States	*Percentage Non-farm workers*	*Head Count Ratio, Rural Poverty*		*Poverty Change*
		2004-05	*2011-12*	*2004-05 and 2011-12*
Andhra Pradesh	30.5	11.2	11.0	–0.2
Bihar	32.4	42.1	34.1	–8.0
Gujarat	25.4	19.1	21.8	2.7
Haryana	42.2	13.6	11.6	–2.0
Himachal Pradesh	36.7	10.7	8.5	–2.2
Karnataka	29.8	20.8	24.5	3.7
Kerala	68.6	13.2	9.1	–4.1
Madhya Pradesh	27.9	36.9	35.7	–1.2
Maharashtra	22.9	29.6	24.2	–5.4
Odisha	37.8	46.8	35.7	–11.1
Punjab	47.6	9.1	7.7	–1.4
Rajasthan	39.2	18.7	16.1	–2.6
Tamil Nadu	48.8	22.8	15.8	–7.0
Uttar Pradesh	36.3	33.4	30.4	–3.0
West Bengal	46.8	28.6	22.5	–6.1
All India	**35.9**	**28.3**	**25.7**	**–2.6**

Note: Based on UPR consumption = Uniform recall period consumption in which the consumer expenditure data for all the items as collected from 30 days recall period.

Source: Government of India, NSSO, 68th Round for non-farm workers Government of India, Planning Commission for Rural poverty data.

value addition, development of infrastructure such as road connectivity, electricity, market, promotion of farm producers group and co-operative enterprises of farmers, appropriate pricing and marketing arrangements for rural non-farm products and finally, quality education, training and skills for farm youth for improving their employability in high tech agriculture and rural non-farm enterprises.

Conclusion

It becomes clear from the foregoing discussion that there is a clear positive trend towards both horizontal and vertical diversification of India's rural economy. As farm sizes decline, farmers will depend more on off-farm and non-farm sources of incomes in rural areas. In other words, substantive diversification of the rural economy will help reduce farm distress and poverty in rural areas. However, accelerating the pace of horizontal diversification would require appropriate risk management mechanisms in terms of either price and marketing support or effective insurance coverage for farmers to switch from traditional low yielding and low income generating crops to high value crops and enterprises. This is important because risk aversion continues to be one of the significant impediments to what would be otherwise considered as rational diversification. Similarly, vertical diversification by way of improving farmers access to non-farm sources of income will help reuce poverty and distress only if farmers have the necessary education and skills to take well paid jobs in the formal sector or have the ability to invest in non-farm enterprises that have potentials for profit and growth. Anyway, the existing poor road connectivity, lack of power, inadequate access to credit and markets continue to slow down the pace of rural non-farm diversification in the organized sector, which could reduce farm distress. Otherwise as we can see, even with diversified income sources, the small and marginal farmers in India do not earn enough to meet their consumption needs and stay in debt trap.

Appendix-1

Percentage Distribution of Persons Employed (Usual Principal Status + Usual Subsidiary Status), by Industry Section, by State, Rural India

States	Year	Agriculture	Mining and Quarrying	Manu-facturing	Electricity, Gas, Water	Construc-tion	Trade	Trans-port	Services	All Sectors	Total non-Agriculture
Andhra Pradesh	1993-94	79.3	0.7	6.9	0.0	1.7	4.5	1.0	5.9	100	20.7
	1999-00	78.8	0.8	5.6	0.1	2.2	4.6	1.7	6.2	100	21.2
	2004-05	71.8	1.2	8.6	0.1	3.3	6.6	2.6	5.8	100	28.2
	2011-12	69.5	0.8	7.8	0.2	6.7	5.2	3.2	6.7	100	30.5
Arunachal Pradesh	1993-94	86.4	0.1	0.8	0.8	2.6	0.2	1.6	7.5	100	13.6
	1999-00	83.4	0.0	2.8	0.3	4.4	0.6	0.1	8.4	100	16.6
	2004-05	81.9	0.0	0.4	0.9	4.0	2.3	0.6	9.9	100	18.1
	2011-12	77.9	0.2	0.8	0.3	4.2	4.9	1.0	10.7	100	22.1
Assam	1993-94	79.2	0.2	3.5	0.3	0.7	6.9	1.3	7.9	100	20.8
	1999-00	67.7	0.3	4.0	0.1	1.8	8.1	2.7	15.3	100	32.3
	2004-05	74.3	0.3	3.1	0.1	2.5	9.0	2.4	8.3	100	25.7
	2011-12	62.0	0.2	5.5	0.0	6.6	12.1	3.2	10.4	100	38.0
Bihar	1993-94	84.3	0.5	3.4	0.3	1.3	4.5	1.0	4.7	100	15.7
	1999-00	80.6	0.5	6.1	0.1	2.2	4.5	1.4	4.6	100	19.4
	2004-05	77.9	0.1	5.7	0.1	2.5	7.5	2.2	4.0	100	22.1
	2011-12	67.6	0.1	4.8	0.0	10.4	7.1	2.4	7.6	100	32.4

(Contd.)

(*Contd.*)

States	Year	Agriculture	Mining and Quarrying	Manu-facturing	Electricity, Gas, Water	Construc-tion	Trade	Trans-port	Services	All Sectors	Total non-Agriculture
Goa	1993-94	45.4	0.5	11.2	2.8	5.0	9.2	6.3	19.6	100	54.6
	1999-00	28.7	5.8	11.4	0.3	12.5	19.0	11.7	10.6	100	71.3
	2004-05	35.3	0.0	6.3	2.1	7.6	20.7	12.7	15.3	100	64.7
	2011-12	4.6	7.8	19.0	0.0	8.8	16.9	7.8	35.1	100	95.4
Gujarat	1993-94	78.7	0.5	9.2	0.2	2.4	2.9	1.5	4.6	100	21.3
	1999-00	79.8	0.4	6.9	0.2	2.7	3.9	2.4	3.7	100	20.2
	2004-05	77.3	0.5	7.8	0.1	3.0	4.7	2.7	3.9	100	22.7
	2011-12	74.6	0.5	8.3	0.0	4.3	4.8	2.8	4.7	100	25.4
Haryana	1993-94	71.9	0.6	4.0	0.4	4.3	5.3	3.5	10.0	100	28.1
	1999-00	68.5	0.4	7.3	0.9	6.5	6.0	3.3	7.1	100	31.5
	2004-05	64.1	0.1	8.9	0.5	8.3	7.5	4.0	6.6	100	35.9
	2011-12	57.9	0.2	9.8	0.2	14.3	6.5	3.6	7.6	100	42.2
Himachal Pradesh	1993-94	80.3	0.1	3.3	0.8	6.3	3.0	0.8	5.4	100	19.7
	1999-00	73.6	0.0	4.0	1.1	9.2	3.4	2.3	6.4	100	26.4
	2004-05	69.6	0.1	4.9	1.1	9.8	4.3	3.2	6.7	100	30.4
	2011-12	63.3	0.1	4.1	1.4	15.3	3.2	3.4	9.2	100	36.7
Jammu & Kashmir	1993-94	75.8	0.1	3.6	1.1	6.0	2.5	2.7	8.2	100	24.2
	1999-00	76.3	0.0	4.2	0.4	6.6	3.6	1.7	7.2	100	23.7
	2004-05	63.9	0.3	9.8	1.1	7.1	5.0	4.0	8.8	100	36.1
	2011-12	50.9	0.1	8.0	1.1	20.3	5.7	2.7	11.3	100	49.1
Karnataka	1993-94	81.2	0.7	6.7	0.2	1.5	3.7	0.7	5.3	100	18.8
	1999-00	82.1	0.8	5.4	0.0	1.5	4.5	1.5	4.2	100	17.9
	2004-05	81.0	0.4	6.2	0.0	2.1	4.5	2.0	3.8	100	19.0
	2011-12	70.2	0.3	9.4	0.2	4.3	4.8	3.3	7.4	100	29.8

(*Contd.*)

(*Contd.*)

States	Year	Agriculture	Mining and Quarrying	Manu-facturing	Electricity, Gas, Water	Construc-tion	Trade	Trans-port	Services	All Sectors	Total non-Agriculture
Kerala	1993-94	56.4	1.5	12.8	0.3	5.7	9.0	4.0	10.3	100	43.6
	1999-00	48.3	1.7	12.6	0.2	9.4	11.9	6.2	9.7	100	51.7
	2004-05	42.0	1.3	13.7	0.2	10.6	12.2	7.0	13.0	100	58.0
	2011-12	31.4	0.7	12.4	0.3	18.5	11.1	8.0	17.6	100	68.6
Madhya Pradesh	1993-94	89.8	1.2	3.2	0.2	0.9	1.6	0.5	2.6	100	10.2
	1999-00	87.1	0.4	4.0	0.1	1.8	2.7	0.7	3.2	100	12.9
	2004-05	82.5	0.7	5.0	0.1	3.6	4.0	0.7	3.4	100	17.5
	2011-12	72.1	0.6	4.8	0.1	12.9	3.8	1.3	4.4	100	27.9
Maharashtra	1993-94	82.6	0.4	5.0	0.2	2.4	3.3	1.1	5.0	100	17.4
	1999-00	82.6	0.1	4.9	0.2	2.3	3.8	1.8	4.3	100	17.4
	2004-05	80.0	0.3	5.6	0.1	2.9	4.7	2.3	4.1	100	20.0
	2011-12	77.1	0.1	5.9	0.1	5.2	4.2	2.0	5.5	100	22.9
Manipur	1993-94	63.8	0.1	12.3	0.3	2.8	4.5	1.5	14.7	100	36.0
	1999-00	75.3	0.6	7.4	0.0	0.9	3.7	1.4	10.7	100	24.7
	2004-05	69.3	0.5	8.6	0.0	3.1	6.6	2.1	9.8	100	30.7
	2011-12	45.5	0.4	7.8	0.3	22.6	7.8	2.8	12.9	100	54.5
Meghalaya	1993-94	86.0	0.5	1.0	0.3	1.6	3.8	1.1	5.7	100	14.0
	1999-00	86.5	0.7	0.7	0.0	1.6	4.2	1.1	5.2	100	13.5
	2004-05	81.8	1.3	3.7	0.4	2.2	5.5	1.1	4.0	100	18.2
	2011-12	66.3	2.5	2.2	0.2	6.2	8.1	3.1	11.4	100	33.7
Mizoram	1993-94	88.9	0.0	0.5	0.1	1.2	1.9	0.1	7.3	100	11.1
	1999-00	85.5	0.3	1.0	0.0	1.1	3.0	0.1	9.0	100	14.5
	2004-05	87.4	0.0	0.9	0.0	1.0	3.5	0.5	6.7	100	12.6
	2011-12	75.8	0.3	1.1	0.2	8.3	4.9	0.9	8.5	100	24.2

(*Contd.*)

(*Contd.*)

States	Year	Agriculture	Mining and Quarrying	Manu-facturing	Electricity, Gas, Water	Construc-tion	Trade	Trans-port	Services	All Sectors	Total non-Agriculture
Nagaland	1993-94	74.9	0.8	0.2	0.2	1.8	4.5	0.6	17.0	100	25.1
	1999-00	79.7	0.1	0.6	0.7	0.8	2.8	1.3	14.0	100	20.3
	2004-05	79.3	0.0	2.1	0.4	1.7	4.9	1.4	10.2	100	20.7
	2011-12	76.7	0.2	0.8	1.1	3.0	5.7	1.9	10.8	100	23.3
Odisha	1993-94	80.9	1.1	6.4	0.1	1.8	4.4	0.7	4.6	100	19.1
	1999-00	78.2	0.4	8.2	0.1	3.2	4.6	1.1	4.2	100	21.8
	2004-05	69.0	0.8	11.1	0.1	5.5	6.7	2.0	4.8	100	31.0
	2011-12	62.3	0.5	9.6	0.2	12.3	6.0	2.4	6.7	100	37.8
Punjab	1993-94	74.7	0.0	4.9	1.2	3.5	4.9	2.6	8.2	100	25.3
	1999-00	72.6	0.0	5.9	0.8	5.3	5.8	3.8	5.8	100	27.4
	2004-05	66.9	0.1	7.4	0.7	8.8	6.4	3.5	6.2	100	33.1
	2011-12	52.4	0.0	11.2	1.1	16.2	5.6	3.6	10.1	100	47.6
Rajasthan	1993-94	79.9	1.8	3.6	0.2	7.0	2.4	1.0	4.1	100	20.1
	1999-00	77.7	1.4	4.3	0.2	7.9	3.5	1.7	3.3	100	22.3
	2004-05	72.9	1.2	5.8	0.2	9.6	4.1	2.1	4.1	100	27.1
	2011-12	60.8	1.0	5.2	0.2	20.3	4.5	2.6	5.5	100	39.2
Sikkim	1993-94	58.6	2.2	4.4	0.7	3.5	5.9	2.0	22.7	100	41.4
	1999-00	60.8	0.8	2.5	2.8	3.8	6.3	2.6	20.4	100	39.2
	2004-05	60.5	0.6	2.6	1.2	8.0	8.6	3.9	14.6	100	39.5
	2011-12	72.8	0.0	2.6	2.0	6.5	3.2	2.6	10.3	100	27.2
Tamil Nadu	1993-94	70.5	0.4	12.9	0.2	2.3	4.8	2.0	6.9	100	29.5
	1999-00	67.9	0.5	13.9	0.2	4.0	5.6	2.6	5.3	100	32.1
	2004-05	65.4	0.3	14.0	0.1	5.7	6.4	2.8	5.3	100	34.6
	2011-12	51.2	0.5	14.6	0.6	15.5	5.2	3.7	8.6	100	48.8

(*Contd.*)

(*Contd.*)

States	Year	Agriculture	Mining and Quarrying	Manu-facturing	Electricity, Gas, Water	Construc-tion	Trade	Trans-port	Services	All Sectors	Total non-Agriculture
Tripura	1993-94	47.6	0.0	5.4	0.2	6.8	10.7	3.2	26.1	100	52.4
	1999-00	45.7	0.0	4.0	0.0	8.0	11.8	2.3	28.2	100	54.3
	2004-05	43.2	0.0	5.1	0.0	11.4	10.9	3.7	25.7	100	57.8
	2011-12	30.8	0.3	6.4	0.0	43.8	8.3	1.8	8.6	100	69.2
Uttar Pradesh	1993-94	80.0	0.2	6.4	0.1	2.0	4.3	1.5	5.5	100	20.0
	1999-00	76.2	0.1	7.8	0.1	3.3	5.4	2.1	5.0	100	23.8
	2004-05	72.8	0.2	8.9	0.1	5.3	6.2	2.1	4.4	100	27.2
	2011-12	63.7	0.4	8.6	0.2	13.4	5.5	2.6	5.6	100	36.3
West Bengal	1993-94	63.3	0.2	16.1	0.1	2.4	7.5	2.7	7.7	100	36.7
	1999-00	63.6	0.3	16.6	0.1	2.2	8.5	3.3	5.4	100	36.4
	2004-05	62.7	0.2	13.5	0.1	3.9	9.5	3.6	6.5	100	37.3
	2011-12	53.2	0.3	19.5	0.1	8.4	6.4	3.5	8.6	100	46.8
A & N Islands	1993-94	63.3	0.4	5.3	1.3	8.1	6.7	2.8	12.1	100	36.7
	1999-00	65.0	0.3	3.2	0.8	11.4	4.6	3.9	10.8	100	35.0
	2004-05	45.4	1.0	3.3	2.6	12.5	11.9	7.4	15.9	100	54.6
	2011-12	38.3	0.0	2.7	1.9	17.2	9.0	8.3	22.7	100	61.8
Chandigarh	1993-94	13.6	0.0	16.4	1.0	20.4	14.5	13.1	21.0	100	86.4
	1999-00	66.1	0.0	10.2	0.0	8.0	6.2	1.6	7.9	100	33.9
	2004-05	7.0	0.0	22.3	0.0	21.3	19.8	11.0	18.6	100	93.0
	2011-12	1.9	0.0	11.7	0.0	1.2	6.8	29.0	49.4	100	98.1
Dadra & Nagar Haveli	1993-94	66.0	1.9	6.9	0.2	6.0	1.7	10.6	6.7	100	34.0
	1999-00	55.2	1.4	18.4	0.0	6.2	2.6	11.0	5.2	100	44.8
	2004-05	48.3	0.0	26.2	0.0	4.4	6.7	6.5	7.9	100	51.7
	2011-12	35.5	2.7	39.1	0.0	8.6	6.1	1.2	6.9	100	64.6

(*Contd.*)

(*Contd.*)

States	Year	Agriculture	Mining and Quarrying	Manu-facturing	Electricity, Gas, Water	Construc-tion	Trade	Trans-port	Services	All Sectors	Total non-Agriculture
Daman & Diu	1993-94	52.4	0.0	26.1	0.0	0.6	6.2	2.9	11.8	100	47.6
	1999-00	34.2	0.0	42.3	0.0	0.8	9.1	4.4	9.2	100	65.8
	2004-05	39.4	0.0	24.7	0.0	4.8	11.6	11.4	8.1	100	60.6
	2011-12	9.2	0.0	81.4	0.0	0.7	0.6	0.6	7.5	100	90.8
Delhi	1993-94	10.3	0.0	51.4	0.0	0.8	10.5	7.1	19.9	100	89.7
	1999-00	7.5	0.0	26.0	0.4	10.0	30.4	8.4	17.3	100	92.5
	2004-05	7.2	0.0	18.5	0.0	9.7	35.4	8.9	20.3	100	92.8
	2011-12	2.0	0.0	7.8	6.9	0.7	12.2	6.3	64.0	100	98.0
Lakshadweep	1993-94	39.9	0.0	5.6	2.6	9.6	5.2	7.2	29.9	100	60.1
	1999-00	53.6	0.0	3.4	0.7	8.4	6.5	7.2	20.2	100	46.4
	2004-05	39.3	0.0	5.5	6.4	31.0	2.1	4.6	11.1	100	60.7
	2011-12	15.0	0.0	4.2	3.2	25.1	3.7	14.7	34.0	100	85.0
Pududicherry	1993-94	64.7	0.0	13.8	0.0	3.5	4.9	2.5	10.6	100	35.3
	1999-00	59.9	0.0	13.8	2.2	4.7	10.6	1.1	7.7	100	40.1
	2004-05	58.6	0.0	15.3	1.0	7.2	10.6	2.0	5.3	100	41.4
	2011-12	31.7	0.0	17.7	0.0	14.9	9.0	8.4	18.4	100	68.3
All India	**1993-94**	**80.2**	**0.6**	**7.0**	**0.2**	**2.1**	**3.8**	**1.2**	**5.2**	**100**	**19.8**
	1999-00	78.4	0.5	7.5	0.1	2.8	4.4	1.7	4.9	100	21.6
	2004-05	74.9	0.3	8.2	0.1	4.2	5.4	2.0	4.9	100	25.1
	2011-12	64.1	0.5	8.6	0.2	11.1	5.6	2.9	7.0	100	35.9

Source: Governmen of India: various Roundes of NSSO, New Delhi.

Appendix-2

Percentage share of various sources in monthly income of agricultural household (2012-13)

States	Average monthly income (Rs.)					% Share from total Income			
	Income from wages (Rs.)	Net receipt from cultivation (Rs.)	Net receipt from farming of animals (Rs.)	Net receipt from non-farming business (Rs.)	Total Income (Rs.)	Income from wages	Net receipt from cultivation	Net receipt from farming of animals	Net receipt from non-farming business
Andhra Pradesh	2482	2022	1075	400	5979	41.51	33.82	17.98	6.69
Arunachal Pradesh	2076	6647	1310	836	10869	19.10	61.16	12.05	7.69
Assam	1430	4211	799	255	6695	21.36	62.90	11.93	3.81
Bihar	1323	1715	279	240	3558	37.18	48.20	7.84	6.75
Chhattisgarh	1848	3347	–19	1	5177	35.70	64.65	–0.37	0.02
Gujarat	2683	2933	1930	380	7926	33.85	37.00	24.35	4.79
Haryana	3491	7867	2645	431	14434	24.19	54.50	18.32	2.99
Himachal Pradesh	4030	2876	1047	824	8777	45.92	32.77	11.93	9.39
Jammu & Kashmir	7336	3063	801	1483	12683	57.84	24.15	6.32	11.69
Jharkhand	1839	1451	1193	238	4721	38.95	30.74	25.27	5.04
Karnataka	2677	4930	600	625	8832	30.31	55.82	6.79	7.08
Kerala	5254	3531	575	2529	11888	44.20	29.70	4.84	21.27
Madhya Pradesh	1332	4016	732	129	6210	21.45	64.67	11.79	2.08

(Contd.)

(*Contd.*)

States	Average monthly income (Rs.)					% Share from total Income			
	Income from wages (Rs.)	Net receipt from cultivation (Rs.)	Net receipt from farming of animals (Rs.)	Net receipt from non-farming business (Rs.)	Total Income (Rs.)	Income from wages	Net receipt from cultivation	Net receipt from farming of animals	Net receipt from non-farming business
Maharashtra	2156	3856	539	834	7386	29.19	52.21	7.30	11.29
Manipur	3815	2924	1563	540	8842	43.15	33.07	17.68	6.11
Meghalaya	3776	6472	657	887	11792	32.02	54.88	5.57	7.52
Mizoram	3655	4561	864	19	9099	40.17	50.13	9.50	0.21
Nagaland	5393	3212	1384	59	10048	53.67	31.97	13.77	0.59
Odisha	1716	1407	1314	539	4976	34.49	28.28	26.41	10.83
Punjab	4779	10862	1658	760	18059	26.46	60.15	9.18	4.21
Rajasthan	2534	3138	967	710	7350	34.48	42.69	13.16	9.66
Sikkim	3113	1696	980	1009	6798	45.79	24.95	14.42	14.84
Tamil Nadu	2902	1917	1100	1061	6980	41.58	27.46	15.76	15.20
Telangana	1450	4227	374	260	6311	22.98	66.98	5.93	4.12
Tripura	2185	2772	311	162	5429	40.25	51.06	5.73	2.98
Uttarakhand	1069	2531	848	253	4701	22.74	53.84	18.04	5.38
Uttar Pradesh	1150	2855	543	376	4923	23.36	57.99	11.03	7.64
West Bengal	2126	979	225	650	3980	53.42	24.60	5.65	16.33
Group of UTs	5179	1864	213	1312	8568	60.45	21.76	2.49	15.31
All India	**2071**	**3081**	**763**	**512**	**6427**	**32.22**	**47.94**	**11.87**	**7.97**

Source: NSS, 70th Round.

References

Commision for Agricultural Costs and Prices (2014), Report on Price Policy for Kharif crops for 2014-15, Government of India, New Delhi.

Dogra, Bharat (2013), Farmers' Crisis in India and Urgency of Remedial Action, *Mainstream*, New Delhi, July 12-18.

Govt. of India (2014), *Key Indicators of Situation of Agricultural Household in India*, 70th Round, NSSO, New Delhi.

Government of India (2013), *Key Indicators of Employment and Unemployment in India*, 2011-12, NSSO 68th Round, New Delhi.

Government of India (2012), *State-wise Estimates of value of output from Agriculture and Allied Activities*, CSO, New Delhi.

Government of India (2007), *Eleventh Five Year Plan, Planning Commission*, New Delhi.

Haque, T. (1996), *Diversification of Small Farms in India*, Policy Paper, NCAP (ICAR), New Delhi.

Indian Council of Agricultural Research (ICAR: 2007), *State Specific Technological Interventions for Higher Agricultural Growth*, New Delhi.

Joshi, P.K., Ashok Gulati and Pratap S. Birthal (2007), Agricultural Diversification in India in Joshi *et al.* edited *Agriculture and Small Holders in South Asia*, Academic Foundation, New Delhi.

Rao, N. Chandrasekhar (2009) *Enhancing Growth and Productive Employment Linkages in the Food Processing Sector*, CESS, Hyderabad.

Rao, P. Narsimha and K.C. Suri (2006): Dimension of Agrarian Distress in Andhra Pradesh, *Economic and Political Weekly*, Mumbai, April 22.

Sheila Bhalla (2005a), Rural Workforce Diversification and Performance of Unorganised Sector Enterprises in Nayyar, Rohini and A.N. Sharma (ed.) *Rural Transformation in India*, IHD, New Delhi.

Sheila Bhalla (2005b), *India's Rural Economy: Issues and Evidence—Working Paper Series No. 25*, IHD, New Delhi.

Sheila Bhalla (1998), Trends in Poverty, Wages and Employment in Rural India, in *Empowering Rural Labour in India*, edited by R. Radhakrishna and Alakh N. Sharma, Institute for Human Development, New Delhi.

Singh, S.P. and B.K. Sahoo (2007), Diversification of Agricultural Economy Towards Horticultural and Livestock, *IJLE*, Vol. 50, No. 4.

Taylor, Donald C. (1994), *Agricultural Diversification in South East Asia*— Paper presented at APO Meeting, January 18-23, Tokyo.

Vaidyanathan, A. (1986), Labour Use in Rural India, *EPW*, Vol. XXI, No. 52.

7

Employment and Income Diversification in Rural India

Shinoj Parappurathu, Pratap Singh Birthal and *Kingsly Immanueraj*

Introduction

Indian agriculture has grown at an annual rate of about 3 per cent over the last two decades, thus contributing to improving farm incomes and reducing rural poverty in the country (Datt and Ravallion, 1996; Warr, 2003). However, the excessive employment pressure and declining landholding size continue to pull down the sector from achieving higher growth, or even to sustain its present performance. In 2009-10, agricultural sector engaged 52 per cent of the country's total workforce, 17 percentage points less than in 1983; while its share in the gross domestic product (GDP) declined at a much faster rate, from 40 per cent to 15 per cent during this period. Further, the average size of landholding declined from 1.84 ha in 1980-81 to 1.16 ha in 2010-11. Given these trends, there arises a basic development question: can the farmers continue to depend solely on such tiny pieces of land for their livelihood? This prompted policy-makers to target their efforts and investment towards developing and supporting non-agricultural income generating activities in rural areas.

Evidences from many developing countries suggest that diversification of rural economy towards non-farm activities has considerable potential to augment income and reduce poverty (Adams and He, 1995; Adams, 2001; Reardon *et al.*, 1998; Barrett *et al.*, 2001; Lanjouw, 1999; de Janvry *et al.*, 2005; Reardon *et al.*, 2007). Non-farm diversification overcomes land constraint to income growth, enables farmers cope with income shocks and improves their capacity to invest in productivity-enhancing agricultural inputs and technologies (Collier *et al.*, 1986; Reardon *et al.*, 1996). At macro-level, a growing rural non-farm economy can absorb surplus labour, slowdown rural-urban migration, reduce rural-urban disparities and promote farm-non-farm linkages.

India's rural economy diversified considerably towards non-farm activities during the last two decades or so. The share of non-farm sector in rural income increased from 35 per cent in 1980-81 to 62 per cent in 2004-05 (Government of India, 2010), and in rural employment from 22.3 per cent to 31.5 per cent (Lanjouw and Murgai, 2008). However, India's rural non-farm sector remains dominated by traditional small-scale, labour-intensive, low-investment, low-return enterprises (Government of India, 2006; Lanjouw and Shariff, 2002; Foster and Rosenzwig, 2004). Nevertheless, in a land-scarce, labour surplus country like India diversification to non-farm activities holds much promise, particularly for the poor and marginal dwellers. Against this backdrop, we examine (i) the changes in the pattern of employment in rural India and the profile of income sources of farm households, and (ii) determinants of farm households' access to different income sources. This paper is organized majorily in three broder headings. The data and methodology used for the study is discribed below. The employment profile of rural households and the changes therein over the past decade is analyzed further. In the last income sources of farm households in rural areas and identifies factors influencing households' access to different income sources discussed. A few conclusions drawn in the light of the study results are included in the final section.

Data and Methodology

This study has used the household level data on employment and unemployment collected by the National Sample Survey Organization (NSSO) to examine the trends in farm as well as non-farm employment of rural households. The analysis pertains to activities which were grouped in to five categories, viz. crops, livestock, other farm (forestry, fisheries, agricultural services etc.), non-farm and others. The first three activities represent the farm sector whereas the fourth activity includes all non-farm activities belonging to the secondary and tertiary sectors. The 'other' category denotes the households which have not reported any particular principal activity as a source of income. Further, to analyze the contribution of different sources to total income of the farm households, we have used household-level data from a unique survey that focused on farm households and was conducted during January-December 2003 by the NSSO (Government of India, 2005). This survey covered 51,770 farm households[1] spread over 6,638 villages throughout the country. The survey reports a number of income-generating farm and non-farm activities, which we have classified into four broad income sources, viz. agriculture (crop production), livestock, wages and salaries and non-farm business.

We have considered 'gross value of output' from an activity as income from that activity. The different income sources covered in the study are defined as follows:

(i) ***Agriculture***: includes income from the cultivation of cereals, pulses, oilseeds, fibres, sugarcane, fruits, vegetables, floriculture, spices, medicinal and aromatic plants and plantation crops.
(ii) ***Livestock***: includes value of output from the dairy, poultry, sheep and goats.
(iii) ***Wages and salaries***: includes farm and non-farm wage earnings, and salaries.
(iv) ***Non-farm business:*** includes income from manufacturing, hotels and restaurants, construction, mining and quarrying, repairing and other services.

The factors influencing households' access to different income sources were determined using a truncated regression (truncated at upper limit) model with share of an income source in total household income (in natural log) as the dependent variable. The model was estimated using Maximum Likelihood Estimation (MLE) method (*See* Birthal, *et al.*, 2014 for more details on methodology).

Employment Profile of Rural Households

To comprehend the process of occupational diversification taking place in the rural areas, the employment profile of rural households corresponding to the years, 2004-05 and 2011-12 was analyzed and compared. Accordingly, the households were first assessed based on their self-reported principal industry of activity. Table 7.1 shows the share of households with their primary industry of activity falling under the five groups defined above. In 2004-05, 48.3 per cent of households were dependent on crop husbandry as their principal industry of activity whose share declined to 34.1 per cent by 2011-12. A similar declining trend was observed in the case of livestock and other farm activities. On the other hand, the share of households dependent on non-farm sector has increased from 44.2 to 58.7 per cent during the period thereby indicating structural shift in occupational pattern of the households. These results underscore the emergence of non-farm sector as a vital sector on which a substantial majority of the households depended primarily for their livelihood during the period under study. However, the above observations are only indicative of the broad process of non-farm diversification as only an assessment based on individual employment that could reveal the actual picture.

Table 7.1: Changes in profile of rural households based on their principal industry of activity, 2004-05 to 2011-12

Sector	*Share of households (%)*		*Change, %*
	2004-05	*2011-12*	
Crops	48.3	34.1	–29.4
Livestock	1.3	0.6	–53.8
Other farm	1.1	0.9	–18.2
Non-farm	44.2	58.7	32.8
Other	5.1	5.7	11.8

Source: National Sample Survey (Govt. of India) 61 and 68 Rounds.

Accordingly, a disaggregate analysis based on the usual status (usual principal status + subsidiary status) of employment of sample population was carried out across expenditure quintiles as well as farm-size groups. Table 7.2 presents the employment of people in farm and non-farm sectors in rural India as share in total workforce across expenditure quintiles. The results suggest that, crops sector provided the highest employment to the rural inhabitants irrespective of their income (expenditure) status in 2004-05. The dependence on crop husbandry was the highest (82.7%) for the people belonging to the lowest quintile, and the share declined as one move towards higher quintile groups. A quite opposite trend was observed in the case of animal husbandry (livestock) wherein, the participation rate of people in this sector was the highest for the highest quintile group. This could be attributed to the relatively higher requirement of resources for livestock farming, the access to which is higher among resource-rich farmers. The employment of people in the non-farm sector remained more or less same across the quintile groups, the rates ranging from 38.6-40.6 per cent. However, over time (between 2004-05 and 2011-12), farm sector employment was observed to decline with a commensurate gain in the non-farm sector. This marks a structural shift in the economy characterized by abandonment of agriculture particularly by the new entrants in the workforce for gainful employment in the non-farm sector. This trend was common across the quintile groups, with the most marked changes being noted in the lowest quintile group. In this quintile, employment participation in crop, livestock and other farm sectors declined by 27.2 per cent, 58.7 per cent

Table 7.2: Changes in employment participation rate (Usual Status) by sectors in rural India across expenditure quintiles, 2004-05 to 2011-12

Sector	*Employment as share of total workforce-Usual Status (%)*		*Change,%*
	2004-05	*2011-12*	
	Lowest quintile		
Crops	82.7	60.2	–27.2
Livestock	4.6	1.9	–58.7
Other farm	2.9	1.6	–44.8
Non-farm	39.5	61.1	54.7
	Second quintile		
Crops	78.7	57.4	–27.1
Livestock	6.8	3.0	–55.9
Other farm	1.9	1.3	–31.6
Non-farm	40.6	58.7	44.6
	Third quintile		
Crops	74.1	55.1	–25.6
Livestock	8.9	3.5	–60.7
Other farm	1.9	1.4	–26.3
Non-farm	39.8	57.8	45.2
	Fourth quintile		
Crops	71.5	54.4	–23.9
Livestock	10.2	4.8	–52.9
Other farm	2.0	1.0	–50.0
Non-farm	38.6	55.9	44.8
	Highest quintile		
Crops	66.7	52.3	–21.6
Livestock	13.2	7.4	–43.9
Other farm	1.0	1.1	10.0
Non-farm	39.1	51.0	30.4
	All groups		
Crops	75.0	56.0	–25.3
Livestock	8.6	4.0	–53.5
Other farm	1.9	1.3	–31.6
Non-farm	39.5	57.1	44.6

Source: National Sample Survey, 61 and 68 Rounds.

and 44.8 per cent respectively, whereas the corresponding gain for non-farm sector was notable at 54.7 per cent. In short, the process of employment diversification from farm to non-farm sector was applicable irrespective of the income (expenditure) profile of the people though the effect was seen to diminish with increase in income. The broad trends shown above were found to be consistent with other recent studies in the field such as that of Lanjow and Murgai, 2008 and Mehrotra *et al.*, 2014.

Table 7.3 presents the above trend across the five categories of farm-size groups categorized based on their status of land ownership. The land-

Table 7.3: Changes in employment participation rate (Usual Status) by sectors in rural India across farm-size groups, 2004-05 to 2011-12

Sector	*Employment as share of total workforce-Usual Status (%)*		*Change,%*
	2004-05	*2011-12*	
	Sub-marginal (<0.5 ha)		
Crops	61.2	41.1	–32.8
Livestock	7.9	3.6	–54.4
Other farm	2.4	1.7	–29.2
Non-farm	53.1	72.1	35.8
	Marginal (0.5-1.0 ha)		
Crops	91.1	71.1	–22.0
Livestock	8.5	4.5	–47.1
Other farm	2.2	1.0	–54.5
Non-farm	30.2	47.6	57.6
	Small (1.0-2.0 ha)		
Crops	91.8	78.1	–14.9
Livestock	9.6	4.4	–54.2
Other farm	1.4	0.9	–35.7
Non-farm	24.9	36.8	47.8
	Medium (2.0-4.0 ha)		
Crops	88.5	80.1	–9.5
Livestock	10.5	5.0	–52.4
Other farm	0.9	0.6	–33.3
Non-farm	21.7	27.9	28.6
	Large (>4.0 ha)		
Crops	87.7	82.0	–6.5
Livestock	9.1	4.8	–47.3
Other farm	0.8	0.5	–37.5
Non-farm	15.9	21.3	34.0

Source: National Sample Survey, 61 and 68 Rounds.

size groups thus defined were sub-marginal (<0.5 ha), marginal (0.5-1. ha), small (1.0-2.0 ha), medium (2.0-4.0 ha) and large (>4.0 ha). The dominance of farm sector as a major employer of people was more apparent across farm-size groups than across quintile groups. Barring the sub-marginal group, more than two-third of the population belonging to rest of the farm-size categories were employed in the farm sector, even in 2011-12. As obvious, the people belonging to the smaller farm-size categories were more inclined to non-farm sector employment. Temporally, the trend of farm to non-farm diversification in employment was pervasive across the farm-size group as well, as with the case of quintile groups. The dynamics were observed more at the lower farm-size groups, with the sub-marginal households exhibiting the highest per cent decline in farm employment followed by marginal houscholds. Evidently, the employment participation in crop husbandry declined from 61.2 per cent to 41.1 per cent in the case of sub-marginal farm-size group between 2004-05 and 2011-12. Similarly, livestock sector participation declined from 7.9 per cent to 3.6 per cent and other farm participation declined from 2.4 per cent to 1.7 per cent in this group. The loss in farm employment was compensated with nearly 36 per cent gain in non-farm employment. With minor case to case disparities, the transition was explicit in other farm-size groups also, thereby paving way for the emergence of the non-farm sector as a major source of employment for the rural masses.

The growth in employment in the non-farm sector was however not even across its various sub-sectors as depicted in Table 7.4. In 2004-05, manufacturing, trade and hospitality and community, social and personal services accounted for nearly three fourth of employment in the non-farm sector. As share of total non-farm workforce, these three sub-sectors respectively catered to 24.8 per cent, 25.0 per cent and 23.3 per cent of sample population for their employment needs in primary or subsidiary status. Nearly 16 per cent of people were employed in the construction sector whereas other sub-sectors such as mining and quarrying, electricity, gas and water supply, transport, storage and communication as well as financing, insurance and real estate made up for the rest. By 2011-12, the scenario changed drastically with the construction sector emerging as the highest employer among others. It is also the only sector which expanded in terms of employment during 2004-05 to 2011-12, whereas all other sectors lost their shares. The finding has several implications, particularly when it comes to a movement of people from tertiary sector to the secondary sector, which is generally not the case. It also has possible connotations in terms of casualization of workforce, marginalization of employment, and so on which forms basis for further in-depth analysis, but, however, not attempted here.

Table 7.4: Changes in employment participation rate (Usual Status) in rural non-farm sector across its subsectors in India, 2004-05 to 2011-12

Non-farm sub-sector	*Employment as share of workforce in non-farm sector – Usual Status (%)*		*Change, %*
	2004-05	*2011-12*	
Mining and quarrying	1.4	0.8	–42.9
Manufacturing	24.8	17.3	–30.2
Electricity, gas and water supply	0.8	0.7	–12.5
Construction	15.9	32.9	106.9
Trade and hospitality	25.0	19.1	–23.6
Transport, storage and communication	8.9	8.0	–10.1
Financing, insurance and real estate	2.1	1.0	–52.4
Community, social and personal services	23.3	20.1	–13.7

Source: National Sample Survey, 61 and 68 Rounds.

Income Sources[2]

Further to the above discussion on broad trends on employment in rural areas, it is worthy to look specifically at the income sources of the farm households in detail. As the usual rounds of the NSSO surveys on employment and unemployment do not provide detailed information on income sources of farm households, the results obtained from the Situation Assessment Survey of the NSSO for the year 2003 was used for onward discussion.

Table 7.5: Income sources of farm households by income quintile

Quintile	*Per capita income (Rs./annum)*	*Share in income (%) from*			
		Agriculture	*Livestock*	*Wages and salaries*	*Non-farm business*
Lowest	2503	49.2	18.7	28.4	3.7
Second	4079	45.2	19.2	31.2	4.4
Third	6162	43.8	19.6	30.1	6.4
Fourth	9834	45.3	19.2	25.2	10.3
Highest	32324	38.4	10.7	13.5	37.4
All	**10411**	**41.4**	**14.4**	**19.8**	**24.4**

Source: Birthal *et al.*, 2014.

There was considerable disparity in the contribution to income by different sources across income quintiles. Among the four occupational categories considered, agriculture (crop production) was the single largest income source for farm households. Close to 91 per cent farm households reported income from crop production and obtained 41.4 per cent of their income from crops (Table 7.5). Non-farm business was the next largest income source (24.4%) after agriculture followed by livestock sector that accounts for 14.4 per cent of the total income. Wages and salaries, mainly that accrues from non-farm sources contributed nearly one-fifth of the total income.

These sources, however, vary across income strata. For the bottom 20 per cent farm households, agriculture was the dominant income source, accounting for nearly half of their total income. Wages and salaries (28.4%) and livestock (18.7%) were the next important income sources for these households. Non-farm business, however, does not constitute an important income source with only a meagre 3.7 per cent of total income. Almost a similar pattern of income diversification was observed for the households in the three middle income quintiles, with some minor variation in contribution of income sources. The income share from non-farm business activities was found to increase with increase in their total income. On the whole, for the bottom 80 per cent of the farm households, the share of non-farm business activities in their total income hardly exceeded 11 per cent. For the top 20 per cent households, non-farm business and agriculture were more important income sources accounting for about three-fourths of their total income and almost in equal proportion. Wages and salaries, and livestock contributed 13.5 per cent and 10.7 per cent to their total income, respectively.

This pattern of distribution clearly reveals that the poor farm households were more diversified towards agriculture and wage labour, while the rich households were more diversified towards non-farm business activities. Further, a comparison of income shares of the non-farm business activities offers an important conclusion that the poor diversify more towards low-paid, low-return non-farm activities, while the rich tend to diversify towards non-farm activities that generate high returns to labour and capital.

Another way of looking at the relative importance of an income source is by landholding size. Table 7.6 presents income shares of farm households from different sources across landholding size. For sub-marginal farm households (=0.5ha), who are at the lowest end of land distribution, wages including salaries, and non-farm business activities were more important income sources, contributing 36 per cent and 33 per cent to their total income, respectively. Agriculture was the dominant income source for marginal farm households (0.5-1.0ha), but non-farm sources also made sizeable contribution to their total income—non-farm business contributed 27 per cent and wages and salaries 21 per cent.

Table 7.6: Income sources of farm households by land holding size

Farm category	*% of total farm households*	*Share in income (%) from*			
		Agriculture	*Livestock*	*Wages and salaries*	*Non-farm business*
Sub-marginal (0.002-0.5ha)	40.0	16.2	14.8	36.1	32.9
Marginal (0.5-1.0ha)	23.3	36.3	15.0	21.4	27.4
Small (1.0-2.0ha)	19.0	49.6	15.7	15.4	19.3
Medium (2.0-4.0ha)	11.5	56.6	13.8	9.7	19.9
Large (>4.0ha)	6.3	67.0	12.2	4.4	16.4
All	100.0	41.4	14.4	19.8	24.4

Source: Birthal *et al.*, 2014.

Nonetheless, as the size of landholding increases, income share of agriculture increases, and shares of livestock, non-farm business activities, and wages and salaries declined. In fact, share of wages and salaries fell steeply with increase in landholding size. This pattern of income diversification is as expected, as the small landholders face acute land constraint and are forced to seek income opportunities in labour market, low-investment non-farm business activities and animal husbandry.

Determinants of Income Sources[3]

Barrett *et al.* (2001) identify a number of push and pull factors that motivate rural households to diversify their income, activity and asset portfolios. Push factors include higher risk, diminishing returns to input use, excess supply of labour in relation to land, and imperfections in land, labour and credit markets. We have used truncated regression model to ascertain the relative importance of some of these factors in farm households' decisions to diversify their income portfolio. The dependent variable is 'the share of a source in total income', and the explanatory variables include land holding size, tenurial status, land productivity, labour availability, access to farm and non-farm credit and age, sex, caste and education of the head of the household. In the regression model we have controlled for location-specific fixed effects by including dummies for states where the sample farm households are located. To improve the efficiency of the parameter estimates we have used seemingly unrelated regression estimation (SURE) by combining parameter estimates and associated var-covariance matrices into a single parameter vector and simultaneous var-covariance matrix.

Households' diversification decisions are largely determined by their land and labour endowments. Probability of participation in non-agricultural activities is expected to be higher for the households having smaller landholdings and larger families and/or more number of workers. Other related factors that motivate farm households to diversify include imperfections in land-lease market and quality of cultivated land or land productivity. If land-lease market is flexible, households may adjust their resources by leasing-in or leasing-out land, although conditional upon availability of alternative income generating opportunities. In the presence of imperfections in land-lease market, smaller farmers would tend to diversify towards non-agricultural income sources. Poor quality of land or alternatively low productivity of land would also push farm households towards non-farm activities. Gender composition of the workforce also influences farm households' decisions to diversify. Women comprise half of the total workers. Many factors such the trade-off between household chores and income opportunity, distance to an income source, nature of work, skills etc. may influence women's participation in non-farm activities. It is thus anticipated that households with more women workers have a higher probability to remain engaged in traditional income-generating activities, e.g. agriculture and animal husbandry, and are less likely to diversify towards non-farm activities. Age may be considered as a proxy for working capacity of a person. Generally, younger persons are more capable to take up strenuous works, such as wage labour, and hence are expected to participate more in non-farm activities that demand physical strength, while older persons are more likely to pursue traditional family occupations that are not as strenuous. Human capital (skills and education) is an important factor in diversification decisions. Unskilled and less educated persons are expected to be more engaged in agriculture, livestock and low-paid wage works, while the skilled and more educated have a higher probability to diversify more towards high-pay off non-farm activities including salaried jobs. The Indian social system is quite heterogeneous and stratified by religion and caste. In order to find out the role of social stratification in farm households' access to different income sources we included caste dummies as explanatory variables in the model. Scheduled caste and tribal households which rank lowest in the social hierarchy together comprised close to one-third of the total sample farm households. The other backward castes comprised 39 per cent of the total households. Lack of capital impedes investment in remunerative non-farm activities requiring higher start-up capital, purchase of lumpy assets and acquisition of livestock (Barrett, *et al.*, 2001). Access to non-farm credit eases liquidity constraint to taking up non-farm business activities. Likewise, access to credit for agricultural purposes improves their capacity to invest in agriculture and therefore a higher share of the agriculture in their total income.

Table 7.7 presents the determinants of income diversification obtained from income diversification equations. Most of the explanatory variables was found to have expected signs and were statistically significant. Landholding size had a significantly positive effect on income from agriculture, and a significantly negative effect on income from livestock, wages and salaries, and non-farm business activities. Effects of agricultural productivity were as expected; the regression coefficient on land productivity was negative and significant in case of wages and salaries, livestock and non-farm business, and a positive and significant in case of agriculture. Further, those households who have access to land through lease-market tend to be more involved in agriculture and labour market, though the estimates were not statistically significant. The parameter estimate of land-lease market was negative and significant in wages and non-farm business equations. These results suggest that small landholdings and low agricultural productivity are important push factors in farm households' decisions to diversify away from agriculture towards non-farm income generating activities.

Excess supply of labour appeared as an important trigger for non-farm diversification. The regression coefficient on workers carried a negative and significant sign in agriculture and livestock income share equations, and a positive and significant sign for wages and salaries. The coefficient on non-farm business income was also positive but insignificant. This implies that households with excess labour look for income opportunities in labour market and non-farm sector. However, households with more number of workers as well as larger landholdings tend to remain engaged in agriculture rather working as wage labour or in the non-farm sector as was indicated by direction of the regression coefficients on the variable that includes interaction between land and

Table 7.7: Determinants of income shares

Explanatory variables	*Agriculture*			*Livestock*			*Wages and salaries*			*Non-farm business*		
	Coefficient	*z-statistic*	*P>|z|*	*Coefficient*	*z-statistic*	*P>|z|*	*Coefficient*	*z-statistic*	*P>|z|*	*Coefficient*	*z-statistic*	*P>|z|*
Proportion of leased in Land	0.0007	0.74	0.461	–0.0016	–2.83	0.005	0.0012	1.34	0.182	–0.0004	–2.32	0.020
Ln Land Productivity	1.0289	94.60	0.000	–0.2894	–14.40	0.000	–1.1532	–18.56	0.000	–0.0622	–12.80	0.000
Ln Landholding size	0.8751	44.59	0.000	–0.0845	–3.53	0.000	–0.9702	–14.17	0.000	–0.0383	–5.75	0.000
Ln Number of workers	–1.1006	–9.94	0.000	–0.5764	–4.00	0.000	0.5192	1.99	0.046	0.0172	0.46	0.643
Number of workers x landholding size	0.1663	10.53	0.000	0.0227	1.24	0.215	–0.1575	–4.75	0.000	–0.0101	–1.95	0.051
Age of the head of household	–0.0025	–3.65	0.000	0.0019	1.75	0.081	–0.1584	–1.39	0.164	–0.0003	–0.99	0.321
Proportion of female workers	0.5310	11.66	0.000	0.3560	4.56	0.000	–0.4840	–3.22	0.001	–0.0131	–0.53	0.595
Availability of credit												
Access to farm credit	0.2769	12.14	0.000	0.1768	5.42	0.000	–0.4236	–7.25	0.000	–0.0402	–3.77	0.000
Access to non–farm credit	–0.8365	–17.24	0.000	–1.0753	–13.44	0.000	–2.0297	–12.53	0.000	0.1232	11.74	0.000
Education level												
Primary	–0.0871	–4.26	0.000	–0.0434	–1.23	0.219	–0.2221	–3.59	0.000	0.0841	8.06	0.000
Middle	–0.2585	–9.65	0.000	0.0509	1.17	0.241	–0.0270	–0.35	0.727	0.0873	7.24	0.000
Secondary	–0.6321	–21.45	0.000	–0.2067	–4.66	0.000	0.6885	7.57	0.000	0.1000	7.86	0.000
Graduation and above	–1.1209	–22.81	0.000	–0.5933	–7.73	0.000	2.0044	10.49	0.000	0.1036	4.89	0.000
Social group												
Scheduled tribe	0.1032	3.61	0.000	–0.2490	–4.79	0.000	0.3534	3.71	0.000	–0.0771	–4.20	0.000
Scheduled caste	0.0510	1.97	0.049	–0.3388	–7.64	0.000	0.2466	3.07	0.002	–0.0648	–5.06	0.000
Other backward castes	–0.0398	–1.79	0.074	0.0258	0.72	0.473	–0.0909	–1.43	0.152	–0.0206	–2.07	0.038
Number of observation		33220			23050			17545			6090	

Source: Birthal *et al.*, 2014.

labour. Lanjouw and Shariff (2002) using data from a large scale survey conducted by the National Council of Applied Economic Research (NCAER) found a strong positive association between household size and farm as well as non-farm income sources excepting agricultural wage employment. Our results contradict this relationship in case of agriculture and animal husbandry.

Parameter estimates of women workers were positive and significant in agriculture and livestock income share equations suggesting that households with more women workers have a higher probability to involve more in agricultural activities. This contrasts finding of Lanjouw and Shariff (2002) which suggest that women are more likely to feature among agricultural labourers than any of other non-farm occupation or cultivation. Age of the head of household was negatively associated with income from agriculture, labour and non-farm business activities indicating that younger persons tend to diversify more towards these activities.

Education was found to have a strong effect on income diversification. Irrespective of the level of education there was tendency among farm households to diversify away from agriculture and livestock towards non-farm activities and wages and salaried employment. Income from non-farm business was positively and significantly related with all levels of educational attainment, which suggests that non-farm sector is quite heterogeneous and has potential to engage individuals with varying degrees of skills and schooling. The relationship of wages and salaries varied with the level of education—those with less years of formal education have a lower share of wages and salaries in their income portfolio. This is not surprising, as more educated persons often seek opportunities in regular salaried activities rather being engaged in low skilled wage opportunities. Income share of agriculture was negatively associated with educational attainment and the relationship becomes stronger with level of education. Likewise, the tendency to diversify away from animal husbandry increased with level of education. These results by and large conform to those reported in Lanjouw and Shariff (2002), Lanjouw and Murgai (2008).

Credit alleviates liquidity constraint to investment in farm and non-farm activities. In our model, credit appears as an important determinant of households' access to different income sources. The households who had access to agricultural credit (includes credit for crop as well as livestock production) were less likely to diversify towards wage employment and non-farm business activities as was indicated by a strongly negative regression coefficient on farm credit in the wage and non-business income equations and a positive and significant coefficient in case of agriculture and livestock. Likewise, access to non-farm credit had a positive and significant influence on non-farm business, and a negative and significant association with other income sources.

Finally, we looked at the empirics of the relationship between social structure and income diversification. The regression coefficients on caste dummies indicate that both scheduled caste and tribal households obtain relatively a smaller share of their income from non-farm business and livestock, and larger share from wage employment and agriculture. These households face greater financial and skill barriers and hence their less involvement in non-farm business and livestock enterprises is expected. Their gravitation towards wage employment including salaried employment was expected as the labour market is heterogeneous and can engage both skilled and unskilled workers. However, our results show that compared to other castes, these households derive a larger share of their income from agriculture which contradicts the common perception that socially disadvantaged groups are less involved in cultivation (Lanjuow and Shariff, 2002), because of their limited access to land. To obtain further insights into this relationship we examined land distribution by social classes and found that scheduled caste and tribal households were not as deprived in land distribution as is often perceived— together they shared about one-fourth of the total arable land in our sample. Lanjouw and Murgai (2008) also found these households were more involved in cultivation and agricultural wage labour than in any other occupation.

Conclusion and Policy Implications

This study brings forth the changes in employment profile as well as income sources of India's rural population during the last decade or so. The findings point to a shift of rural inhabitants from the farm sector to non-farm sector for employment and income thereby contributing to the emergence of the latter as an important source of livelihood in rural India. Both in terms of primary industry of activity of the households as well as usual status of employment of population, the farm sector has been gradually being replaced by the non-farm sector. It was also noted that, as against the common perception that agriculture is the main source of income for farm households, non-agricultural sources make a sizable proportion of their total income. Non-farm business, wages and salaries accounted for as much as 44 per cent of their household income, and were more important to the households at lower-end of land distribution. For small landholders, wages and salaries were an important income source besides livestock and related activities. Access to different income sources was, however, determined by a number of factors. Small size of landholding, surplus labour and low agricultural productivity were important motivations for farmers to diversify towards non-farm activities, while skills, education and access to financial resources emerged as important pull factors in diversification towards non-farm activities.

These results have some important implications for policy. First, agriculture still constitutes a major source of employment and incomes of rural households, and hence the the development policies should emphasize on intensification as well as diversification of small farms, which dominate the Indian agriculture, and make immense contributions to national food-security. Simultaneously, there is a need to develop markets, infrastructure and institutions that may enable agriculture to tread on a sustainable growth path. The policy should also emphasize development of animal husbandry, which generates a regular stream of income and can be taken up with small initial investment and scaled up through reproduction. Second, poor households depend more on wages for their livelihood. Labour market, however, is characterized by seasonality which necessitates generating sustainable employment opportunities in the rural non-farm sector. Finally, the rural non-farm sector has considerable potential to enhance income of the poor small landholders if they are facilitated to overcome some of the financial and market barriers they face in their entry into non-farm sector.

Notes

1. The survey defines a farm household as the one engaged in one or another agricultural activity during the survey period.
2. This section is heavily drawn from Birthal *et al.* (2014).
3. This section is heavily drawn from Birthal *et al.* (2014).

References

Adams, Jr. R.H. (2001). *Non-farm income, inequality and poverty in rural Egypt and Jordan*. Policy Research Working Paper 2572. World Bank, Washington, D.C.

Adams, Jr. R.H. and He, Jane J. (1995). *Sources of income inequality and poverty in rural Pakistan*. Research Report No. 102. International Food Policy Research Institute, Washington, D.C.

Barrett, C.B., Reardon, T. and Webb, P. (2001). Non-farm income diversification and household livelihood strategies in rural Africa: Concepts, Dynamics and Policy Implications. *Food Policy* 26(4): 315-332.

Birthal, P.S., Negi, D.S., Jha, A.K. and Singh, Dhiraj (2014). Income sources of farm households in India: Determinants, distributional consequences and policy implications, *Agricultural Economics Research Review*, 27(1): 37-48.

Collier, P., Radwan, S. and Wangwe, S. (1986). *Labour and Poverty in Rural Tanzania*. Clarendon Press, Oxford, U.K.

Coppard, D. (2001). *The Rural Non-farm Economy of India: A Review of the Literature*. NRI Report No. 2662. Natural Resources Institute, London, U.K.

Datt, G. and Ravallion, M. (1996). *Why have some Indian states done Better Than others at Reducing Poverty*. Policy Research Working Paper 1594. World Bank, Washington, D.C.

de Janvry, A., Sadoulet, E. and Zhu, N. (2005). *The Role of non-farm Incomes in Reducing Rural Poverty and Inequality in China*. Working Paper 1001. Department of Agricultural and Resource Economics, University of California, Berkeley. Available at http://repositories.cdlib.org/are_ucb.

Foster, A.D. and Rosenzweig, M.R. (2004). Agricultural Productivity Growth, Rural Economic Diversity and Economic Reforms: India, 1970-2000. *Economic Development and Cultural Change* 52(3): 509-542.

GoI (Government of India). (2005). Income, expenditure and productive assets of farmer households. *NSS Report No. 497* (59/33/5). National Sample Survey Organization, Ministry of Statistics and Programme Implementation, Government of India, New Delhi.

GoI. (2006). Provisional Results of the Fifth Economic Census. Central Statistical Organization, Ministry of Statistics and Programme Implementation, Government of India, New Delhi.

GoI. (2010). National Accounts Statistics, 2007. Central Statistical Organization, Ministry of Statistics and Programme Implementation, Government of India, New Delhi.

Lanjouw, P. (1999). Rural Non-agricultural Employment and Poverty in Ecuador. *Economic Development and Cultural Change* 48(1): 91-122.

Lanjouw, P., and Murgai, R. (2008). Poverty Decline, Agricultural Wages and Non-Farm Employment in Rural India: 1983-2004. Working Paper 437. Stanford Centre for International Development, Stanford University, Stanford.

Lanjouw, P. and Shariff, A. (2002). *Rural Non-farm Employment in India: Access, Income and Poverty Impact*. Working Paper 81. National Council of Applied Economic Research, New Delhi.

Mehrotra, S., Parida, J., Sinha, S. and Gandhi, A. (2014). Explaining Employment Trends in the Indian economy: 1993-94 to 2011-12. *Economic and Political Weekly*, 49(32): 49-52.

Reardon, T., Berdegue, J., Barrett, C.B. and Stamoulis, K. (2007). Household Income Diversification into Rural Non-farm Activities. In: Haggblade, S., Hazell, P., Reardon, T. (Eds.), *Transforming the Rural Non-farm Economy: Opportunities and Threats in the Developing World*. Johns Hopkins University Press, Baltimore.

Reardon, T. and Taylor, J.E. (1996). Agro-climatic Shock, Income Inequality and Poverty: Evidence from Burkina Faso. *World Development* 24(4): 901-914.

Reardon, T., Stamoulis, K., Cruz, M.E., Balaiscan, A., Berdegue, J. and Banks, B. (1998). *Rural Non-farm Income in Developing Countries. The State of Food and Agriculture 1998* (Part III). Food and Agriculture Organization, Rome.

Warr, P. (2003). Poverty and Economic Growth in India. In: Kalirajan, K., Shankar, U. (Eds.), *Economic Reform and the Liberalization of the Indian Economy*. Edward Elgar, Cheltenham, UK and Northampton, MA, USA.

8
Non-Farm Employment in Rural India
Trends and Determinants

Sharmistha Sinha

A marked increase in India's GDP growth is witnessed in the new millennium. During the period 2002-2007, GDP grew by 7.7 per cent per annum, which increased further to a growth rate of 7.9 per cent per annum during 2007-2012. These growth rates were much higher than those experienced in the last quarter of the previous century, let alone the first three decades after Independence (the latter being characterized by the Hindu rate of growth of 3.5 per cent per annum). With the economy growing at such an unprecedented rate, it is also witnessing a structural change since the 2000s. The contributions to GDP over 2000-01 to 2011-12 changed for agriculture from 24 to 14 per cent, for industry from 27 to 28 per cent, and for services from 49 to 58 per cent. However, change in output has not been translated into commensurate change in employment. The diverging trend between the structure of output on the one hand, and the structure of employment on the other, in the last decade since the GDP growth rate increased since the beginning of the new millennium, is a matter of concern among both academia and policy-makers. This is more so, since the share of the working age population in total population has been growing—the so-called demographic dividend.

Employment in India grew from 399 million in 2000 to 459 million in 2005, and then in the next five years, when the output growth reached its peak, employment, practically did not increase at all. It rose by only 1 million during 2005 to 2010. In the next two years, however, it increased by 14 million. Though the structural change in employment is taking place more slowly than in output, but it is notable that for the first time in the post-Independence economy, agricultural workforce declined in absolute terms since 2005, particularly among rural females.

The structural shifts in employment, particularly in rural India coupled with significant increase in rural wages since 2004-05 had raised the consumption expenditure per capita, thus leading to a decline in absolute numbers of the poor on a scale unprecedented in the post-Independent history of India. Post 2004-05, when a revised (Tendulkar) poverty line raised the absolute and relative poverty estimates, the absolute number of poor in rural areas in 2004-05 which was 326 million fell by over 110 million to 216 million in 2011. Now, the question is whether this transition out of agriculture, fall in the incidence of poverty is sustainable.

One way that the increase in labour force is gainfully engaged, the structural change in the labour force is sustained and strengthened and the incidence of poverty is reduced further is through sectoral diversification, particularly in rural areas. However, mere growth of non-farm employment is not by itself sufficient as this may occur due to people seeking shelter in petty non-farm activities as their last resort. Sectors having high employment potential and linked with other productive sectors should be encouraged to bring about sectoral diversification by creating jobs at higher levels of productivity.

This chapter has three sections. First section looks at the type of non-farm employment opportunities available in the rural areas in India and how did it change over time 1993-94 to 2011-12, computed from the Unit level data of various rounds of the Employment-Unemployment Surveys of the National Sample Survey Organization. In the second section, the chapter tries to look at the factors affecting non-farm employment growth. The factors discussed here are always not specific to rural areas but they are macro-economic policy factors that have linkage effect impacting rural non-farm employment growth. The last section concludes the chapter with policy recommendations.

Employment in Rural Areas: Nature and Trends

Workforce in rural India increased from 292 million in 1993-94 to 304 million in 1999-2000, and then it accelerated significantly to reach 345 million in 2004-05 (Table 8.1). Nevertheless, out of this 41 million increase in the first half of the decade, 18 million increase was in agriculture sector itself. This is to be noted that the first half of the decade, that is 2000 to 2005 was a period of agricultural distress—when the agricultural growth rate was very low at about 1.7 per cent per annum as compared to the 3.2 per cent per annum, since the mid 2000s. Therefore, the significant rise in agricultural employment during this period was clearly a retrograde development. Post 2005, rural employment reduced sharp to reach 336.3 million in 2009-10 and further to 335.9 million in 2011-12. And this decline is primarily attributed to fall in agricultural employment. 22 million and then another 13 million rural people left agriculture since 2005. The question is: if there was commensurate increase in non-farm employment (NFE) opportunities? During 2005 to 2010, non-farm employment in rural areas increased by only 13 million, mainly driven by the huge increase in employment in the construction sector. Employment in the manufacturing sector fell and that in services sector increased marginally. Post 2010, in the next two years, employment opportunities grew significantly in all non-farm sectors.

Table 8.1: Employment in rural areas

(*in millions*)

Sectors	*1993-94*	*1999-00*	*2004-05*	*2009-10*	*2011-12*
Manufacturing	20.5	22.6	27.9	24.1	29
Non-manufacturing	9.1	11.9	19.1	34.3	39.5
Services	33.6	37.7	47.5	49.4	52.1
Total non-farm employment	**63.2**	**72.2**	**94.5**	**107.8**	**120.6**
Agriculture	227.3	230.6	248.3	226.7	213.6
Total	**292.5**	**304.7**	**345.1**	**336.3**	**335.9**

Source: Author's estimates based on EUS NSS Unit level data, various Rounds.

Non-farm sectors employ 36 per cent of total rural employment. There is no doubt that rural economy has undergone a transformation with the share of non-farm sector in rural employment growing over time, increasing from 22 per cent in 1994 to 27 per cent in 2005 to 36 per cent in 2012. The share of non-manufacturing (which mainly comprises the construction sector) sector in rural non-farm employment rose from 14 per cent in 1994 to 20 per cent in 2005 to 32 per cent in 2012. Increase in investment, both public and private, in infrastructure, real estate, and public programmes like Indira Awaas Yojana, Pradhan Mantri Gram Sadak Yojana and Mahatma Gandhi National Rural Employment Guarantee Act (MGNREGA) have raised the employment in this sector. This alternative employment opportunity at relatively higher wages has actually acted as a demand factor inducing people to move out of agriculture.

The capacity of the manufacturing and services sectors have not expanded sufficiently, so as to provide enough employment opportunities to the rural labour force. Even though employment increased in absolute terms, the share of manufacturing employment in rural non-farm employment fell from 32 per cent in 1994 to 29 per cent in 2005 to 24 per cent in 2012. Similarly, for services, it declined from 53 to 50 to 43 in the corresponding periods.

Within the manufacturing sector, employment increased mostly in the labour-intensive sectors like wearing apparel, textiles, furniture, non-metallic mineral products and wood products etc., which again had been fluctuating over the period. Also the share of output in these sectors in total manufacturing value-added has been stagnant implying that these are low-productive small scale enterprises that are driving employment in these sectors (Mehrotra *et al.*, 2014). In services also, it is retail trade and transport that engages most of the rural population.

Rural Non-Farm Employment—Is it Vulnerable Employment?

What type of work is seen in the non-farm sectors? In the manufacturing sector, workers are mainly self-employed; a significant share of them works as unpaid family workers in family enterprises, which they neither own nor get any salary or wages for their work (Figure 8.1). In services also, more than 50 per cent of the workers are self-employed. They are either own account workers or family helpers, working without any regular wage, any social security, compromising on work conditions, mostly in

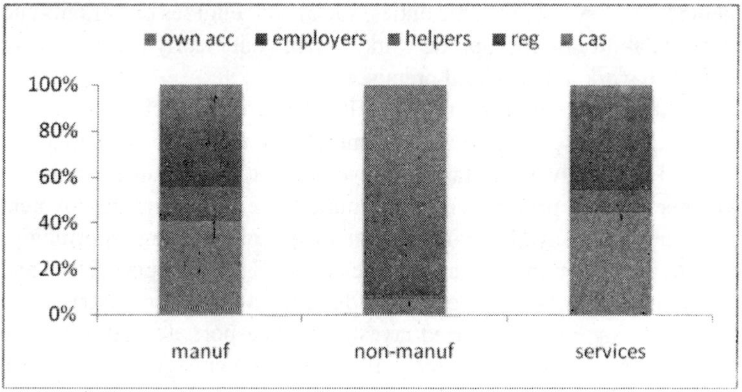

Fig. 8.1: Classification by type of employment in broad sectoral categories in 2011-12

Source: Author's estimates based on EUS NSS Unit level data, various Rounds.

petty trade. Regular work, mainly in education, health, public administration etc. is also prevalent. The non-manufacturing sector, which practically means construction, employs manly casual workers.

Gender-wise Distribution of Workers

Gender based sectoral and occupational segregation of work is widespread. Also there is segregation by type of employment, type of enterprise in which they work, mode of payment etc. More than three quarters of rural female workers are engaged in the agricultural sector in 2011-12 (Table 8.2). There is certainly a massive decline, with 27 million women coming out of agriculture in absolute numbers since 2005. But there is not adequate absorption of females in non-farm sectors. During 1994 to 2005, female employment in rural areas in non-farm sectors grew at par with that of men. In the next seven years, there was practically no growth in female employment in manufacturing and services sectors.

Increment in educational participation among females is considered as a key reason behind falling female labour force participation rate since 2005, but female labour force participation rate had fallen in all age groups, even for those who are out of the educational age cohort (Sinha 2014). The NSS 68st Round Report (NSS 2012) shows that a significant share of females is moving out of labour force and engaged in domestic activities (care work, household chores) and domestic duties with specified activities (collecting fuel, water, dairy, poultry for household use, tailoring for household use). And the reasons were mainly absence of any other member to carry out domestic duties, social and religious constraints and non-availability of appropriate work. It came out clearly that they were willing to work at household premises.

If it is considered that female labour force participation rate is declining with rise in household income, now reaching a bottom of a 'U', and will rise at advanced stages of development, then the question arises whether enough productive and remunerative non-farm employment opportunities are available in the rural areas. Employment opportunities need to be created in the vicinity. Focus should be laid upon Tier-II and Tier-III cities also, in terms of development of infrastructure, manufacturing hubs, increased investment, transport etc. this will raise the domestic economy, which will further generate more non-farm employment opportunities. The educated females likely to join the labour force in the next 5 to 7 years would surely not like to join agriculture or low productive non-agricultural work like *zari*, *zardosi*, *bidi* rolling and fish processing.

Table 8.2: Gender-wise distribution of workers in different sectors in Rural India, 1993-2012

(in millions)

Sectors	Male			Female		
	1993-94	*2004-05*	*2011-12*	*1993-94*	*2004-05*	*2011-12*
Manufacturing	13.1	17.5	19.1	7.4	10.5	10
Non-manufacturing	7.8	17	32.5	1.23	2.2	7.1
Services	27.9	39.3	43.7	5.9	8.2	8.5
Non-farm employment	48.8	73.8	95.3	14.5	20.91	25.6
Agriculture	139.1	146.7	139.2	90.4	104.3	76.2
Total	**187.8**	**220.7**	**234.6**	**104.7**	**125.2**	**101.8**

Source: Author's estimates based on EUS NSS Unit level data, various Rounds.

Number of male workers has continuously risen in every non-farm sector. But for women, post 2005, employment in manufacturing, services has remained constant. It has increased only in construction. There is also gendered diversity in the type of work that one does. In manufacturing, for women, self-employment is very high (74%); for men the share of regular wage worker is higher (43%) (Mehrotra and Sinha, 2014). The classification of self-employed makes the gendered nature of work all the more stark. Among male self-employed, own account work is high, but for females unpaid family work is more. So women really work under the aegis of men. With contractualisation, and sub-contract there had been an increasing trend of shifting work from factory to floor, mostly piece rated work, where one do not have any bargaining power, any social security or job security. And typically more women than men work as such home-based workers.

Female employment is higher in specific industries like tobacco, textile, apparel, non-metallic mineral, jewellery and toys. But for male, it is much more diversified, spread in all sub-sectors. Female employment in services increased from 8.2 million to just 8.5 million. But for males, it rose from 39 million to 43 million. Share of regular work among female in services is high at 51 per cent. But self-employment is also high (42%) and in that 85 per cent are unpaid family workers. They work mostly in retail trade, education, domestic workers and health sector. In 2011-12 there had been a rise in personal services (Mehrotra and Sinha, 2014).

Reasons behind Insufficient Generation of Non-farm Employment in Rural Areas

A number of exogenous factors, both from demand side and supply side, affect the expansion of rural non-farm activities and generation of more employment. Price of a product produced, the wage received in the sector, the prices of the array of inputs used in the production process or employment, market linkage (Haggblade, Hazell and Reardon, 2007, Jha, 2006) etc. determine the profitability and/or risk associated with it. Agricultural profitability, land holding, cropping pattern, credit facilities, agro-based tradable etc. often link to the expansion of employment in non-agricultural sectors, be it upstream or downstream. Along with that availability of employment opportunities in the nearby areas, distance, road connectivity, presence of industrial centres, power supply, sub-contracting of work etc. also decides on taking up work outside agriculture. Relative poverty, family size, basic service facilities, skill level (Mishra, 2013) has impact on the type of work taken up by rural people. Some of the important macro-economic factors in terms of expenditure, investment, deployment of credit etc. which has significant implications on growth of employment are discussed to understand the insufficient generation of rural non-farm employment in India.

Deployment of Credit

Deployment of credit by Scheduled Commercial Banks (SCB) shows that the share of credit to industries over total non-food credit is 45.3 per cent in March 2014 (Figure 8.2), which is also constant since 2012. In fact, growth of credit to manufacturing by SCBs fell from 19 per cent in 2011-12 to 18 per cent in 2012-13 to 14 per cent in 2013-14. This credit is largely directed to large enterprises (Figure 8.3a), which comprises of 81 per cent of total industry credit, showing an increasing trend. For small and medium enterprises the share of credit has declined since 2011.

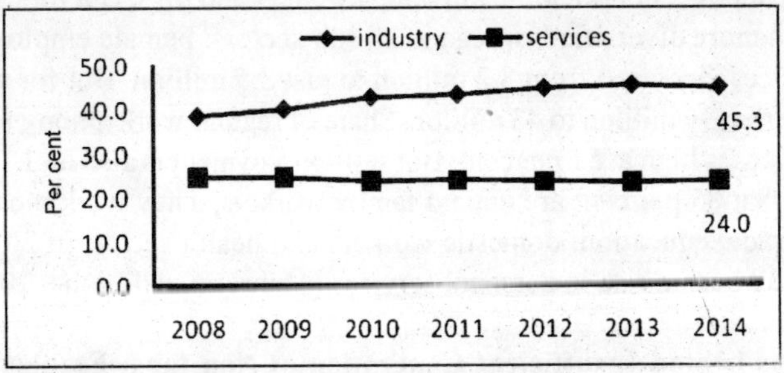

Fig. 8.2: Credit to industries over total non-food credit

Source: Selected Macroeconomic Indicators 17/05/2014, EPW Research Foundation.

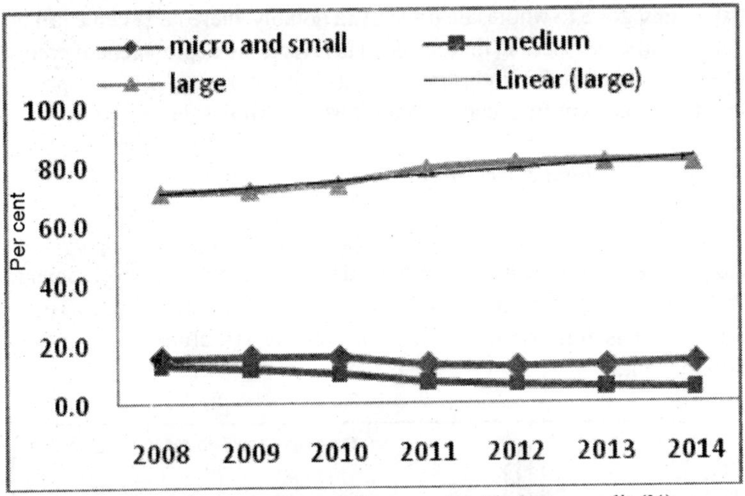

Fig. 8.3 (a): Credit by type of enterprises over total industry credit (%)

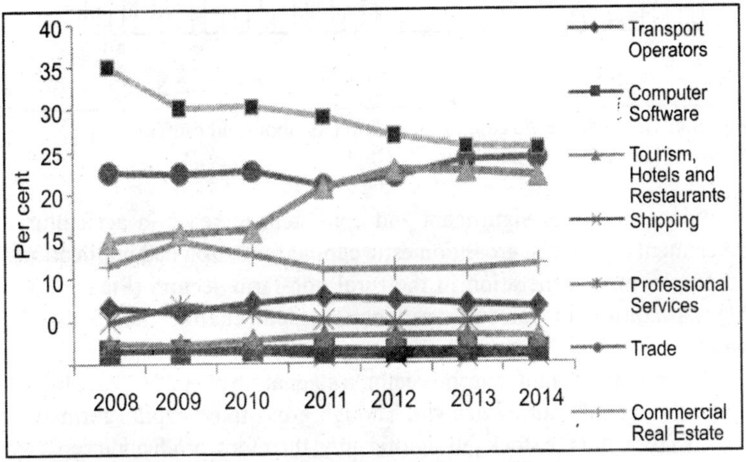

Fig. 8.3 (b): Sector-wise credit over total services credit

Source: Selected Macroeconomic Indicators 17/05/2014, EPW Research Foundation.

In services, sector, credit mainly goes to trade, other services and non-banking financial companies (Figure 8.3b). There had been a significant rise in credit to non-banking financial companies since 2010; trade also had seen a rise in its share in total services credit, however, it had mainly gone to wholesale trade. And notably, there had been a fall in credit to other services in this period. Therefore, the deployment of credit is highly skewed both in industries and services, favouring large enterprises and in services the financial corporations and wholesale dealers.

Changes in Capital Formation

Capital formation is an important factor to raise productivity, employment and income and thus can be instrumental towards economic development. The structure and trend of gross domestic capital formation (GDCF) over the years has important implications on the structure and trend in employment.

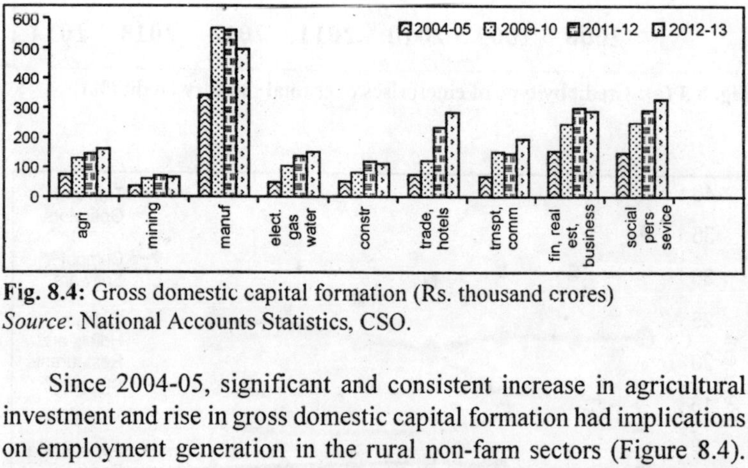

Fig. 8.4: Gross domestic capital formation (Rs. thousand crores)
Source: National Accounts Statistics, CSO.

Since 2004-05, significant and consistent increase in agricultural investment and rise in gross domestic capital formation had implications on employment generation in the rural non-farm sectors (Figure 8.4). Mechanization in agriculture, technological change increased the productivity of this sector and it released the excess workers in this sector. However, in the manufacturing sector, a stagnation post 2010 has slowed down the manufacturing growth. Though gross fixed capital formation (GFCF) rose, but the stock fell significantly; therefore, production capacity declined. Employment increased post 2010, but it had been mainly in low capital intensive sectors. Similarly in construction, there had been a significant increase in Gross Domestic Capital Formation (GDCF) during 2005-2010 followed by a fall post 2010; GFCF rose but stock fell. This is visible from the slowing down of the growth of employment in this sector post 2010. The rise in GDCF in electricity and other utilities, transport, trade and hotels have had impact in employment generation in these sectors. But since the incidence of employment in electricity, water, transport are less, so the impact does not look significant. To sustain and improve the growth of employment in non-farm sectors, more capital formation is required in every sector.

Capital formation helps in the growth of human capital. Expenditure on education, health, social service, and social welfare is an investment on human capital which then raises its productivity. Figure 8.4 shows an increase in Gross Domestic Capital Formation in community social and personal services during 2004-05 to 2009-10. However, it is to be noted that public administration and defence comprises the major share in it. And second, the increase had been mainly due to rise in gross fixed capital formation in this public administration and defence. Community services, education, health have seen only a marginal rise in GFCF. Investments, from both public and private quarters, in these sectors, mobilization of savings can step up productivity, employment and income and thus can be instrumental towards economic development.

Low Public Expenditure on Education and Health—Implications on Employment

India's total (Union + State governments) expenditure on each of education and health as a proportion of GDP is one of the lowest proportions recorded anywhere in the world. The share of combined expenditure on education over GDP increased from 3.0 per cent in 2004-05 to just 3.72 per cent in 2009-10 (Figure 8.5), which is much below the 6 per cent of GDP, as recommended in 1966 by the Kothari Commission. Combined Public expenditure on health as a proportion of GDP was just 0.91 per

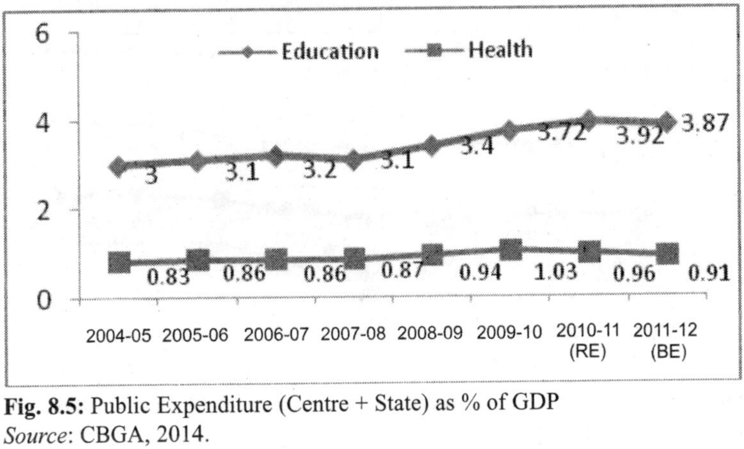

Fig. 8.5: Public Expenditure (Centre + State) as % of GDP
Source: CBGA, 2014.

cent in 2011-12. The combined expenditure on health has increased in absolute terms in the last ten years, since 2004-05, but its share to GDP had remained constant during the period 2004 to 2008; increased marginally till 2010 and reduced since then till 2011-12. Low public expenditure on these sectors which absorb a significant share of the workforce, is one of the reasons behind low non-farm employment.

A sectoral break-up shows that majority of the allocation on education goes for elementary education (Figure 8.6). Though a marginal improvement in allocation is observed in elementary and secondary level of education since 2007-08, however, 2008-09 onwards, allocation for University and Higher Education has gone down and continuously decreasing in the following years. Higher education and technical education is very important to generate productive employment. And since skill development is cited as a priority of the government, higher and technical education, vocational education needs financial provisioning. To reach to the level of higher education, spending on primary and secondary education is needed. With the Sarva Shiksha Abhiyan, Mid-Day meal scheme, RMSA, Right to Education Act and others, enrollment and retention in educational institutions have increased, particularly of females, which is one of the key reasons behind declining female work participation. However, these schemes have opened avenues for employment as well, which is much more productive, remunerative and decent than agricultural or low productive home based manufacturing or petty trade. Strengthening and expanding the capacity of these schemes and the education sector as such will absorb more people much more suitably than many more sectors.

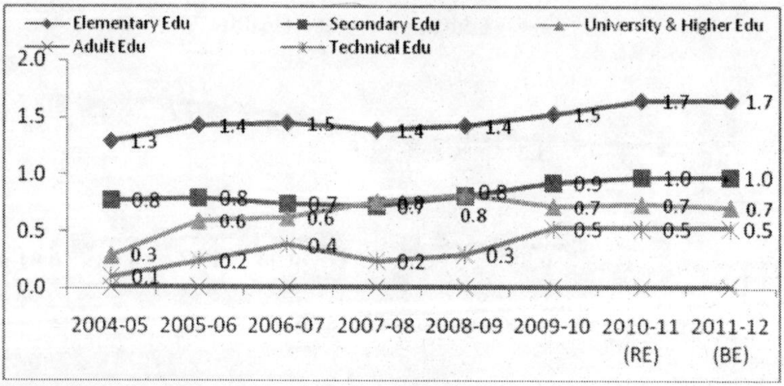

Fig. 8.6: Sector-wise combined public expenditure (Centre+State) on education as % of GDP

Source: CBGA, 2014.

The health sector comprises a significant share of workforce in non-farm sectors, particularly females in both rural and urban areas. With constrictions on public expenditure on health, it would definitely restrict job creation. The National Rural Health Mission (NRHM), launched in 2004-05 in rural areas was a landmark programme to provide universal quality healthcare. This generated employment opportunities in rural areas with the creation of a cadre of Accredited Social Health Activities (ASHA) and other public health workers. It is another matter that government is bequeathing a lot of responsibilities by delegating work to volunteers rather than recruiting regular workers, but it is certain that NRHM created job opportunities in rural areas in large numbers, particularly to women. Later in 2012-13, the NRHM and the NHM (and also many other programmes[1] of the Government) were together converted into the National Health Mission (NHM). But the share of NHM in total expenditure had been fluctuating with its budget estimates in 2013-14 reduced to only 56 per cent (Figure 8.7). To meet the objective of providing essential primary healthcare needs of households, there is a need to create job opportunities, both professional and semi-skilled workers, at individual local places.

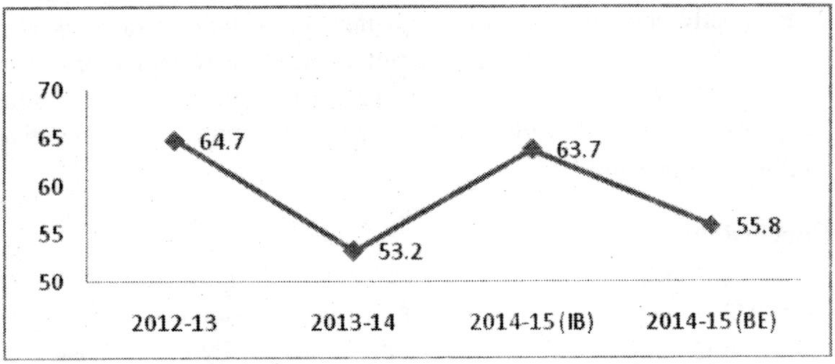

Fig. 8.7: Share of NHM in Total Health Expenditure by Union Government
Source: CBGA, 2014.

More public expenditure can address the shortage of human resources in the health sector. Not just health sector, but increased public expenditure in general (education, drinking water, sanitation, waste management etc.) can generate more employment, therefore, raise earnings, lift up domestic and regional demand (for goods and services) and therefore, more production with a strong and sustained market, therefore can pave an effective mode for growth.

Though these arguments do not really pertain to rural areas, but it has significant impact on rural non-farm employment generation, migration, earnings and therefore a cycle progresses with more multiplier effect. Looking at the sectoral classification of the rural non-farm workers, they are concentrated in a few sectors like manufacturing, construction, retail trade, education and health. Therefore, increase in investment and expenditure will certainly give an optimistic boost to generate more employment.

Is Growth of Rural Non-farm Employment Enough?

Growth of non-farm employment opportunities is not sufficient to ensure equity in access to and distribution of returns from such prospects. It varies by relative endowment with financial and human capital, by social group, gender and other vulnerable groups and depends on the socio-cultural context in a region. Those with higher level of skills, education and landholding/capital are better equipped to take advantage of growth in the high-productivity sectors, both as entrepreneurs and as wage employees. On the other side, the poor with labour as their only asset is renounced with only low productive employment avenues left for them. They mostly engage in small-scale, often one-person, rural non-farm enterprises and/or often home based piece rated work which brings in only subsistence level of earning. Information asymmetry also arises from the power asymmetry which then impacts engagement in productive employment possibilities.

Conclusion

This chapter shows, that non-farm employment opportunities in rural India are growing but at a slow pace. It is driven mainly by the expansion of the construction sector as in the recent years investment in infrastructure, real estate has increased. In manufacturing and services it is concentrated only in some subsectors like textile, wearing apparel, retail trade etc. in small scale low productive enterprises, mostly as self-employed and there are fluctuations in the level of employment over the period. Female employment had been declining in rural areas since 2005, primarily from the agricultural sector, because of various reasons like increase in educational participation, rise in household income, mechanization in agriculture etc. There is not adequate absorption of women in the non-farm sectors; thus calling for more productive, remunerative and diversified employment generation in the non-farm sectors in the rural areas.

Public policies must focus on improvement in rural education and health in order to improve the human capital stock of the poor. There is increased participation in educational institutions in the recent period with the Sarva Shiksha Abhiyan, Right to Education Act etc. this transition will earn a dividend if vocational training, technical education is encouraged, linking it with the need of the industry. Any training should, however, be supported with adequate hand holding so that it comes to its logical conclusion. Focus should be on increasing job opportunities in the vicinity in rural and semi-urban areas with Infrastructural development like public transport, road connectivity, electricity etc. Public support is needed on Initial finance-working finance, which needs more banking facilities. Banking should relate to the local needs and can be manned by local people whose skill can be developed though appropriate training. Development of cooperative banks in villages should be promoted. Different insurance schemes should be introduced to cater to the rural people. There should be opportunities for agricultural households to moderate seasonal income swings, finance agricultural input purchases etc.

A system also needs to be developed to assess the quality of a product, improve market linkage, reduction of information asymmetry and credit facilities particularly relevant for the home based workers. Government must focus on providing basic service facilities, like child care etc. which can release the female labour force for work. Public expenditure on health, education etc. can, on the one hand, improve the outcome indicators and on the other has significant implications on employment generation. It should not be looked at with adhocism, but a consistent and significant allocation of resources is needed to address the issue of shortage of human resources. Rise in credit extension to transport, tourism, hotel, professional services and other services can expand these sectors, which again can open employment avenues to a large extent in rural areas. Consistent rise in investment in non-agricultural sectors and the agro-based manufacturing focusing on the domestic market can expand the non-agricultural sectors and can link the rural poor to growing income. It is not just robust growth in non-farm sectors that can reduce the poverty level, but developing an environment wherein the access to and returns from such growth is equally distributed can lead to a sustained growth with development and reduce farm distress as well.

Note

1. Major interventions of the MOHFW, like the NRHM-RCH Flexi pool, the NUHM Flexipool, the National Vector Borne Diseases Control Programme (NVBDCP), the Routine Immunisation, Pulse Polio Immunisation and others, have been brought under the ambit of NHM.

References

EPW Research Foundation (2014). *Selected Macro-economic Indicators* 17/05/2014.

Haggblade S., Hazell P. and Reardon T. (2007). *Transforming the Rural Non-farm Economy Opportunities and Threats in the Developing World*, Published for the International Food Policy Research Institute, The Johns Hopkins University Press, Baltimore.

Jha, B. (2006). *Rural non-farm employment in India: Macro-trends, Micro-evidences and Policy options,* Institute of Economic Growth Working Paper.

Mehrotra, S. Parida, Sharmistha Sinha and A. Gandhi (2014), 'Explaining Employment Trends in the Indian Economy: 1993-94 to 2011-12', *Economic and Political Weekly*, Vol. XLIX, No. 32, August 9.

Mehrotra and Sinha (2014), "Low and Declining Female Labour Force Participation in India: the way forward", presented at Centre for Informal Sector and Labour Studies (JNU)-UNWOMEN Conference on "Labouring Women: Some Major Concerns at the Current Juncture", 31 July - 1 August, 2014, Jawaharlal Nehru University, New Delhi.

Misra, Shuchi Benara (2013), Growth of Rural Non-Farm Employment in India: Pre- and Post-Reform Trends and Patterns, *Journal of Land and Rural Studies*, Vol. I, No. 2, 99-112, July.

National Account Statistics, Central Statistical Organisation, Different Rounds.

Employment-Unemployment Survey, National Sample Survey Organisation, Different Rounds.

Sinha, Sharmistha (2014) "Female Employment and the Planning Process in India"—in Sagade, Jivan and Forster (ed.) *Feminism in the Subcontinent and Beyond: Challenging Laws, Changing Laws*, EBC, Lucknow.

The Centre for Budget and Governance Accountability (2014), *Response to Union Budget*.

9
Investments and Subsidies in Indian Agriculture
Recent Trends and Implications

Seema Bathla

Introduction

Among several challenges that Indian agriculture faces, the biggest has been stagnation in the growth rate which hovers between two to three per cent per annum. Of late, some states have experienced an upsurge in the rate of agricultural growth but it has not turned around the persisting slowdown in incomes and the consequent agrarian distress in the country. The National Agriculture Policy announced in 2000 laid down several measures, with an aim to achieve over 4 per cent annual rate of growth by 2020, the majority being hike in the resource allocation towards agriculture, irrigation, research and education. Concerted efforts were initiated during the early 2000s to reverse the declining trend in public investment and stimulate agricultural growth by better targeting of subsidies, increased investment in productive assets such as power, credit and development of rural infrastructure. Public and private capital formation in agriculture grew at 6.3 per cent and 8.6 per cent per annum respectively from 2000-2011, much faster than the previous decades. The share of public expenditure in agriculture and irrigation in agricultural income averaged 12 per cent during 2000-2011. Concurrently, subsidies also grew at a far rapid pace at 5.43 per cent per annum and reached 13 per cent of agricultural income by the end of the Eleventh Five Year Plan (2006-2011). An increased quantum of investment seems to have paid off as agriculture income has lately surpassed 3.5 per cent rate of growth.

The government intends to inject more funds into research, extension, education and rural energy during the 12th Five-Year Plan (2012-2017), which may defy the notion of 'neglect of agriculture' built up during the 1990s and early 2000s and uphold growth too. But achieving growth with equity, as proposed in the National Agriculture Policy, seems to be a distant dream in view of the lopsided budgetary allocations to agriculture, major and medium irrigation works and also input subsidies across the states. An unevenly regional spread and growth of public investment and subsidies has implications for farm productivity which calls for deliberation and prescription.

This chapter aims to evaluate the broad trends and patterns in public and private capital formation and input subsidies in agriculture for the recent period upto 2011-12, and their impact on farm production and growth. The following key questions have been addressed. Firstly, what has been the magnitude of allocation of public expenditure towards agriculture and other economic services and input subsidies during the 2000s compared to the earlier decades? Secondly, has government increased public investment in agriculture at the expense of input subsidies? Lastly, to what extent has a big push in public investment in 2000s influenced private investment and in turn agricultural output? The inter-linkages between public and private investments, subsidies and agricultural income are examined from 1981-82 to 2011-12 using a simultaneous equation model. Based on the findings, some broad implications are drawn.

The main sources of data on public expenditure are Finances Accounts, Government of India. The magnitude of fertilizer subsidy is taken from Fertilizer Statistics of India. The estimates for irrigation, power, and credit subsidies are arrived at using data given in the Finance Accounts, Annual Reports of the State Electricity Board, and National Federation of State Co-Operative Banks. Irrigation subsidy is calculated as the difference between the total operation and maintenance costs and the total revenue in the irrigation sector plus 1 per cent of the cumulative capital expenditure on major and medium irrigation and command area development programmes. Electricity (power) subsidy is estimated as total power supplied to agriculture multiplied by the difference between unit cost of power supply and unit revenue from agriculture. Credit subsidy consists of interest subsidy and default subsidy. Only the former is estimated due to non-availability of time series data on non-performing assets of the commercial banks. Interest subsidy is the difference between commercial interest rates and the subsidized interest rate for agriculture multiplied by the amount of outstanding loan. Time series on gross domestic product (GDP) and GDP in agriculture and allied activities

(GDPA), public and private GCFA are extracted from National Accounts Statistics (NAS), CSO. Production of foodgrain and non-foodgrain, procurement of wheat and rice are taken from Agricultural Statistics at a Glance, Ministry of Agriculture. The nominal series on public expenditure on capital and revenue accounts are converted into real price at base 2004-05 using GDP deflator.

Inter-linkages between Agricultural Investments, Input Subsidy and Growth—Literature Survey

Past studies indicate that the behaviour of public investment in agriculture[1] is explained by multiple factors, both from the supply and the demand sides. Some of the factors have acted as constraints and hence dissuaded the government to allocate funds to irrigation and agriculture during the 1980s as well as the 1990s. These included growing deficit in the revenue account, input subsidies and other grants and less priority of the government towards the agricultural sector (Mishra and Chand, 1995; Misra and Hazell, 1996; Dev 2000, and 2008). Empirical evidence also confirms that net market borrowings and funds by the State Governments, and the savings on revenue account and input subsidies have affected investments in agriculture. The demand side factors, viz. population density, procurement of foodgrains and share of agricultural workforce have influenced the government to undertake investments, (Chand, 2000; Roy and Pal, 2001; Bathla and Thorat, 2006). The analysis carried out during the pre- and post-reform periods suggests that public GCFA is largely governed by the size of government spending, availability of funds and need to sustain agricultural growth by the respective states (Bathla, 2014).

A deceleration in the foodgrain output during the 1980s became a cause for concern due to well-founded linkages of public spending with agricultural growth and reduction in rural poverty through the increase in productivity and employment, and decline in prices (Fan *et al.,* 1999; Fan, 2008; Bisaliah *et al.,* 2013). Agricultural growth was also stated to be affected by partial implementation of reforms, inadequate technological upgradation and lower returns due to high input cost (Landes and Gulati, 2004). Literature abounds on the positive impact of investments in irrigation, agriculture, education and infrastructure such as roads and power on farm productivity and poverty (Hazell and Haggblade, 1991; Ravallion and Dutt, 1995). The impact may be direct as well as indirect through the 'crowding in' effect of public investment on private investment (Misra and Hazell 1996; Dhawan, 1998; Gulati and Bathla 2002).

Many studies have favoured strong inter-dependency between the public and private GCFA. Private GCFA is also influenced positively by a favourable incentive structure, technology and credit and negatively by rural poverty and incidence of marginal holdings. A deceleration in rate of growth in public GCFA during the eighties and the nineties did affect growth in the private GCFA. A recovery in private investment during the early 2000s was explained by an increase in the number of holdings and flow of institutional credit (Chand and Kumar 2004; Bisaliah *et al.* 2013), external trade, crop diversification towards high value crops (Joshi *et al.* 2004 and 2006) and backward linkages with the growing organized food industry (Bathlam 2014).

While agricultural investments hold key to higher farm growth and productivity, input subsidies have been widely used not only in India but across the world, primarily to address market failures, support agricultural production and boost productivity. Input subsidy—direct and indirect—may reduce capital requirements and costs of inputs viz. fertilizer, irrigation, seeds and power for the small and marginal farmers (Lopez, 2005). While subsidy on inputs may influence their application, low rate of interest on credit enables farmers to purchase inputs, machinery and other capital. A renewed interest in the evaluation of farm subsidy programmes has transpired across several developing countries due to stagnant agriculture productivity growth, concerns on food security and poverty reduction, and hike in commodity prices worldwide (Jayne and Rashid, 2013). It is argued that subsidy was given to provide incentives to farmers and address their constraints on account of market failures. But over a period of time, input subsidy has become costlier, ineffective, inequitable and might have suggested a tradeoff between consumer and producer gains. It is, therefore, important to rework on its implementation keeping in view its costs and benefits, and targeting, effectiveness and sustainability (Chirwa and Dorward, 2013).

In India, the input subsidy programme was implemented during the sixties to enable a wider adoption of new technology under the Green Revolution programme aimed to raise foodgrain production. Presently, nearly 13 per cent of agricultural income is devoted to the provision of indirect input subsidy, the maximum going to fertilizer and power subsidies. Many studies viz., Gulati and Sharma (1997), Gulati and Narayanan (2003), Fan, Gulati and Thorat (2008) reiterated that subsidies enabled farmers to adopt new technology at the initial stage of Green

Revolution. An increased level of subsidies leaves much less resources for investment in agriculture. Besides, there has been a decline in the returns from subsidies during the eighties and the nineties whereas the same from public investments in agriculture, research and infrastructure remained strong and kept growing. Input subsidies have also been under attack on environmental grounds due to over pumping of groundwater and adverse soil and water quality. On equity grounds, the argument is that a handful of subsidies actually reach the farmers that too in select regions without much contribution to productivity.

Controversy remains on the effects of input subsidies including a lack of evaluation on the impact of input subsidy on farmers' incentives, agricultural productivity and profitability. A few suggested that a deceleration in public investment in agriculture was caused by a decline in the expenditure on agriculture and irrigation relative to the aggregate national expenditure (Shetty, 1990; Mitra, 1996; Mishra and Chand, 2005). Recent literature indicates that fertilizer subsidy has accelerated productivity of certain crops in certain regions, albeit imbalance in the use of NPK (Chand and Pandey, 2008; Sharma, 2013). Bathla, Bing, Thorat and Joshi (2014) have found productivity enhancing effects of fertilizer and irrigation subsidies in the preceding decade.

Stagnancy in the rate of growth in agriculture and need to build up foodgrain stocks to ensure food security to population at large might have impelled the government to accelerate spending on both investments and subsidies during 2000s, which may possibly, explain a spurt in the private investment being witnessed from that time. Notwithstanding significant acceleration in both investment and subsidy, farm productivity and income have marginally improved in some states, which require probing. It is also imperative to address the issue of public resource allocation towards various types of investments *vis-à-vis* subsidies and their impact on private investment, input use and productivity in agriculture during the 2000s.

Trends in Public and Private Capital Formation, Subsidies and Agricultural Growth

Fig. 9.1 and Table 9.1 furnish the temporal trends in gross capital formation in agriculture and subsidies in the last three decades. The behaviour of GCFA is mapped as per the type of institution and its share in overall gross capital formation of the economy (GDCF) and gross domestic product (GDP). The share of public GCFA has always been lower than that of private GCFA, at nearly one-fourth of the total investment in agriculture. It reveals periodic ups and downs, a higher rate till the 1970s, a significant decline during the 1980s, marginal recovery during the 1990s and a significant improvement during the 2000s.

The GCFA is found to be on an increasing trend in the early 1980s, somewhat downwards during the 1990s, followed by a significant acceleration from 2000 onwards. The GCFA as per the type of institution showed peaking on public account during the late-1970s, followed by a gradual decline from early-1980s to 1990s and then a sharp recovery from 2003. In contrast, a declining trend in the private GCFA is visible only in some years during the 1980s and the 1990s, followed by a steady increase. Private GCFA has increased from Rs. 205 billion to Rs. 245 billion from 1981-82 to 1991-92 and to Rs. 631 billion in 2001-02 and then jumped to Rs. 1244 billion in 2011-12. Public GCFA was Rs. 110 billion in 1981-82 and fell to Rs. 76 billion in 1991-92, remained at the same level for many years and then shot up to Rs. 199 billion in 2005-06 and reached Rs. 220 billion in 2011-12.

Accordingly, the annual rate of growth in public GCFA which was negative during the 1980s became positive at 2.7 per cent during the nineties and went up by 6.3 per cent in the subsequent decade. In contrast, private investment grew at a much faster rate from 1.81 per cent during the eighties to 3.11 per cent during the 1990s followed by 8.6 per cent during the 2000s. A bifurcation of the period into pre- and post-liberalization reveals a higher rate of growth in private GCFA from 1991. However, the same in public GCFA, which has always been lower than that of the private GCFA, has exceeded it only during the 2000s.

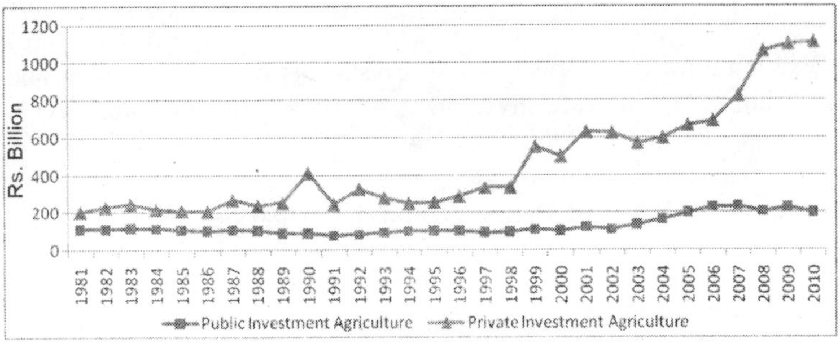

Fig. 9.1: Gross Capital Formation on Public and Private Accounts in India at 2004-05 Price

Notably, acceleration in the private GCFA has been responsible for recovery in GCFA over time. The share of private GCFA in total has also been increasing steadily from 60 per cent during the 1960s to 67 per cent during the 1980s, and finally touching a peak of 82 per cent in 2011. The public GCFA share in total investment has gradually gone down and went up slightly above 20 per cent from 2004. But it is the private GCFA that continues to hold the maximum share in total GCFA.

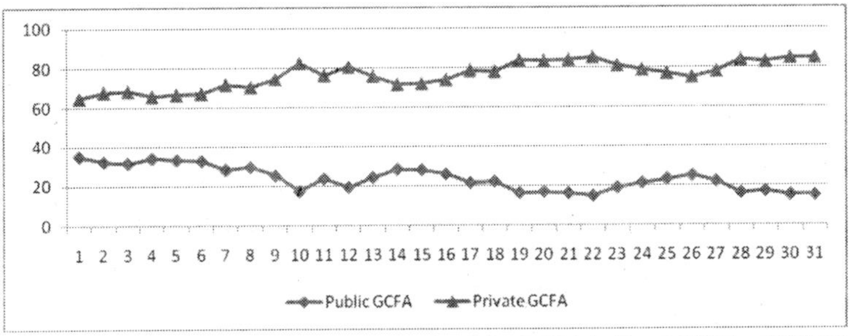

Fig. 9.2: Percentage Share of Private and Public GCFA in Total GCFA: 1981-82 to 2011-12

Fig. 9.3 depicts the share of GCFA in relation to gross domestic capital formation in the whole economy (GDCF) and net domestic product in agriculture and allied activities. The share of GCFA in GDCF was about 17 per cent during the 1980s, touched a peak of 14 per cent in the subsequent decade, fell thereafter touching a trough of 7 per cent in 1990s and then rose to 11 per cent during 2000s and again fell to 6.7 per cent in 2011-12. The fall in share was relatively more in the private sector (from 25 to 9%) than the public (from 14 to 7%) sector. Similarly, in relation to GDPA, the share of GCFA has become almost doubled, from 6 per cent during the 1960s to 12 per cent during the 1980s, fell to 7 per cent during the 1990s and rose to a maximum 22 per cent during the 2000s. On an average, 17.5 per cent of an increase in farm income was ploughed back for investment during 2000s compared to an average of 10 per cent during the 1990s.

An important point to note is that the share of both GCFA in overall GDCF and GDPA in overall GDP fell till the 1990s, implying that agriculture was perpetually losing out to the rest of the economy. The downward trend in the share of investment in farm income has got arrested during the 2000s, especially in the allied activities which may imply alterations in the sources of agricultural growth from crops to allied activities (Bathla, 2014). Notably, a fall in the share of GCFA in GDCF was much faster than that of GDPA in GDP.

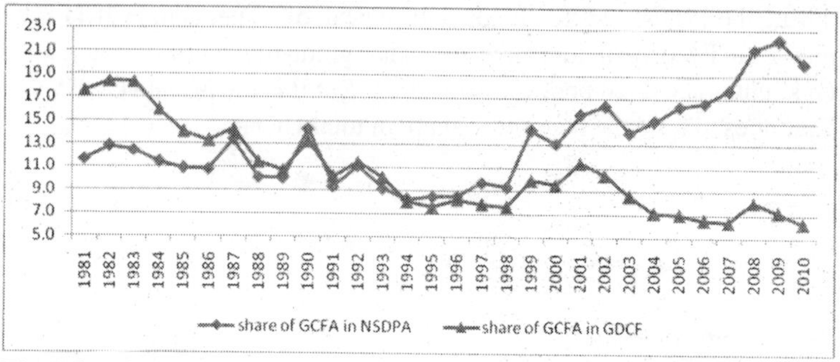

Fig. 9.3: Percentage Share of Agricultural Investment in Agriculture Income and Total Investment

Like public investment, input subsidy in agriculture on account of fertilizer, credit, power and irrigation has also expanded rapidly. Both were about the same size at nearly Rs. 100 billion in 1981 but the gap quickly blown up over time, may be due to stagnancy in public investment in the 1990s (Fig. 9.4). It is posited that expenditure on subsidy leaves much less resources for financing of public investment in irrigation and for that matter agriculture along with other factors such as rise in irrigation cost, impact of environmentalist movement (Shetty, 1990; Mishra and

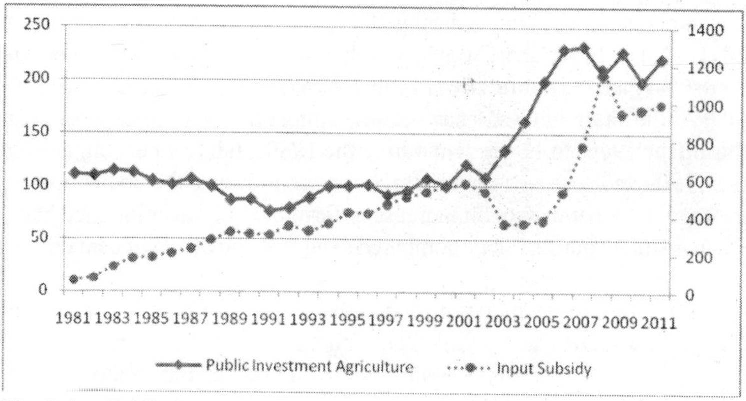

Fig. 9.4: Public investment and Input Subsidy in Agriculture, in Billion Rupees at 2004-05 Prices

Source: Bathla *et al.* (2014).

Chand, 1995; Gulati and Sharma, 1997; Gulati and Narayanan, 2003). The scenario seems to have changed from 2004 as both started increasing and reached Rs. 1000 billion. A trivial downward trend in public GCFA and an equivalent increase in subsidy are observed in recent years.

Given the magnitude of investments, production and income in agriculture has grown substantially over the period. Table 9.1 elicits statistics on foodgrain and total production and GDPA from 1981 to 2011. Average production has shot up from 146 million tons to 196 million tons during the nineties and subsequently to 217 million tons. But the rate of growth has decelerated from 2.7 to 2.1 per cent in the nineties and picked up to 2.9 per cent. The rate of growth in foodgrain together with oilseeds and sugarcane has shown signs of deceleration. Besides, GDPA has hovered around average 3 per cent annually during the last three decades. Perhaps growth has been maintained in the 2000s due to hike in public and private investments and also favourable terms of trade to agriculture. Average gross irrigated area has increased only by nearly 15 million ha during 2000s, at an average rate of 1.59 per cent per year. Only 46 per cent of gross area is irrigated so far, necessitating more expenditure on irrigation works. In what follows, the share of public investment in major heads of economic and social expenditure is investigated to gauge the priority of the government in allocation of resources.

Public Investment in Social and Economic Services and Share of Agriculture

The total public expenditure has expanded more than 7 times from 1980 to 2011, growing at a rate of nearly 7 per cent per year. Of this, the development expenditure has increased at a much slower rate than the non-development expenditure. As a result, the share of development expenditure declined continuously from 75 per cent in 1981 to below 60 per cent in 2004 and then recovered slightly. In 2011-12, nearly 57 per cent of development expenditure went to social services[2], mostly education and health compared to economic services share at 43 per cent (Table 9.2). It is evident that public spending (revenue and capital) on social services has acquired prominence over the period. Within economic services, the share of agriculture, irrigation, rural development and village industry has gone down and that of energy, transport and communications have gone up. Within agriculture and allied sector, spending on food

Table 9.1: Agricultural Investment, Input Subsidy, and Production: 1981-82 to 2011-12

(Rs. billion at 2004-05 Price)

Average	Public GCFA	Private GCFA	GCFA	GDPA	Input Subsidy	Irrigation	Credit	Power	Fertilizer
1981-1989	105.13	231.83	336.97	3343.27	182.3*	48.15*	20.12	56.75	67.67
1990-1999	92.89	329.61	422.50	4534.11	397.42	63.86	14.87	201.75	116.93
2000-2011	174.91	763.31	937.82	6151.14	680.83	86.06	47.23	254.71	292.84
1981-2011	126.49	459.30	585.79	4814.33	444.67	64.87	28.92	180.15	170.72
Annual Rate of Growth (%)									
1981-1989	–2.49	1.81	0.49	2.90	14.4*	8.9*	9.38	20.71	23.05
1990-1999	2.78	3.11	3.06	3.34	6.79	2.40	–10.4	11.80	3.97
2000-2011	6.26	8.65	8.15	3.31	5.38	3.11	9.89	2.25	8.71
1981-2011	2.71	5.96	5.16	3.01	5.99	3.18	3.52	7.50	7.93

Average	Fertilizer Consumption (Million Tons)	Foodgrain Production	Production Million Tons	%of Gross Area Irrigated	Road Density (Km/1000 sq.km)	Power Consumption in Agriculture (%)	TOT
1981-1989	8.54	146.77	348.97	33.24	747.62	24.89	90.85
1990-1999	14.26	186.50	473.29	39.06	890.89	32.99	103.19
2000-2011	21.43	217.42	541.71	46.73	1231.22	26.01	112.54
1981-2011	15.37	186.93	463.68	40.34	981.03	27.94	103.23
Annual Rate of Growth (%)							
1981-1989	7.85	2.77	2.53				
1990-1999	4.41	2.10	2.46				
2000-2011	4.15	2.86	2.35				
1981-2011	4.68	1.96	2.12				

Note: Production include food grain, oilseeds, sugarcane. *estimated from 1983-84.
Source: Finance Accounts, Agriculture Statistics at a Glance, Fertilizer Statistics of India.

storage, agricultural research and cooperation has acquired prominence in terms of relatively higher allocation of resources.

Notably, the share of capital expenditure (mainly for investment) in total expenditure is relatively more in irrigation, transport and rural energy due the nature of services. But resources seem to have shifted from capital formation to current expenditure on government's day-to-day operations, subsidies and losses as the share of capital expenditure in total has fallen over time. This could explain mounting bureaucracy and inefficiency in public expenditure, the brunt of which is felt more on agriculture and rural development sectors.

Table 9.2: Public Spending on Various Social and Economic Services and Share: Rs. Billion at 2004-05 Price

Major Expenditure Heads and annual rate of growth	*1981-1989*	*1990-1999*	*2000-2011*	*Growth rate (%) 1981-2011*	*Share of capital exp.(%)*
Total expenditure	1412.2	2427.3	5245.0	6.6	12.3
Development expenditure	1034.9	1589.8	3283.4	5.8	17.9
Social expenditure	516.9	827.2	1763.1	6.3	5.56
Economic expenditure	517.9	762.6	1519.7	5.4	31.0
Economic services (% share)	100	100	100	5.4	
Agriculture and allied services	21.8	22.0	17.3	4.2	9.1
Rural development	15.0	14.9	12.9	4.6	6.7
Irrigation and flood control	33.1	26.8	23.7	3.9	55.7
Rural energy	1.3	4.3	5.4	12.2	28.9
Rural road and transport	9.9	8.6	11.4	1.8	9.3
Village and small industry	2.84	2.35	1.37	1.76	45.6
Agriculture and allied services	100	100	100	4.2	
Crop husbandry	23.1	27.9	27.8	5.8	2.5
Soil and water conservation	5.9	6.9	5.7	3.9	20.9
Animal husbandry	11.6	11.4	10.9	3.9	2.3
Dairy development	13.6	8.5	4.6	–1.9	2.1
Fisheries	2.3	2.5	2.5	4.7	17.7
Forestry and wildlife	22.3	19.7	19.1	3.8	10.6
Plantations	0.4	0.1	0.9	—	71.2
Food, storage warehousing	2.5	5.4	10.2	—	—
Agri. research and education	6.3	6.2	7.3	4.8	1.8
Agri. financial institutions	0.8	0.7	0.02	—	—
Cooperation	11.1	10.4	11.1	4.0	21.7
Other agri. Programmes	0.8	0.9	0.7	3.7	17.1

Note: Growth rate and share of capital expenditure refer to average annual growth rate and average from 1981 to 2011.

Source: Finance Accounts, Government of India.

Relationship between Public and Private GCFA, Input Subsidy and GDPA

For examining the relationship between public and private GCFA and their impact on agriculture, the following equations are estimated.

Equation 1: Public GCFA = f {Size of government spending, savings, funds (loans from centre+ net market borrowings), savings (surplus/deficit on revenue account), input subsidy, foodgrain procurement/production, NIA, dummy-reforms}.

Equation 2: Private GCFA = f (Public GCFA, terms of trade, institutional credit, public infrastructure, industrial demand for raw material, world price, infrastructure, dummy-reforms).

Equation 3: GDPA = f {GCFA (public+private), lagged terms of trade, institutional credit, NSA, NIA, dummy-reforms}.

Equation (1) suggests that public GCFA is influenced by factors originating from both supply and demand sides such as size of public spending, denoted by total revenue and capital expenditure per GDP, input subsidy, availability of funds through loans and net market borrowings, savings on revenue account. The demand side factors that may push public GCFA include population pressure and food shortages, deceleration in crop and land productivity. For capturing the impact, the proxy variables used are: (a) share of wheat and paddy/rice procurement in total wheat-rice production. An increase in the procurement of cereals indicates preference and/or compulsion on the part of the government to allocate more funds towards agriculture; (b) net irrigated area as proportion to net sown area (NIA/NSA); and (c) land productivity (GDPA/NSA).

Equation (2) exhibits private investment behaviour to be determined by public GCFA, terms of trade (TOT), institutional credit by regional rural banks, commercial banks and primary agricultural cooperative societies, public expenditure on infrastructure—denoted by road density or capital expenditure on rural roads-bridges-transport, world price, and industrial demand for raw material, proxied by gross value added per food factory. As mentioned in Mitra *et al.* (2002) and Bathla and Gautam (2013), private investment in food processing may incite farmers to invest owing to a high correlation between industrial productivity and land and labour productivity in agriculture[3].

Finally, Equation (3) shows agriculture income as a function of investment (public and private GCFA), lagged TOT, net sown area, irrigation intensity and institutional farm credit. A dummy variable is included in each equation to explore the impact of reforms on investments and income.

The model was analysed from 1981-82 to 2011-12 first using OLS and then 3SLS. Public GCFA is taken in absolute as well as in stock/cumulative basis due to time lag between investment in major and medium irrigation works. A high multicollinearity among the explanatory variables was identified due to which a few variables had to be dropped and alternate equations were estimated. The model was tested for robustness. Multiple

Table 9.3: Relationship between Public and Private GCFA and Agriculture GDP: Three-stage least-squares regression in double log form: 1981-82 to 2011-12

Endogenous/Exogenous variables	*I*	*II*
Eq.1: Public GCFA		
Size of government spending	0.91 (0.69)	1.18**(0.66)
NIA/NSA	2.13* (0.90)	2.02*(0.85)
Foodgrains procured/production	0.86*(0.17)	0.78*(0.16)
Funds (loans + borrowings from market)	–0.16** (0.07)	–0.13***(0.08)
Input subsidy/NSA	–0.02 (0.15)	–0.001(0.14)
Dummy—reforms 1991	–0.46*(0.16)	–0.45*(0.16)
Constant	1.39 (3.19)	0.64(3.01)
R^2	0.70	0.69
Eq.2: Private GCFA		
Public GCFA	0.41***(0.25)	1.03*(0.24)
Lagged TOT	0.67 (0.67)	—
Institutional credit	0.58*(0.23)	—
Industrial demand	0.001(0.17)	0.35***(0.21)
World price index	—	0.23(0.39)
Dummy reforms 1995	0.11(0.09)	0.23**(0.12)
Constant	–3.02 (2.94)	–1.87(1.75)
R^2	0.90	0.79
Eq.3: GDPA		
GCFA (public+private)	0.11*(0.02)	0.098* (0.02)
Lagged TOT	0.59*(0.09)	0.63*(0.09)
NSA	0.67*(0.14)	0.68*(0.15)
NIA	0.89*(0.08)	0.89*(0.08)
Dummy reforms	0.02 (0.01)	0.02(0.01)
Constant	–6.48*(1.58)	–6.8*(1.64)
R^2	0.98	0.99

*, ** and *** indicate statistical significance at 1%, 5% and 10% levels respectively.

Note: Figures within the parentheses are standard errors.

goodness-of-fit tests are performed to check the acceptability of the model. The coefficient of determination shows that the model is adequately fit. The test results indicates that the 3SLS has goodness of fit based on the combined rule of low RMSEA and SRMR and high stability values. The Hausman test is to check the presence of endogeniety in the explanatory variables.

Table 9.3 furnishes the empirical results. The coefficient of multiple determination (R^2) is reasonably high at 0.70. It shows that public GCFA is determined positively by the size of public spending and negatively by the funds (market borrowings+loans) and input subsidy. The value of coefficient of funds is estimated to be 0.12, which implies that a 10 per cent increase in resources by the government may reduce public investment in agriculture by 1.2 per cent. The value of the coefficient of subsidy is negative but insignificant (elasticity = 0.02) suggesting that spending on subsidies may not deter the government to allocate more towards agriculture. The policy variables, i.e. cereals procurement as proportion to production and irrigation intensity have exerted relatively stronger influence in explaining variations in agricultural investment. The impact of reforms captured through dummy appears to have put pressure on public resources.

As regards the behaviour of private GCFA, results indicate that public GCFA, lagged TOT, institutional credit and raw material demand by the food industry are propelling factors for farmers to undertake investments. The elasticity is the highest for public GCFA and TOT (1.0), institutional credit (0.58) and industrial demand (0.35). A positive and significant value of the coefficient of public GCFA validates the 'crowding in' effect of public GCFA on private GCFA, as has been argued in the literature. A positive and significant influence of food processing industry on private farm investment is in conformity with the economic theory and empirical findings on the positive backward agri-industry linkages. Terms of trade and public investment in road-bridges infrastructure exerted positive bearing on the private GCFA but were dropped due to the problem of multicollinearity with credit (equation 1). World price did not favour private investment much.

The effect of public and private GCFA on agriculture income is positive and significant. An estimated value of elasticity at 0.11 suggests that a 10 per cent increase in capital formation in agriculture would lead to an increase in income by nearly 1 per cent. When GCFA is replaced by another variable in physical terms, viz. NIA, it shows a higher value of elasticity at 0.89. Lagged terms of trade turns out to be a significant factor with elasticity at 0.59. NSA and reforms have also influenced GDPA positively and significantly.

The findings clearly indicate that both public and private investments along with an incentive structure are imperative for augmenting agriculture income. A major push in public investment and favourable TOT during 2000s seem to have instigated private investment and agricultural growth, which must be maintained. An increased public GCFA has also coincided with a spurt in the input subsidy, and both remained at nearly a same level in the recent period. The contention that rising subsidy bill could be diverted from investments in agriculture may hold true but not in a significant way. This may imply that cutting down of subsidies could slightly help fiscal resources, but the bigger issue that needs deliberation is whether it would impinge input use and thus farm production. This aspect is examined through estimation of the following equations simultaneously. The equations are run separately for fertilizer, irrigation and power used in agriculture.

Equation 1: Production = f {TOT, GIA, Input use (fertilizer, irrigation, power), Rainfall, Dummy-reforms}

Equation 2: Input use = f {Subsidy (fertilizer, irrigation, power), GIA, Credit}

Equation 1 is the aggregate production function which is explained by price (lagged terms of trade), annual rainfall, gross irrigated area, fertilizer and power consumption in agriculture. Lagged price is taken as farmers respond to last years' price. A dummy variable is taken to capture the impact of reforms, initiated from 1991 and picked up from 1995. Equation 2 evaluates the impact of each input subsidy on the consumption of respective input. While agriculture production and input use are taken to be endogenous variables, other variables in the equations are treated as exogenous. Theoretically, input use is determined by its respective price. But due to unavailability of time series data on irrigation and power prices, input use is considered to be a function of its respective subsidy.

Table 9.4 presents the estimated parameters. The explanatory variables explain 95 per cent of the variations in the endogenous variable in each equation. Farm output is highly responsive to price structure, rainfall and input use. An elasticity of lagged TOT at 0.48 suggests a 10 per cent increase in price would raise output by 4.8 per cent. The elasticity estimates of irrigation, fertilizer and power with respect to output varied between 0.18 to 0.29, each indicating a high level of statistical significance. World price was also tried as an exogenous variable but was dropped due to its high correlation with TOT. Results from the second equation on fertilizer, power and irrigation use are again robust. The use of each input is highly responsive to institutional credit and the elasticity estimates vary between 0.15 to 0.24. Dummy variable is positive and significant, suggesting that reforms did make difference in the input use.

As regards the impact of subsidy on input use, the results are positive and significant in each case but irrigation subsidy tends to have insignificant impact on irrigation use. The elasticity of fertilizer consumption with respect to subsidy turns out to be 0.19 and for power consumption to be 0.45. This suggests that a ten per cent increase in input subsidy would increase usage of fertilizer and power by 1.9 and 4.5 per cent respectively. The insignificance of irrigation subsidy can be explained by the fact that it is given primarily on surface water (major and medium) and farmers' dependence is more on the groundwater sources.

The extent to which input subsidy can influence crop output is examined by taking the product of elasticity of output with respect to

Table 9.4: Impact of Input Subsidy on Input Use and Agricultural Output: Three-stage least-squares regression in log form 1981-82 to 2011-12

Equation 1: Output		*Coefficients*	
Terms of Trade (lagged)	0.48* (0.19)	0.43* (0.18)	0.44** (0.19)
Gross Irrigated Area	0.28* (0.11)	0.20** (0.10)	0.29* (0.22)
Fertilizer Consumption	0.19*(0.08)	0.24*(0.07)	0.20***(0.12)
Electricity Consumption (%share)	0.18 * (0.05)	0.22 * (0.05)	0.19 * (0.05)
Annual Rainfall	0.19*(0.08)	0.21*(0.08)	0.20*(0.09)
Constant	3.89*(1.13)	4.27*(1.01)	3.79*(1.81)
R^2	0.95	0.95	0.95
Equation 2: Input Use	*Fertilizer*	*Agriculture Electricity*	*Irrigation*
Input Subsidy	0.19*(0.05)	0.45*(0.05)	0.01 (0.02)
Institutional Credit	0.24* (0.05)	0.22* (0.06)	0.15* (0.03)
Gross Irrigated Area	0.28** (0.17)	0.06 (0.28)	—
Dummy (Reforms) 1991	0.23* (0.05)	0.38*(0.09)	0.23*(0.04)
Constant	0.72 (1.6)	1.66(2.5)	8.74*(0.37)
R^2	0.96	0.97	0.86

Notes: *, ** and *** indicate statistical significance at 1%, 5% and 10% levels respectively.

Figures in the parentheses are standard errors. Output relates to foodgrain, oilseeds and sugarcane.

input use and elasticity of input use with respect to its subsidy. In other words, it gives elasticity of production with respect to input subsidy, which is estimated to be 0.036 for fertilizer, 0.013 for electricity and 0.003 for irrigation respectively. This suggests that a 10 per cent increase in subsidy on account of fertilizer and power may result in 0.1 to 0.3 per cent increase in output and *vice versa*. Chand and Pandey (2008) estimated the impact of withdrawal of fertilizer subsidy on production by taking into account the price change. Their estimates reveal a 9 per cent decline in foodgrain production if fertilizer price is increased from a subsidised rate to actual market price. The results obtained point towards a tradeoff in the provision of input subsidy by the government. Its continuation may negatively affect public resources and hence farm investment whereas its withdrawal would impinge upon input use, asset creation by farmers from subsidised credit and finally production.

Conclusion and Implication

The recent policy mandate has been to accelerate public investment in agriculture, which in tandem with private investment may lead to a targeted four per cent annual rate growth in agriculture. This study has mapped the temporal trends in public and private investment in agriculture and input subsidies for the recent period and empirically analysed their impact on agricultural growth. The analysis has been undertaken from 1981-82 to 2011-12 with an aim to capture different phases of growth in investments, subsidies, production and income.

Broad findings indicate a significant increase in the magnitude of both public and private investments during 2000s, especially private GCFA which has also captured a major share in total farm investment at 82 per cent. Though an increasing share of agriculture income is now ploughed back into the sector for investment, the share of agricultural sector investment in total investment in the economy is way below (7%) compared to that in the non-agricultural sectors. Furthermore, a bias against agriculture is apparent as the share of agriculture along with irrigation and rural development in public spending has gone down and that on energy, transport and social services has gone up in 2000s. Another disturbing aspect is a decline in the share of capital expenditure in total expenditure on most of the economic services, implying more spending on revenue account for government's operation, other expenditures such as on subsidies. The estimated total input subsidy on account of irrigation, power, fertilizer and credit showed a declining trend during the nineties followed by an increase in 2000s to nearly the same magnitude as that of public GCFA at Rs. 1,000 billion.

An increase in public investment in agriculture and irrigation is, however a reflection of effective public policy in favour of capital deepening in agriculture more so in view of bridging the growth deficit in agriculture and mounting pressure to procure grain stocks to ensure food security to the masses. Public GCFA along with other factors such as better prices, increase in institutional credit, infrastructure and productivity in agro-processing industry has helped in raising private investment. The empirical results obtained verify an inducement effect of public GCFA on private GCFA and their positive impact on agricultural growth. The findings suggest that input subsidy may negatively impinge on resource allocation towards agriculture but the impact may not be significant. Besides, it would not be feasible to cut down subsidy in view of its positive and significant impact on input consumption, especially of fertilizer and power. The withdrawl of farm subsidy would disable and discourage small and marginal farmers to stay in agriculture, even though there is not much scope for their upward mobility outside agriculture because of lack of education and skill. As a result their agrarian distress may result in distress migration, not by choice, but by adverse circumstances. As argued in Jayne and Rashid (2013) and Chirwa and Dorward (2013), subsidy and investment are complimentary as small farmers who are capital constrained may need subsidy to purchase inputs. The Sub-Saharan African countries' experience suggests an increased input use in an efficient and effective way by 'smart subsidies' through proper targeting and rationing. Determining the appropriate subsidy and making it reach the needy farmers and states in a cost effective manner could be the steps forward. It is equally imperative to assess the costs and benefits of each subsidy in terms of constraining public resources *vis-à-vis* accelerating the input use and hence production.

Notes

1. The estimates on public capital formation in agriculture, given in the NAS, CSO refer to major and medium irrigation works mainly and not agriculture (for details see Gulati and Bathla, 2002).
2. Social services include four broad categories, viz. education and sports, medical and public health, welfare of SC, ST and backward classes, social welfare and nutrition and others.

3. The latter study revealed investment in organised food manufacturing to have increased more in agriculturally dominant states which could be attributed to economies of scale due to easy procurement of raw material. The empirical analysis also testifies strong backward linkages of the industry, which may have provided impetus to farmers to undertake investments.

References

Bathla, Seema (2014). Public and Private Capital Formation and Agricultural Growth: State-wise Analysis of Inter-linkages during Pre- and Post-reform Periods, *Agriculture Economics Research Review*, 27(1), January-June.

Bathla, Seema, Bing Yu, S.K. Thorat and P.K. Joshi (2014). *Accelerating Agricultural Growth and Poverty Alleviation through Public Expenditure: The Experience of India*. Report submitted to IFPRI, New Delhi.

Bathla, Seema and Madhur Gautam (2013). *Accelerating Private Investment and Competitiveness in Food Processing Industry in India: A State Level Analysis*, Mimeo, World Bank Washington D.C.

Bathla, Seema and Thorat, S.K. (2006). Trends and Determinants of Public Investment in Agriculture at the Regional Level. *Man and Development*, 28 (2), June: 115-136.

Bisaliah, S., Mahendra Dev, S. and Saifullah, S. (2013). *Investment in Indian Agriculture: Macro and Micro Evidences*. Academic Foundation, New Delhi.

Chand, R. (2000). *Emerging Trends and Regional Variations in Agricultural Investments and their Implications for Growth and Equity*, Policy Paper, National Centre for Agricultural Economics and Policy Research, New Delhi.

Chand, R. and Kumar, P. (2004). Determinants of Capital Formation and Agricultural Growth: Some New Explorations. *Economic and Political Weekly*, 39 (52): 5611-16.

Chand, R. and L.M. Pandey (2008). *Fertiliser Growth, Imbalances and Subsidies: Trends and Implications*, Policy Paper, National Centre for Agricultural Economics and Policy Research, New Delhi.

Chirwa, E. and A. Dorward. (2013). *Agricultural Input Subsidies: The Malawi Experience*. Oxford University Press. Available online at: http://fdslive.oup.com/www.oup.com/academic/pdf/openaccess/9780199683529.pdf

Dev, S.M. (2000). Economic Reforms, Poverty, Income Distribution and Employment, *Economic & Political Weekly*. 35 (10): 823-35.

Dev, S.M. (2008). *Challenges for Revival of Indian Agriculture*. National Centre for Agricultural Economics and Policy Research, New Delhi.

Dhawan, B.D. (1998). *Studies in Agricultural Investments and Rural Savings*. Commonwealth Publishers, New Delhi.

Fan, S. (Ed.) (2008). *Public Expenditures, Growth and Poverty: Lessons from Developing Countries*. Oxford University Press, New Delhi.

Fan, S., A. Gulati and S. Thorat. (2008). "Investment, Subsidies and Pro-poor Growth in Rural India." *Agricultural Economics*. 39: 163-170.

Fan, S. Hazell, P. and Thorat, S. (1999). Linkages between Government Spending, Growth and Poverty in Rural India. Report submitted to International Food Policy Research Institute, Washington D.C.

Fertilizer Association of India, *Fertilizer Statistics of India* (various issues), New Delhi.

Government of India, *Estimates of National Accounts Statistics* (various issues), Central Statistical Organisation, Ministry of Statistics, Planning and Implementation.

Government of India (various issues), *Estimates of National Accounts Statistics*, Central Statistical Organisation, New Delhi.

Government of India (various issues). *Agriculture Statistics at a Glance*, Ministry of Agriculture.

Gulati, Ashok and Seema Bathla (2002). Capital Formation in Indian Agriculture: Trends, Composition and Implications for Growth. NABARD Occasional Paper No. 24. Also published in *Economic and Political Weekly*, (2001):1697-1708.

Gulati, Ashok and Anil Sharma (1997). "Subsidies and Investments in Indian Agriculture", Conference Papers, Rajiv Gandhi Institute for Contemporary Studies, Paper No. 21, July.

Gulati, A. and S. Narayanan. (2003). *Subsidy Syndrome in Indian Agriculture*. New Delhi. Oxford University Press.

Hazell Peter B.R. and Haggablade Steven (1991). "Rural-Urban Growth Linkages in India", *Indian Journal of Agricultural Economics*, Vol. 46 (4), October-December.

Jayne, T. and S. Rashid (2013). Input Subsidy Programmes in Sub-Saharan Africa: A Synthesis of Recent Evidence, *Agricultural Economics*, 44, pp. 547-562.

Joshi, P.K., Gulati, A., Birthal, P.S. and Tewari, L. (2004). Agricultural Diversification in South Asia: Patterns, Determinants and Policy Implications. *Economic and Political Weekly*, 39 (24): 2457-68.

Joshi, P.K., Birthal, P.S. and Minot, Nicholas (2006). *Sources of Agricultural Growth in India: Role of Diversification towards High Value Crops*. MTID Discussion Paper No 98, International Food and Policy Research Institute. Washington DC.

Landes R. and Gulati, Ashok (2004). Farm Sector Performance and Reform Agenda. *Economic and Political Weekly*, 39 (32), August 7 -1: 3611-3619.

Lopez, R. (2005). Under investing in Public Goods: Evidences, Causes and Consequences for Agricultural Development, Equity and the Environment, *Agricultural Economics*, 32, Issue Supplement 1, pp. 211-224, January.

Mishra S.N. and Chand, R. (1995). Public and Private Capital Formation in Indian Agriculture: Comments on the Complementarity Hypothesis and Others. *Economic and Political Weekly*, 30, (25), June 24: A64-A79.

Misra, V.N. and Hazell, B.R. (1996). Terms of Trade, Rural Poverty, Technology and Investment: The Indian Experience, 1952-53 to 1990-91. *Economic and Political Weekly*, 31, (43), March 30: A2-A14.

Mitra A. (1996). Public and Private Investments in Agriculture. In: *Agricultural Development Paradigm for the Ninth Plan under New Economic Environment*. Eds: B.M. Desai, Oxford and IBH, New Delhi.

Mitra, Arup, A. Varoudakis and M.A.V. Varoudakis (2002). Productivity and Technical Efficiency in Indian States' Manufacturing: The Role of Infrastructure. *Economic Development and Cultural Change,* 50 (2): 395-426.

Planning Commission (2013). Twelfth Five Year Plan (2012/2017): *Economic Sectors*, Volume II. Government of India.

Roy, B.C. and Pal, Suresh (2001). Investment, Agricultural Productivity and Rural Poverty in India: A State-level Analysis. In: *Indian Agriculture in the New Millennium: Changing Perspectives and Development Policy*. (Eds): N.A. Majumdar and Uma Kapila, Academic Foundation, New Delhi.

Ravallion M. and G. Dutt (1995). *Growth and Poverty in Rural India*, World Bank Policy Research Working Paper No. 1405, World Bank.

Sharma, V.P. (2013). *The Role of Fertilizer in Transforming Agriculture in Asia: A Case Study of the Indian Fertilizer Sector. Centre for Management in Agriculture (CMA),* Indian Institute of Management, Ahmedabad, India.

Shetty, S.L. (1990). Investment in Agriculture: Brief Review of Recent Trends, *Economic and Political Weekly*, February 17-42, 1990.

10

Minimising Agrarian Distress by Select Strategies and Priorities

Investment in Indian Agriculture

C. Ramasamy

Introduction

During the past two and half decades which can be construed as post-green revolution period, there has been a growing agrarian distress in India. The agrarian distress is well evidenced by farmers' suicides, poverty among small and marginal farmers and other agriculturally dependent population such as farm and other rural labourers, migration of rural population to urban and semi-urban areas, a shift from farm to non-farm occupations and overall poor quality of life among agriculturally dependent population. This chapter analyses how agrarian distress can be minimised by select strategies and prioritised investment in Indian Agriculture.

Causes and Distress

Monsoon rain failures and consequent drought situations, market risks resulting in unremunerative prices for farm products, crop losses due to pests and diseases, damage to crops by cyclones, floods etc., death of livestock because of infections, high cost of and adulterated inputs such as fertilizers and pesticides, indebtedness and lack of capital to opt for application of yield increasing modern inputs, fragmented unviable farm holding, little knowledge and skills to migrate to up non-farm and non-agricultural activities, ill-conceived and poorly implemented agricultural development policies and programmes, decline in public investment, individualisation of agriculture and decline of the traditional social order, egoism and anomie (types of suicides) and import liberalisation are often considered as primary causes and components of agrarian distress in India (Hebous and Klonner, 2013; Mohanty, 2013; Harper, 2011; Deshpande, 2011; Sainath, 2006; Vaidyanathan, 2006; Vyas, 2005).

Inability to bear debt has led to farmers' suicides on an unprecedented scale and suicides are concentrated mostly in low rainfall and poorly irrigated regions. This has caused considerable distress in regions where they are prominent in the farm economy. This has also led to growing economic and social disparities between agriculture and the rest of the economy and between rural and urban sectors. One of the concerns is the marked slowing down of agricultural growth in States that are highly irrigated, have shown the greatest dynamism in adopting new technology, and have been a significant factor in sustaining the national growth rate. The slowing down of growth is widely attributed to the slowing down of investment in agriculture, especially public investment. The Central Planning Commission seems to share this perception and sees increased investments in irrigation and watershed development as the means to achieve the projected growth. But, what is relevant is not the magnitude of investment but its contribution to increasing production capacity. For a variety of reasons peculiar to agriculture, production capacity has not increased in proportion to the quantum of investment. Hence agrarian distress continues that all developmental and safety measures to alleviate it have not shown expected results.

Causes not Unprecedented

Agrarian distress most often caused by drought is a common recurring feature of the climate in most parts of the world causing distress to those who rely on agriculture for livelihood. It is among the earliest documented climatic events, present in the *Epic of Gilgamesh* and tied to the *Biblical* story of Joseph's arrival in and the later *Exodus from Ancient Egypt*. Hunter-gatherer migrations in 9,500 BC Chile have been linked to the drought phenomenon (Drought-Wikipedia, the free encyclopaedia). Drought in India has resulted in tens of millions of deaths over the course of the eighteenth, nineteenth, and twentieth centuries. Indian agriculture is heavily dependent on climate: a favourable southwest monsoon is critical in securing water for irrigating crops. Droughts, combined with policy failures, have periodically led to major Indian famines, including the Bengal famine of 1770, the Chalisa famine, the Dojibara famine, the Great Famine of 1876-78, and the Bengal famine of 1943 (Iqbal and

You, 2001). Government of India has identified 90 drought prone districts of India (Government of India, 1999).

Similarly, the indebtedness of farm households is not the recent phenomenon and has remained with for long. Farmers' indebtedness in India is not merely an economic problem. It is embedded in social structures and this embedding adds a historical dimension to the problem. This problem assumes particularimportance in low rainfall dry land areas with less irrigation resulting in indebtedness due to crop failures leading subsequently to suicides. Various credit schemes implemented have not extricated the farmers from spectre of indebtedness since Independence of the country. Out of total farm households in the country, 48.6 per cent are reported to be indebted. The provision of credit through cooperatives and commercial banks were to the extent of about 4 per cent of the total outstanding debt as at end-June 1951. This finding of Report of the *All India Rural Credit survey* (RBI, 1954) had laid the foundation stone for furthering the role of institutional credit to rural sector through formal channel of cooperatives and commercial banks. But Report of the 'Task Force on Credit Related Issues of Farmers' (MoA, GOI, 2010) has revealed that, in recent years, policy interventions have led to doubling of agricultural credit, but the limited access of small and marginal farmers to institutional credit continues to be a matter of concern. What is worrying is that the proportion of such farmers is increasing and they form more than four-fifths of the operational holdings.

Another major factor which adds to agrarian distress is market and price risks which were much higher during earlier decades. The market risks result from fluctuations in the prices of inputs and outputs, outside competition, changing supply and demand, market imperfections, and changing consumer preferences. Various measures were taken since 1950s to correct or moderate undesirable trends. Prices, however, fluctuated widely in the First Plan period and they have shown a rising trend through the Second Plan period. At the commencement of the Third Plan, the levels of wholesale prices and cost-of-living were already high and it was essential to ensure that there was no accentuation of inflationary pressures in the course of the Third Plan and that the levels of living of the more vulnerable classes in society were safeguarded (Plan Document, Third Plan, 1961). Volatility in agricultural markets seems to have increased extreme price movements of agricultural commodities in recent periods posing a threat to world food security. Policy measures should improve market functioning and increase countries' resilience to shocks (FAO, 2010).

The number of holdings, which was 71.0 million in 1970-71, went up 138.35 million in 2010-11 (Agricultural Census, 2010-11). The average size of operational holding has declined to 1.15 ha in 2010-11 as compared to 2.28 ha in 1970-71. The proportion of marginal and small farmers which stood at 70 per cent in 1970-71 increased to 85 per cent in 2010-11. The message is that proportion of marginal and small farmers remained 70 per cent and above during the last four and half decades signalling the poor economic viability of farming. Fragmentation of holdings which was considered as a constraint for improving farm efficiency continues to be a discouraging factor for improving farm income resulting in agrarian distress all along the past decades.

Before 1970s, production risks such as crop losses due pests and diseases and weeds were in much larger measure but still continue to add to the distress of the farmers. Damage because floods occur in high rainfall areas throughout history. A recent study released by industry body Assocham and Yes Bank said that pest and disease infestation resulted in crop losses worth Rs. 50,000 crore annually in India. According to a very recent study in West Bengal, Sarkar *et al.* (2013) reported crop losses due to pests 17.72 per cent in rice and 18.05 per cent in wheat. In some quarters, it is argued that trade liberalisation leading to integration with world market has added to more market risks. As regards trade liberalization, share of agricultural exports in the total exports remains at 10 per cent and the respective figure for imports about 5 per cent. Hence trade liberalization is important not only to improve farmers' income but for the import of essential commodities such as edible oils and pulses which are approximately 40 per cent and 25 per cent respectively short of total demand which cannot be met from domestic production. And as such, trade liberalization has caused agrarian distress needs deeper investigation.

In implementation of agricultural development programmes, there is widespread corruption which leads to huge leakages of funds and also unfair processes followed at different levels of programme implementation. Farmers' *interface* discussions brought out that the inputs are often low quality and there is absence of timely availability (CCS HAU, 2007). In India, around 70 per cent of quality problems in modern farm inputs are due to adulteration or misbranding, 20 per cent due to deliberate manufacture of poor quality products and the remaining 10 per cent to the difference in quality sold on the official and black markets (Mujeri *et al.*, 2012).

Approach to Minimise Distress

With independence, a number of policy measures addressing it have been implemented. The impact of those measures has not solved agrarian problems to the farming dependent population who aim for a decent living. Alongside the relief obtained by the suffering farming population, new problems arose while existing problems only partially solved. The causes also lie in poor implementation of agricultural development projects, improper governance, and corruption at all levels of development and welfare projects, little awareness among beneficiary population, absence of rigorous monitoring and evaluation, lack of accountability of public sector employees and managers, vested interests influencing elected representatives and bureaucrats, silencing of people's voices and caste discrimination, among many other factors. Administrative reforms, technology induction and exposure to rest of the world have not paid proportionate dividends in improving efficiency of governance.

A huge amount of literature examining the effects of various agrarian policies implemented spatially and temporarily is available and has come out with varied results. The diverse results and conclusions of the studies have naturally the shadows of respective disciplines which authors belong to such as economists, sociologists, agricultural experts, political scientists, activists etc. carrying a wide ranging policy prescriptions and suggestions. Enough evidences have been generated to frame policies and programmes aimed at minimising the agrarian distress. The basic argument of this chapter that there is persistence of agrarian distress in all the times in India and short-term reliefs and crying wolf and subjective assessment underpinning of ideological tones will not solve the problem. Only a long-term strategies—many of them on-going but implemented poorly—addressing the problem objectively which will deliver solid outcomes are the need of the hour. The select strategies will be in the form of resource development, enhancing production capacity through technological and efficiency factors, creating better marketing environment for farm products, ensuring safety nets against risks and better governance of developmental projects. These strategies must have top priority in public investment and guiding the private sector including the farm operators in such kind of investment is critical to minimise the agrarian distress.

Select Strategies

1. Bolster Efficiency in All Sources of Irrigation

Canal Water

Given land resources, water is fundamental for crop growth and farming. If water is assured, everything follows. Having realised this, huge investment has gone into development of irrigation since launch of First Five Year Plan. It is common knowledge that irrigation development schemes fall under major, medium and minor irrigation projects. By 2011-12, irrigation potential created is 113.24 m.ha of which potential utilised is 79.42 per cent leaving 23.30 mha unutilised (12th Plan Document, 2012). The ultimate irrigation potential in the country is estimate data bout 140m. Ha. Of this, about 58.5 m ha is from major and medium irrigation sources, and 81.5 m. ha is from minor irrigation sources and of which 64.1 m. ha from groundwater irrigation and 17.4 m. ha from surface water (*State of Agriculture* 2012-13, Government of India). Thus we have created a potential which is only 80 per cent of ultimate potential and again what is created has not been fully utilised. Factors responsible for low utilization of irrigation water in major and medium projects are lack of proper operation and maintenance, incomplete distribution systems, non-completion of CAD works, changes from the initially designed cropping pattern and diversion of irrigable land for other purpose, among others. Inadequate provision of budget for operation and maintenance of the irrigation system is significantly responsible for under-utilization followed by non-completion of distributaries, minors, field channels and on-farm development. Thus outlay on surface irrigation, which accounts for the bulk of public investment in agriculture, is concentrated on projects that have taken an unconscionably long time to complete. Costs have risen and outlays have not resulted in any significant additions to irrigated area.

There has been a massive increase in plan expenditure on Major and Medium Irrigation (MMI) projects in the last 60 years. These projects have suffered from massive time and cost over runs with some with average cost run being 1382 per cent for major projects. There is a spill over of 337 projects—154 major, 148 medium and 35 Extension, Renovation, Modernination (ERM) projects into the Twelfth Plan from previous Plan periods (12th Plan Document, 2012). Major factors responsible for this include, *inter alia* improper synchronization of project components and delayed tendering and contract management, land acquisition, delays in construction of railway/highway crossing. In Major and Medium irrigation projects, the conveyance efficiency is 69 per cent, on-farm application efficiency 52 per cent and overall project water use efficiency 38 per cent. In the years ahead, shifting public investment towards modernization of surface irrigation works to facilitate higher yield per unit of water used and prioritising incomplete irrigation projects after strict cost-benefit analysis and monitoring implementation will yield immediate results benefiting agriculture. Whatever surplus water which could be extracted by addressing the above problems will bring substantial un-irrigated areas under cultivation, a measure expected to reduce agrarian distress in these areas.

Ground Water (GW)

The unplanned and non-scientific development of ground water resources, mostly driven by individual initiatives has led to an increasing stress on the available resources in many parts of the country resulting in agrarian crisis in the areas which are classified as dark zones. On the other hand, there are areas in the country, where ground water development is still at low key not benefiting the farmers in spite of the availability of sufficient resources. Ground water potential created is 49.40 m.ha but what is utilised is 41.82 m. ha. In minor irrigation, surface water projects (largely irrigation tanks) potential created is 15.72 m.ha and utilized is 12.43 m.ha. Private investment has been concentrated on groundwater exploitation deepening wells and installing more powerful pumps causing distress due to over extraction of water. And these investments do not increase the volume of groundwater available for irrigation and therefore, production capacity. Water table in many States has been falling at an alarming rate. India is by far the largest and fastest growing consumer of groundwater in the world. But, groundwater is being exploited beyond sustainable levels and with an estimated 30 million groundwater structures in place. India may be hurtling towards a serious crisis of groundwater over extraction and quality deterioration which must be arrested; otherwise it will strengthen the pathway for agrarian distress. The measures which are required for sustained use of ground water for irrigation will include:

- Re-mapping India's aquifers and planning to use them rationally
- Addressing areas with sub-optimal exploitation of GW in Indo-Gangetic plains

- Formulating suitable GW development plans for areas in the country with artesian conditions.
- Assessing deeper aquifers in alluvial areas and using them.
- Exploiting GW in flood plains along river courses.
- Augmenting GW supplies in shallow aquifer areas.
- Coordination by Central Ground Water Board (CGWB) the various Ministries of Central Government and State Governments in implementing various schemes for addressing GW issues in a holistic manner.
- Introducing participatory regulatory (and flexible) legal mechanisms for optimal use of GW with involving Water Users Groups and Panchayati Raj Institutions.
- Wherever feasible, large scale use of drip and sprinkler technologies.
- Dissemination of System Rice Intensification method for well irrigated rice areas and promotion of Precision Farming in other crops for efficient use of GW.

Tank Irrigation

Tank irrigation is popular in the peninsular plateau area where Andhra Pradesh, Tamil Nadu and Karnataka are the leading States exploiting tank irrigation for crop production. Undivided Andhra Pradesh is State with largest area Tamil Nadu has the second largest area of 589 thousand hectares under tank irrigation. Years of poor management in three states has reduced the area under tank irrigation. Problems above the outlet such as siltation, foreshore encroachments and poor tank structures have reduced the 15-20 per cent of the tank capacity (Palanisamy, 1991). Tank irrigation today is a neglected resource, and irrigation tanks are underutilized due to mismanagement. However, suitable management strategies and organization would allow the better utilization of existing tanks. An analysis of five tanks in Northern Tamil Nadu shows that more number of tanks yield a benefit-cost ratio of more than one in all sensitivity combinations, so that modernization would yield significant benefits over costs. The studies show that tank performance can be improved significantly by periodical investment in repair works such as repairing sluice structures and surplus weirs, desilting tanks to increase their storage capacity and cleaning supply channels (Govindasamy and Balsubramanian, 1990).

In the case of tanks, the following approaches will be rewarding:

- Undertaking tank system improvement through public spending in a large scale;
- Establishing an enabling environment for the sustainable, decentralized management of tank systems;
- Strengthening community-based institutions to assume responsibility for tank system development and management; and
- Large scale public investment with participatory management involving all stakeholders critical to improve tank irrigation systems.

Overall, India's water future is worrying even without climate change. Despite daunting seasonal and regional variations, it should have ample water for agricultural, industrial and household use. But most of it falls, in a remarkably short time, in the wrong places. India's vast task is, therefore, to trap and store enough water; to channel it to where it is most needed; and, above all, to use it there as efficiently as possible. And on all three counts, India fares badly. Without huge improvements, according to a decade-old official estimate, by 2050, when its population will be a shade under 1.7 billion, India will run short of water. All sources of irrigation from pond development to huge dams must be supported and managed most efficiently for alleviating agrarian distress. The emphasis in this chapter is to treat 'irrigation development of all sorts' as number one priority and that expanding irrigation will greatly minimise production risks, a source of agrarian distress.

2. Make Rainfed/Dry Land (RF/DL) Agriculturemore Resilient

Rain fed/dryland agriculture holds an important position in Indian agriculture as it provides livelihood for half of agriculturally dependent population and also substantially adds to food security of the nation. Still 55 per cent of net cultivated area of 140 m. ha comes under RF/DL farming contributing closer to 40 per cent of food production. But risks are very high due to total dependence on rainfall with little support of irrigation water. They are also hot spots of suicide, highly diverse, ranging from resource-rich areas with good agricultural potential to resource-poor areas with much more restricted potential. Some resource-rich rainfed areas potentially are highly productive and already have experienced widespread adoption of improved seeds. In drier, less favourable areas, on the other hand, productivity growth has lagged behind, and there is widespread poverty and degradation of natural resources. On an all-India scale, for example, rainfed systems includes high-rainfall agriculture in the east and northeast as well as the drought-prone areas of the Deccan Plateau. Other agro climatic characteristics such as soil types also vary (Chand, 2012). The rainfed agricultural areas with below 600 mm of rainfall defined as dryland agriculture in this chapter as do infrastructure development, human capital, and other socio-economic factors. Local-level diversity of RF/DL agricultural systems, meanwhile, implies that planners must recognize that changes in policy, technology or infrastructure may have varying impacts across small areas, and that there is a limit to the extent to which external interventions can induce finely-tuned responses. Regional or district-level planning must be complemented by local initiatives that can be more responsive to specific needs. Planning development for rain fed farming needs defining rain fed typologies, infrastructure development, human capital, and other socio-economic factors which can be effectively utilised for planning developing rain fed agriculture (NBSS&LUP, 1992). The Crop Division of Ministry of Agriculture has also classified total area into five regions as cold northern states, arid western states, semi-arid to arid central and southern states, sub-humid to humid eastern states and sub-humid to humid north-eastern states. Hence districts in those regions are to be carefully identified and plans must be formulated and implemented to address the agrarian distress. Poor infrastructure development, human capital, and other socio-economic factors cause agrarian distress affecting the livelihood of farming population in these areas. Nearly 54.6 million farm holdings are affected. This leads to loss of crop, reduced productivity and loss of income, loss of wages to labourers and unemployment, increased drought relief, migration of farm labourers and farmers, high indebtedness and debt trap for farming community, migration and loss of livestock etc. The efficacy of public investments in nearly 54.6 million farm holdings are affected. They have had hardly any impact on raising product potential or productivity of rainfed lands.

Strategies

There is a need to re-invent country's monsoon agriculture through change in policy and approach. The rain fed agriculture must take off completely redefining the areas at Block/Taluka level to implement rain fed area development schemes given their varying natural resource endowments.

Then investment must be substantially increased to improve *in-situ* moisture conservation which will support current cultivation and ground water recharge. Available mechanical technologies can be leveraged to improve rain water harvesting and *in situ* moisture conservation. Construction of farm ponds and water storage will help to provide life/ support irrigation. Mobile sprinklers can be an option for giving supportive irrigation. Multipurpose seed drill can place coated seeds; fertilizers and mulch in one go which is cost effective and preventive shield against early drought. Raised-bed, ridge furrow, zero tillage, and mulching will minimize risks of crop loss. Dry land areas, instead of mono cropping where largely farmers suicides occur, must be targeted for livestock, horticulture, silviculture, grassland, and fodder in keeping with natural resource availability. Watershed development programmes have demonstrated their potential to contribute to increased farm incomes in rain fed areas. Participatory and knowledge-based watershed development programmes in Andhra Pradesh, Madhya Pradesh, Rajasthan and Gujarat have shown that farmers' and public investments can provide attractive social returns, which inturn contribute to poverty reduction and ecological sustainability. There is an urgent need to evaluate, re-energize and scale up such initiatives with a plan to cover all the dryland areas in the country. The programme requires substantial public and private investments and technical support to create the desired impacts in a reasonably short period (Bantilan *et al.*, 2004). As a long term measure, multiple watershed programmes with different Agencies/Ministries must be integrated to insulate farmers from the risks in rainfed farming. Ensuring food security through provision of wage-paid employment, organized women's self-help groups (SHG), higher inflow of institutional credit to dryland agriculture, covering more households under crop and livestock insurance, covering more crops under the minimum support prices scheme, dryland farmers receiving a higher priority in the allocation of funds for subsidies on farm inputs, facilitating migration, and better management of wastelands and common pool lands are other strategies which need serious policy considerations in minimising distress in dry land areas. Studies confirm that prices, roads, markets and primary schools have all been found to have relatively strong impacts on agricultural growth (Kerr *et al.*, 1996). Farm/non-farm incomes in the dryland regions are constrained by deficient infrastructure comprising roads, markets, hospitals, electricity, irrigation, means of transport and communication. Constraints in seed availability and other input supply also emphasize the importance of an effective public and private sector in reaching the rural poor. Results show that the marginal returns to investment in infrastructure and technology in dry areas are higher than those in irrigated areas. Investment in rural infrastructure particularly in dryland areas will have a direct impact on food security through income, employment and wage impacts (Bantilan *et al.*, 2004). The recent policy of creating Agricultural Producer Companies must be given more thrust in dry land areas. Recent years witness bourgeoning imports of edible oils and pulses and renewed emphasis on oilseeds and pulse production can help reduce the dependence on imports since these are predominantly grown in dry lands. Hence special schemes could be launched for these crops in RF/DL areas to address agrarian distress.

3. Leverage Technology

Past research has unequivocally proved that science based agricultural tools hold great promise for tackling India's farmers' distress and food demands of growing population. Technology offered improved seeds, modern crop protection solutions, use of mobile technology for farmers in the fields, making of foods fresher, safer, and healthier along the food chain. Again technology holds the future of the agricultural and food systems which have to be more productive, more sustainable, more efficient, and more interconnected. Impact of modern crop related technologies were responsible for substantial productivity hikes in crops such as rice, wheat, maize and other crops to the farmers and thereby increased farm incomes. Green revolution technologies in Asia provide ample evidences of benefits due to higher crop productivity increases due to technology. Very recent evidence is the introduction of Bt cotton and its enormous success. So also hybrid seeds of different crops delivered productivity hikes and contributed for the growth of agri-businesses in India. Improved crop varieties with different traits offered resistance to pests and diseases and prevented crop losses to the farmers. Consumer preferences were addressed through improved varieties of crops.

Irrigation technologies came as a handy tool to efficiently use the scarce water resources in the form of micro irrigation tools. Mechanical technologies in the form of machinery and tools have become useful to manage labour scarcity and large scale farming increasing the efficiency of farm operations. Recent experience suggests that even small farmers confront labour scarcity and they require tractor, harvester, rice planter and seed drill services. These services are readily available at the village level from the private sector, and to a limited extent from the government agencies. The provision of services can be encouraged with tax exemption to the purchase of machinery for agricultural purposes by the service providers who are mainly local small scale rural entrepreneurs. In the most recent decades, hybridization, molecular-marker assisted selection and agricultural biotechnology opened up opportunities in improving the traits of different crops. GM crops are already delivering benefits to the farmers of the world. Government must take very decisive policy positions on introduction of GM crops as its benefits are proven beyond doubt. On the crop management technologies, System Rice Intensification (SRI), precision farming, IPM, INM, and IWM which have saved resources to a significant extent. CGIAR and NARS and private sector have generated location specific technologies thus improving farm incomes cutting across countries.

Despite this, for some time now, there is stagnation in productivity increases in major crops in India. Though an appreciable number of technologies have been adopted by the farmers, still a large number of them remain in the shelf. A committed public extension system must work hard to transfer these technologies to the field. A new policy agenda for a demand-driven extension system is currently needed by providing farmers with access to linkage mechanisms through which they would be provided all relevant information to help them articulate their problems and needs with reference to their production and marketing plans. A key factor in improving these feedback systems is to organise farmers into functional groups, such as Self-Help Groups (SHGs), Farmer Interest Groups (FIGs), Commodity Associations (CAs), and/or other types of Farmer Organisations (FOs). These FOs can provide an effective channel for both the dissemination of technology to large number of small and marginal farmers and feedback to research and extension (DoA and C, Government of India 2012). Various ministrics of the government dealing with credit, fertilizers, irrigation, marketing, electricity, etc. must converge to provide synergized support in delivering knowledge and technologies at the farmers' fields. One of the difficulties farmers face is the use of spurious inputs. If public extension is careful and guide the farmers, this difficulty can be easily circumvented.

With major thrust of extension agencies on production techniques, marketing extension so far has not received the attention it deserves. This assumes greater significance in the light of the new international trading regime under the WTO and the export opportunities being opened up. Public extension functionaries are presently ill-equipped to deal with market extension. The multi-agency extension service will need to address these issues through strengthening capacity of the public agency, supporting private sector in marketing extension and making extensive use of media and IT in knowledge and technology dissemination. Encouraging private sector involvement in technology transfer will certainly complement public sector and yield higher dividends. Hostile nature of public sector personnel to private sector participation must disappear. Subsidy distribution is often quoted as a problem in order to involve private sector. One must, however, note the useful role played by the private sector in providing micro-irrigation subsidies. It is easier to provide subsidies for agricultural machineries and equipments and tools involving private sector. Healthy participation of both public and private sector will stimulate emergence of a competitive input supply network to provide hybrid seeds, artificial insemination services, fertilisers, agro-chemicals, animal feed, machinery and equipment, and other agricultural supplies and services to farmers. Promotion of private extension needs to be matched by corresponding shifts in the allocation of public resources. Public funds would be made available to NGOs, Farmer Associations, Para-Professionals or private foundations for extension work as monopoly of public sector extension has led to stagnation in knowledge and delivery. Also an environment in which private investment in technology generation and transfer is more attractive, will, therefore, have to be created.

Capacity building of extension functionaries must be a continuous process as science and technology is fast changing. Unless we keep them updated, latest technologies may not be able to reach the field. They must be trained in accessing information through the use of ICT platforms. The extension agencies must train farmers in accessing information on technology, marketing, inputs supply through mobile communications and from information portals of agricultural universities and research centres. Strengthening Farmers' Associations could be another strategy. By ensuring that wherever conducive, programmes in the field can be planned and implemented through farmer user groups, such as watershed associations, fruit and vegetable growers cooperative societies, agricultural produce marketing societies etc. as farmers' participation could be ensured in administrative and financial decisions. Mainstreaming women in agriculture has not been achieved to the extent needed despite past efforts. With agriculture getting transformed faster, efforts will need to be made by governments to improve extension services to reach farm women through recognizing them as agricultural extension clientele, equipping men and women extension staff on women's role in agriculture, and training of women on decision-making in the context of farm management, agricultural marketing, particularly with respect to post-harvest processing, on farm value addition and market demand.

4. Capital on Easy Terms

With no consistent flow of farm incomes, capital/credit supplies are critical to invest in modern inputs, provision of irrigation, hire machinery and adopt good management practices. One of the reasons attributed for farmers' suicides is sourcing loans from moneylenders at an unreasonably high interest rates which often adds to the repayment burden. It is reported that institutional credit has fulfilled only 61 per cent of the credit demand leaving rest of it to the traditional sources. The focal point is the neglect of the small and marginal farmers by the credit institutions. Provision of institutional credit to agriculture in 2012-13 was to the tune of Rs. 6,07,375 crore (MoA, Government of India, 2013). Various credit schemes such as (i) Farm credit package (2004), (ii) Interest subvention to farmers, (iii) Extension of interest subvention to post harvest activity loans, (iv) Collateral free loans, (v) Credit relief in event of occurrence of natural calamities, (vi) Interest subvention for loan restructured in the drought affected states and (vii) Kisan Credit Card and its conversion to a Smart Card-*cum*-Debit Card. However, these schemes only served in a limited way and improving their performance is important to realise the full potential of these schemes. Still transaction costs of getting institutional credit ishigh and credit flow is also skewed across states. Addressing the needs of left out categories of farmers and focusing on the low performing regions in institutional credit use must get priority in credit delivery. A cursory look at the official statistics mirrors the slow growth rate and share of long term loan affecting capital formation which needs correction. Appraisal and disbursement of credit for rainfed and dryland agriculture requires a different orientation and approach and there is a need to design innovative credit products.

5. *Broaden and Modernise Agricultural Markets*

Agricultural markets in India suffer from many deficiencies which are well listed in the literature. Considerable income during selling agricultural produce is lost due to lack right marketing practices and poor marketing environment. Linking farmers to markets is of greater importance as getting right prices for their products will improve their income. Given the poor bargaining power of small primary producers, integrating them with markets assumes greater emphasis. The small and marginal farmers are subject to more risks. Farmers must be trained to reorient their crop production decisions keeping market demand for their products in mind. They must be made aware of commercialization and globalization of agriculture and facilitated to understand the supply chains for their products. ICT has become handy in the present times to sensitize the farmers on prices and market situations so as to make better marketing decisions and to stimulate the interactions between farmers and agro-industries. Contract farming and other arrangements for vertical coordination are emerging as alternatives to open markets.

All types and models of markets must be encouraged as they add to competitiveness. For example, many oppose FDI and organized retailing. India has more than 138.5 million farm holdings and variety of crops grown; there is ample space for all types of market channels. The emphasis should be to eliminate all unethical practices and introduce transparency in all operations.

Regulatory barriers have constrained investments in development of storage and processing, hampered the development of effective market institutions, and lowered the capacity of agricultural producers to be internationally competitive. Regulatory measures must be made investor friendly to attract private investment to address inadequate infrastructure for storage, sorting, grading, transport and overall post-harvest management. Primary and Periodic markets numbering 22221 and cattle markets are most neglected and special schemes must be implemented to improve the basic amenities as these markets provide opportunities for marginal and small farmers and rural poor. In 7157 Regulated markets operating in the country, there is no professional management. At present, around seven per cent of the total quantity sold by farmers is graded before sale for only 3049 grading units are functioning in the organized marketing. Most farmers still have only little knowledge about the need for grading and the graded products which could fetch premium prices. Reports prepared for Planning Commission indicates that available scientific storage capacity is only 30 per cent of required capacity and existing cold storage capacity could store only 10 per cent of fruits and vegetables. Unless modern supply chains are created with investment by both public and private sector and the services taken to the farm gates, it will take much more time for the farmers to make use of post-harvest infrastructure. Hence, it is imperative that producers are empowered with knowledge, information and capability to undertake market-driven production. Agricultural markets policy environment should be such that more competitive marketing channels are opened to farmers' choice.

Alternative Market Channels

Recently many initiatives have been taken by NABARD and other organizations to promote and involve Self-Help Groups, Joint Liability Groups, Farmer Clubs, Farmer Federations, SHG Federations, Producer Organizations such as Producer Companies, Producer Cooperatives, etc. in direct marketing of the farmers' produce for better price realization. These new initiatives need close monitoring and strengthened further wherever there are lacunae. Another type of market which has come into existence is the Farmers' Markets wherein producers directly market their produce to the consumers. This is named as *Apni Mandis* in Punjab, *Rythu Bazaar* in Andhra Pradesh, *Uzhavar Santhai* in Tamil Nadu and *Shetkari Bazaar* in Maharashtra which promote direct interactions of producers with consumers in fresh produce. Farmers' markets have helped participating farmers to become aware of the products required by the end users and created awareness among farmers to improve product quality and diversify their product portfolios, besides bringing about resource use maximization. However, farmers' markets have really a positive impact on small, marginal and women managed small farms. Other significant government initiatives include Horticultural Producers' Cooperative Marketing and Processing Society (HOPCOMS—a Cooperative) in Karnataka and SAFAL F&V project of National Dairy Development Board in Bangalore. More number of market channels can reduce market gluts which often lead to fall in prices of agricultural produce.

Food Corporation of India, National Dairy Development Board and National Agricultural Cooperative Marketing Federation of India and markets organized by Commodity Boards set-up earlier have performed well. But there is a decline in the functioning of many marketing institutions and cooperatives established by State Governments. Country needs to revive these institutions and build them on business lines as they act as a balance to the private sector and serve the purpose of attaining social objectives (Chand, 2012). Strong policy decisions on allowing FDIs and large scale domestic investment in promoting organized retail is the need of the hour which will create more opportunities to the farmers to realise higher product prices and will catalyse farmers to adopt good agricultural practices (GAP) to meet demands of quality conscious consumers. Though, it is argued that organised retail can cover 10-15 per cent of the produce, it will still promote competitive market environment and force other market channels to adopt good marketing practices and ensure quality of the products marketed.

Amendment to the Companies Act in 2004 allows formation of agricultural producer companies. Registrations gathered steam only recently when farmers realised the value of economies of scale and collective negotiation in a more market-driven agriculture. Currently, Maharashtra and Madhya Pradesh lead the country in formation of producer companies and other States are also making progress. Producer companies offer farmers several benefits allowing growers to formulate a collective production and marketing strategy in line with customer requirement, negotiate better terms with buyers, buy inputs such as seeds and chemicals in bulk, and receive technical guidance at their door step. An added attraction of the corporate structure is that it protects farmers from the influence of local politicians and the State government, which has been the bane of farmer cooperatives. Government must come forward to extend all support to promote producer companies.

Ensuring a free, competitive marketing system is need of the hour as the present system that is heavily weighted in favour of the merchants. Providing e-markets and electronic auction halls which are connected to the national/international market place in each district for the sale of farm produce will increase the competitive markets. The use of such market places would considerably improve prices for produce; they would also improve traceability, tagging systems which are so important for supply chain management. Another option could be to set-up such modern market places in select urban centres and to connect these with collection centres in each district. Use of information and communication technology tools is important to strengthen the knowledge of market trends and to facilitate access to market information for farmers. A combination of newspapers, radio, television, telephone and the internet could be used for effective dissemination of market information to increase competition in the markets. The only word of caution needed is about avoiding a one-size-fits-all solution. For States with good infrastructure and resources, internet kiosks may be a viable option. How to provide these information to the farmer should take into account specific local conditions instead of the details being dictated by some centralised norms.

Corporate-led contract farming has made a positive impact on crop productivity and farm income. According to a study by Sharma (2008) the socio-economic factors that influenced participation in contract farming were education, age, farm size, and access to institutional credit, source of off-farm income and membership to an organization. The conclusions from the study have wider significance in connection with the question of how successful approaches to contract farming can be developed. First, there is a need to assist farmers to have better education, access to timely and quality inputs such as extension services, institutional credit, and better opportunities of off-farm income to improve financial status. The results have shown that membership to farmers' organization was positively related to the likelihood of being a contract farmer. Thus, there is a need to promote non-political farmers' organizations to improve small holders' bargaining power as well as reduce transaction costs to agri-business companies. This central theme of discussion on markets in this chapter is that *all kinds of markets add to competition and increase the market opportunities to some section of the farmers or others and hence must be encouraged through policy reforms and their effective implementation as a strategy to minimise agrarian distres.*

6. *Cushion Risks in Agriculture*

Increasingly, agriculture is not preferred as a profession as there are more risks and uncertainties involved and there is no assurance of stable income because of risks associated production, markets, credit supply, institutions (such as government policy changes in fixing input prices and subsidies) personal (such as cyclones, snake biting etc.), input supplies (untimely supplies and spurious ones, supply of water in the canals), owning assets—such as auctioning by creditors, sale of livestock due to drought—and technology such as poor performance of a new variety and so on. Hence farming becomes an unviable proposition compounded by income not sufficient to meet educational needs of wards, medical requirements of family members and other social obligations, repayment of previous loans, increase in market induced consumerism resulting in distress. Uncontrollable risks are deficit or excess rainfall, distribution of rainfall, extreme temperature conditions, hail storms, wind speeds, humidity and fog etc. Technology, effective monitoring and usage of inputs mitigate controllable risk. The uncertainties in global markets also cause wide swings in agricultural income. Rational actions at individual farmers' and policy level and ensuring proper governance of developmental schemes can help to minimise the risks.

Continuous droughts paralyse the farmers often leading to suicides. In a study on suicides in Karnataka and Maharashtra by Hebous and Klonner (2014) supported this conclusion. According to their econometric estimates, a poverty increase of one per cent increases male suicide mortality by 0.57. Further, they also find that the area under irrigation is negatively associated with the Suicide Mortality Rate (SMR) of male farmers. Some argue that Bt cotton has contributed for suicides but this study results refute showing that cotton cultivation decreases the suicide rate of male farmers. In India, more than 60 per cent of land is vulnerable to droughts which lead to economic losses resulting from low agricultural production, loss of animal wealth, reduced nutritionand loss of health of workers.

Greater investments in assets like bore wells particularly in Southern States led to tragedies when the investments failed and not even one in three bore wells is successful (Palanisamy, 2002). It is common that farmers get into debt trap because of well failures in States like Tamil Nadu. When rainfall fails consecutively for two to three years, the farmers' debt increases and there are no non-farm jobs available for them. This leads to distress. The inevitability of groundwater extraction has been politically exploited too. Successive governments have even given free electricity to farmers most of Indian States. Water tables have fallen at alarming rates in many places across the country during the last few decades. Injudicious use of canal-irrigation water without regard to soil conditions and inadequate attention to drainage have led to water-logging and salinity in many areas, resulting in valuable agricultural land going out of use. The report by Expert Group on Ground Water Management and Ownership (2007) has, however, revealed that in many States, the irrigation potential created has exceeded the ultimate potential, showing that mining of ground water, that is exploitation beyond the dynamic resource, is already taking place. This situation is intensifying the agrarian distress.

Risks and agrarian distress are closely correlated particularly in areas where farming totally depends on rainfall. The diversities in the sources of risks require avariety of instruments for protecting the farmers. In India, these include crop insurance, rainfall insurance, farm income insurance and a calamity relief fund. Most of these measures other than crop insurance are not effectively implemented. If these measures are successful, it is in only selective districts and states for specific crops. Technological, entrepreneurial and policy solutions are expected to provide protection against risks. For example, growing varieties which are drought, pest and disease tolerant; diversification into dryland horticulture and livestock and right kind of crop insurance schemes, investment in '*in- situ*' moisture conservation and adoption of moisture conserving agronomic practices such as mulching, seed coating, etc. are some of the specific solutions which must be taken up. Besides schemes like MNREGA, providing incentives for locating industries in the distressed areas can be considered to minimise the impact of distress.

To address the income losses of farmers due to risks, Government of India introduced in 1999-2000, a new scheme titled 'National Agricultural Insurance Scheme' (NAIS). It envisaged coverage of all food crops (cereals and pulses), oilseeds, horticultural and commercial crops. It covered all farmers, both loanees and non-loanees, under the scheme. However, this scheme was closed and a new a Central Sector Scheme of National Crop Insurance Programme (NCIP) has been formulated by merging earlier schemes, viz. NAIS, WBCIS and CPIS and has come into effect from *Rabi* 2013-14. This scheme has been based on the recommendations of evaluation committees, experience gained through the implementation of various crop insurance schemes and views of the stakeholders. Various improvements and changes have been incorporated in NCIP to make them more farmer friendly. It is critical that Central and State Governments must allot more resources for insurance schemes of this kind and carry out effective monitoring as crop insurance is the single major recovery measure against agrarian distress arising due to risks in farming.

7. *Absolve Misgovernance and Revamp Institutions*

Good governance is perhaps the single most important factor in eradicating poverty and promoting development (Kofi Annan, Former Secretary General of the United Nations). Good governance is epitomized by predictable, open and enlightened policy making; a bureaucracy imbued with a professional ethos; an executive arm of government accountable for its actions; and a strong civil society participating in public affairs; and all behaving under the rule of law (World Bank, 1994). It is the exercise of economic, political, and administrative authority to manage a country's affairs at all levels and comprises mechanisms, processes, and institutions, through which citizens and groups articulate their interests, exercise their legal rights, meet their obligations, and mediate their differences (UNDP). Good governance creates a sound, ethical and sustainable strategy and facilitates decision-making that is rational, informed and transparent which leads to organizational efficiency and effectiveness (World Bank, 2010).

Since India's Independence, poor governance of development

schemes has seriously affected the delivery of benefits to the people. Huge leakages of resources, financial or otherwise, along the channels of delivery are not uncommon. Those who are in the helm of affairs whether elected representatives, bureaucrats, public sector employees at different hierarchies and grassroutes workers had their share of mismanagement of public sector programmes. The people are put into untold sufferings in getting the scheme benefits. The agricultural programmes could not be an exception to this melody. Agricultural developmental efforts were marked by imprudence in the use of key inputs due to failures of governance, poor management of public systems catering to agriculture and pricing, and subsidy policies inimical to the prudent use of resources and effective exploitation of the potential of available technology. With limiting cultivation frontier and scope for further expansion of irrigation getting diminished, the institutional barrier comprising governance and the quality of public systems has become the most serious impediment to agricultural growth.

There is a policy failure of Five Year Plan resource allocation leading to a decline of public investments in irrigation and other related infrastructure and inadequate supply of credit from formal sources to the agricultural sector resulting in greater reliance on informal sources at higher interest burden. Legal frameworks were enforced impartially, particularly in case of abuse of office and financial powers. Many of the agricultural programmes did not fulfil people's aspirations, using resources efficiently. Concerns and commitment towards poor farmers and performance reviews and evaluation of economic outcomes of schemes were not taken up seriously. This kind of situation continues at present and the change towards better governance environment is not insight. Good governance requires that farmers and their interest groups have the opportunity to participate during the formulation of development strategies and in the design and implementation of programmes and projects. Stakeholders should exercise influence over policy decisions and share control of resources and institutions that affect their lives, thereby applying check and balance on the power of government. Transparency and information are critical components of good governance and to reinforce accountability and small and marginal farmers should have easy and reliable access to institutions, processes and information through various authentic sources to realise the benefits of various schemes meant for them. Governance reforms are needed to improve existing legal and institutional framework by promoting effective participation of farmers and their organizations.

A large number of agricultural development schemes framed by the Central Government and are being implemented by State Governments. Decentralization process, as a part of good governance for agriculture, brings government closer to farmers. Important features of decentralization include managing professional staff, administrative functions and financial resources to lower levels of government at district, block and villages and this political decentralization has to be accompanied by fiscal and administrative decentralization. Several of subsidies and institutional credit schemes are not easily accessible to farmers and they need their participation and consultation to design hassle-free and speedy procedure, methods and system to access them. Farmers should have opportunity to voice their views on important legislations being drafted or passed by Parliament, viz. Seed Act, Land Bill, Commodity Spot Exchange Regulation Act, Groundwater Extraction Regulation Bill, National Water Policy, Policy on Genetically Modified Crops, Contract Farming, and Warehousing Receipt, etc. Implementing agencies and their employees should be held accountable for their actions to the entities from which they derive their authority. Farmers as users should have access to transparent information on policy-decisions and programmes of investment and to ensure efficient resource use, minimize waste, avoid diversion and unproductive use. A pro-development and pro-governance legal and judicial system should be in place in which the laws are unambiguous, clear and are uniformly applied by an independent judiciary. Legal frameworks should be fair and enforced impartially, particularly in case of abuse of office and financial powers (Patel, 2013). Unless we have the right kind of governance, the aim to reduce agrarian distress will remain as adream.

Conclusion

Agrarian distress is not a new phenomenon for India. Historically, Indian farmers have been subjected to all forms of uncertainties and risks and had experienced hunger and famine. Occurrence of varying degrees of drought has been cyclical to an extent and caused distress. Preventive measures have been taken in the earlier centuries to promote agriculture in the form of channelling and storing water in large irrigation tanks and irrigating lands by constructing canals. For example, this practice is 2000 years old in Peninsular India. Much has been written and discussed on India's agrarian distress. Basic premise of this chapter is that unless strong foundations are laid in the form of expanding and sustaining irrigation, transfer of available crop production technologies and best agricultural practices to the farmers' fields, insulating them against risks in farming through insurance schemes, integrating farmers in the existing marketing and modern supply chains and ensuring good governance in delivery of agriculture related services by both public and private sector, alleviating agrarian distress will be far a cry. Any amount of debates and discussions will be of only academic in nature and futile. Nothing will happen at the ground level. Hence solid implementation of some of the prioritised proposals made in this chapter is critical to minimise agrarian distress and need of the hour.

References

Bantilan, M.C.S., K.P.C. Rao, K. Singh, P. Parthasarathy Rao, B. Shiferaw, and R. Padmaja (2004). *Overcoming Poverty through Dry land Agriculture: A Strategy for India*, Policy Brief No. 6, ICRISAT, Patancheru, Andhra Pradesh, India, June.

Chand, Ramesh, (2012). "Development Policies and Agricultural Markets", *Economic & Political Weekly,* Vol. XLVIII, No. 52, December 29.

Central Statistical Organization, Ministry of Statistics and Programme Implementation, Government of India (1999). *Compendium of Environment Statistics.*

Department of Agriculture and Cooperation, (2012). *Policy Framework for Agricultural Extension*, Ministry of Agriculture, Government of India, New Delhi.

Deshpande, R.S. and N. Prabhu, (2005). "Farmer's Distress: Proof Beyond Question", *Economic and Political Weekly*, Vol. XL, No. 44, p. 4.

Deshpande, R.S. and Saroj Arora, (2011). *Agrarian Crisis and Farmer Suicides*, Sage Publishers Pvt. Ltd., New Delhi.

en.wikipedia.org/wiki/Drought

FAO, (2010). "Price Volatility in Agricultural Markets -Evidence, Impact on Food Security and Policy Responses", *Economic and Social Perspectives*, Policy Brief, 12 December.

Government of India (2014). *Agriculture Census 2010-11, All India Report on Number and Area of Operational Holdings,* Agriculture Census Division, Department of Agriculture and Cooperation, Ministry of Agriculture.

Government of India, (2010). *Report of the Task Force on Credit Related Issues of Farmers* (Chairman: U.C. Sarangi), Submitted to the Ministry of Agriculture, Government of India, June.

Haryana Kisan Ayog, (2007). *Report on Policy Issues and Options (Based on Interface with Farmers),* CCS Haryana Agricultural University Campus, Hisar, India.

Govindasamy, Ramu and R. Balasubramanian, (1990). "Water Resources Development: Problems and Prospects", *International Journal of Water Resources Development*, Volume 6, Issue 3.

Hebous, Sarah and Stefan Klonner, (2013). *Sources of Agrarian Distress and Farmers' Suicides in Southern India: An Econometric Investigation*, University of Heidelberg, Germany, June.

Iqbal, F. and You, J.I. (2001). "Ideas of Justice", *Democracy, Market Economics, and Development: An Asian Perspective*, Other World Bank Books, Washington, D.C.: World Bank, pp. 9-24.

Kerr, John M. with contributions by Derek Byerlee, Kumaresan Govindan, Peter Hazell, Behjat Hojjati, S. Thorat and Satya Yadav, (1996) *Sustainable Development of Rainfed Agriculture in India*, Environment and Production Technology Division, International Food Policy Research Institute, Washington, D.C., November.

Meeta and Rajivlochan (2007). *Countercurrents.org*, 19 December.

Malcolm Harper, (2011). "Taking Stock: Agrarian Distress in India—Poor Indian Farmers, Suicides and Government", *Enterprise Development and Microfinance*, Vol. 2, No.1, February.

Ministry of Agriculture, (2013). Government of India, Pocket Book of Agricultural Statistics, New Delhi.

Mohanty, B.B. (2013). Farmer "Suicides in India—Durkheim's Types" *Economic & Political Weekly*, Vol. XLVIII, No. 21, May 253.

Mohanty, B.B., and Sangeeta Shroff, (2004). "Farmers' Suicides in Maharashtra", *Economic & Political Weekly*, Vol. XXXIX, No. 52, December 25, 2004.

Mujeri, M.K., S. Shahana, T.T. Chowdhury, and K.T. Haiderm (2012). *Improving the Effectiveness, Efficiency and Sustainability of Fertilizer Use in South Asia*, Briefing Paper Number 8, GDN Agriculture Policy Series, 23.

NSSO, (2003). *Indebtedness of Farmer Households*, NSS Report No. 498, 2003.

Palanisamy, K. (1991). "Conjunctive Use of Tank and Well Water in Tank Irrigation Systems" in *Future Directions for Indian Irrigation: Research and Policy Issues*, edited by Ruth Meinzen-Dick and Mark Svendsen Published by International Food Policy Research Institute (IFPRI), Washington, D.C., January.

Patel, Amrit. (2013). "Good Governance: A Key to Enhance Agricultural Growth", *Microfinance*, March 18.

Planning Commission, Government of India, (1961). *Third FYP Document*, New Delhi.

Planning Commission, Government of India (2007), *Report of the Expert Group on Ground Water Management and Ownership*, New Delhi.

Reserve Bank of India (1954). *All-India Rural Credit Survey*,, Bombay, 1954.

Sainath P., (2006). "It's Official: Distress Up, Suicides Appalling", *The Hindu*, Madras, 22 November.

Sarkar, D., V. Datta and K.S. Chattopadhyay, (2013). *Assessment of Pre and Post-Harvest Losses in Rice and Wheat in West Bengal*, Agro-Economic Research Centre, Visva Bharati, Santiniketan, Study No. 172.

Sharma, Vijay Paul (2008). "India's Agrarian Crisis and Corporate-Led Contract Farming: Socio-economic Implications for Smallholder Producers", *International Food and Agribusiness Management Review,* Volume 11, Issue 4.

Vaidyanathan, A. (2006). "Farmers' Suicides and Agrarian Crisis", *Economic & Political Weekly*, Vol. XLI, No. 38, September 23-29.

Vyas, V.S. (2005). "Agrarian Distress: Strategies to Protect Vulnerable Sections", *The Indian Journal of Labour Economics*, Vol. 48, No. 1, 2005.

World Bank, (2014). *Accelerating Agricultural Productivity Growth, Agriculture, Irrigation and Natural Resources Sustainable Development, South Asia,* May 21, Washington, D.C.

World Bank, (1994). *Governance: The World Bank's Experience.*

World Bank (2010). *Governance of Technical Education in India,* Working Paper 190, Washington D.C., March.

11
Institutional Reforms for Inclusive Agricultural Development
Role of Financial Institutions and Innovations

Gyanendra Mani

Background

Land resources play an important role in determination of man's economic, social and cultural progress, especially in the rural sector where it is regarded as the nucleus of all farm activities. Unlike in many developed countries who followed the concept of intensive cultivation by using less land for agriculture purposes and managing it more efficiently; India resorted to rather extensive cultivation extending to even marginal and sub-marginal lands, particularly during pre-green revolution period (i.e. up to late sixties). It is only when the limit to extending cultivated area reached in the mid-seventies, the efforts towards intensive land use in agriculture began sincerely. The land use patterns over a period of time has gone under changes on account of demand of land for non-agricultural uses (industry and housing), reclamation of waste land etc. There have been several institutions in the recent years to strengthen credit supply for agricultural and non-agricultural development in rural India. This chapter analyses the trends and impact of such innovations.

Changes in Land Use Pattern

It would seen from Table 11.1 that the area under forest and non-agricultural uses in the country has increased over time. However, the loss in area available for cultivation due to the above has been compensated by bringing-in some culturable waste and fallow land and thereby maintaining the net sown area at around 140 million hectare. In the process, the quality of total stock of agricultural land has deteriorated as good quality agricultural land in the urban fringe areas have gone out of cultivation and comparatively inferior quality land in the form of waste land etc., have been added to the available stock of net sown area. Transfer of good quality agricultural land to non-agricultural uses is to be viewed seriously and some solution to this problem is required to be suggested. Some countries have resorted to the policy of agricultural zoning which seems to have worked well in protecting the agricultural land from going to non-agricultural uses.

Table 11.1: Land use pattern in India

(*Area in Million Ha*)

Sl. No.	*Area under different land uses*	*1950-51*	*1970-71*	*1990-91*	*1995-96*	*2011-12*
1.	Area Reported	284.32	303.75	305.02	304.88	305.82
2.	Forests	40.48	63.83	67.70	68.82	70.02
3.	Non-agricultural uses	9.36	16.48	21.22	22.36	26.29
4.	Barren and Unculturable land	38.16	28.13	19.51	19.01	17.23
5.	Culturalble Waste	22.94	17.50	15.00	14.10	12.64
6.	Other Uncultivated land (excl. fallow)	26.50	17.63	15.22	14.55	13.46
7.	Fallow land	28.12	19.33	23.50	23.85	25.38
8.	Net sown area	118.75	140.86	142.87	142.20	140.80
	Total cropped area	**131.89**	**165.79**	**185.74**	**187.47**	**195.25**

Source: www. Agricoop.nic.in.

At the time of Independence, the land distribution was widely inequitable, with three-fourth of the arable land under the ownership of only seven per cent of the rural population, and the remaining one-fourth pathetically subdivided and fragmented under the ownership or quasi-ownership of about 48 per cent of the rural population. The remaining 45 per cent of the rural population were landless or tenant labourers of which approximately 25 per cent were working on land owned by other for daily wages in cash or kind, and the rest worked in miscellaneous occupations as daily wage earners or were unemployed. In addition to the above, the National Sample Survey in 1953-54 (8th round) made an estimate of the land held under tenancy and sub-tenancy in different part of the country (occupancy tenants were excluded). The percentage of area leased out varied from 21 to 26 per cent. It showed that about one-fifth of the total area was held under tenancy and, thus, it was not possible to ignore the problem affecting such a wide area. Besides this open tenancy, there was considerable amount of land leased-out on the basis of oral or hidden tenancy which accounted for between 35 to 40 per cent of total cultivated area (Datt and Sundharam, 1989). The land reform measures concentrated mainly on abolition of intermediaries, security of tenancy and regulation of rents, consolidation of holdings, imposition of ceiling and redistribution of surplus land beyond ceiling among the landless. The re-distribution of land resources as a result of land reform measures resulted into a more equitable distribution and removal of *benami* holdings to a great extent. A brief account of availability of land resources for agricultural operations is given as under:

Table 11.2: Size class-wise land holding pattern in India

Land size class/ category	*1970-71*			*1990-91*			*2010-11*		
	No. of OprHol (million)	*Area (million Ha)*	*AvgHol Size (Ha)*	*No. of OprHol (million)*	*Area (million Ha)*	*AvgHol Size (Ha)*	*No. of OprHol (million)*	*Area (million Ha)*	*AvgHol Size (Ha)*
Marginal	36.20	14.60	0.40	63.39	24.89	0.39	92.83	35.91	0.39
Small	13.43	19.28	1.44	20.09	28.83	1.43	24.78	35.24	1.42
Semi-Medium	10.68	29.30	2.81	13.92	38.38	2.76	13.90	37.71	2.71
Medium	7.93	48.23	6.08	7.58	44.75	5.90	5.88	33.83	5.76
Large	2.77	50.06	18.10	1.65	28.66	17.33	9.73	16.91	17.38
All Sizes	**71.01**	**162.32**	**2.28**	**106.64**	**165.51**	**1.55**	**138.35**	**159.59**	**1.15**

Sources: Agricultural Census of various years.

The pressure of rising population and labour force on the limited and non-expanding land base, division in the family and thereby division of total land available with the family has resulted into a continuous decline in the availability of land per agricultural holding. This has also resulted into an increased marginalization of farm holdings. As may be seen from Table 14.2, during the period from 1970-71 to 2010-11, the number of marginal holdings has increased by 2.6 times from 36.23 million to 92.83 million. The share of marginal and small size together in the number of operational holdings has increased from 70.0 per cent to 85.0 per cent during the latest decade; their share in the area operated has risen from 20.9 per cent to 44.6 per cent during the same period. The average size of holding for all size groups has fallen from 2.28 hectare in 1970-71 to 1.15 hectare in 2010-11.

Table 11.3: Steeper Marginalization of Marginal Farmers: Declines in Average Landholdings

Year	*Average Landholdings in Hectares* by Size Groups			
	Marginal Farmers (Up to 1 hectares)	*Small Farmers (1 to 2 hectares)*	*All Other Size Groups*	*Total*
1970-71	0.403	1.436	6.001	2.28
1976-77	0.393	1.419	5.598	2.00
1980-81	0.394	1.442	5.328	1.84
1985-86	0.393	1.434	5.060	1.69
1990-91	0.393	1.435	4.827	1.55
1995-96	0.395	1.419	4.592	1.41
2000-01	0.395	1.416	4.466	1.33
2005-06	0.383	1.383	4.315	1.23
2010-11	0.383	1.422	4.283	1.16

Source: NABARD Occasional Paper No. 59.

Apart from the phenomena of marginalisation and fragmentation of landholdings, a distinct feature of the agricultural scene in India has been the steady decline in the area cultivated. In a different context, Singh (2006), who has made an incisive study of agrarian changes, has brought out how, with acute distress amongst the farm community, it is likely that a good section are withdrawing from land cultivation:

> "... the proportion of marginal holdings and the area owned by them has increased sharply over the years, while it has declined sharply in the case of large holdings in all the states. This could have happened due to many factors including agrarian distress, non-viability of a large section of holdings, urge to search for alternative avenues of employment, indebtedness etc. It may be mentioned that the agrarian distress in certain parts of the country has been so severe that a large number of farmers have committed suicide" (NABARD Occasional Paper No. 59).

Institutional Response for Credit Support to Farm Sector

The setting up of an Agricultural Credit Department (ACD) simultaneously with the establishment of the Reserve Bank of India in 1935 can be considered as the first step towards recognising the importance of institutional finance, both for reducing the dependency of farmers on moneylenders and for providing crop loans as well as investment credit for agriculture and allied activities. Starting with the creation of the State Bank of India (SBI) in 1955 with the specific target of opening 400 new branches in rural and semi-urban areas and starting agricultural lending; nationalization of major commercial banks in July 1969 and April 1980; establishing Regional Rural Banks (RRBs) in 1976 to provide more focused attention to the rural areas, particularly in underdeveloped and under-banked regions; and finally setting up the National Bank for Agriculture and Rural Development (NABARD) on July 12, 1982 as an apex body for the rural credit institutions to undertake supervisory functions in respect of cooperative banks (other than urban/primary cooperative banks) and Regional Rural Banks; India completed the process of development of institutional framework for credit support to rural sector, more specifically to agriculture production system in the country. Therefore, the period from 1935 (establishment of RBI) to 1982 (setting up of NABARD) can be defined as stage of 'Financial Institution Development'. During this period, the financial institutions mainly focused on expanding their service base (branch expansion) in rural and semi-urban areas as well as their clientele base.

The period from 1982 onwards can be seen as period of 'Product Innovation' with NABARD taking the lead in product development, particularly for those looking for avenues for their survival in rural areas and more specifically for those deriving their livelihood from agriculture and allied activities. Introduction of Kisan Credit Card (KCC) by NABARD in 1989 allowing farmers to draw credit for all production needs through the production cycle can be considered a major milestone in the rural financial sector. The KCC has now emerged as a most important banking mechanism for cutting down transaction costs both for the farmer and the bank. Launching of Self-Help Group (SHG)—Bank Linkage Programme (SBLP) was another major product by NABARD which changed the life of many belonging to poor and vulnerable section of the society. Setting up of Rural Infrastructure Development Fund (RIDF) with NABARD in 1995 and directing Commercial Banks to deposit the shortfall amount of their target of priority sector/agriculture/weaker sections lending helped creation of infrastructure in rural areas aimed at supporting the agriculture sector. Although, the financial innovations like financing farmers through 'Joint Liability Groups' (JLGs) and supporting promotion of 'Producers Organisation' (POs) can be considered relatively young products as compared to KCC and SBLP but have been found to have great acceptance amongst bankers as well as farmers. Many other innovations like developing Primary Agricultural Credit Societies (PACS) *into multiservice centres especially to* meet the credit and non-credit services required by small and marginal farmers, banking through the business correspondents (BCs) and business facilitators (BFs), particularly in the remote areas have helped financial institutions to reach the unreached in a faster way.

Agricultural Credit Flow in India

The policy on agricultural credit in India has been guided mainly by the considerations of ensuring adequate and timely availability of credit at reasonable rates through the expansion of institutional financial network, its outreach and scale as also by way of directed lending. Over time, significant progress has been made in terms of the scale and outreach of institutional structure, particularly by commercial banks, for credit flow to agriculture sector. The Agricultural Credit Policy during post-nationalization (of banks) period laid greater emphasis on augmenting credit flow at the ground level through credit planning, adoption of region-specific strategies, rationalization of lending policies and procedures and bringing down the cost of borrowing. Bank credit has been made available to the farmers in the form of short-term credit for financing crop production programmes and also in the form of medium-term/long-term credit for financing capital investment in agriculture and allied activities. This policy of having two formats for credit support gave the choice to farmers/rural people to avail the credit as per his requirement and convenience both in terms of availment and repayment of loans.

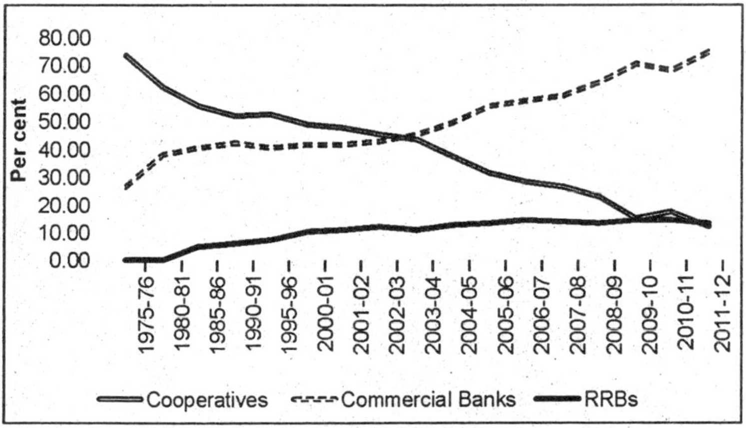

Fig. 11.1: Share of Agencies in Credit for Agriculture (short-term)

Table 11.4: Direct Institutional Credit for Agriculture (Short-term)—Loan Outstanding

(Rs. Billions)

Year	Direct Institutional Credit for Agriculture (Short + Long Term) – Loan Outstanding				Direct Institutional Credit for Agriculture (Short-Term) – Loan Outstanding			
	Cooperatives	Comml. Banks	RRBs	Total GLC	Cooperatives	Comml. Banks	RRBs	Total GLC
1975-76	23.57	7.90	—	31.47	10.12	3.64	0.02	13.78
1980-81	43.15	30.43	1.80	75.38	19.08	11.62	—	30.7
1985-86	69.47	84.16	8.71	162.34	32.37	23.55	2.65	58.57
1990-91	105.31	170.32	17.53	293.16	51.78	42.35	5.90	100.03
1995-96	234.27	34.67	460.20	234.27	93.12	71.73	13.08	177.93
2000-01	461.35	382.70	72.49	916.54	181.68	154.42	36.92	373.02
2003-04	714.03	681.03	117.21	1512.27	308.08	319.82	76.64	704.54
2004-05	788.22	955.19	167.09	1910.50	324.81	427.98	109.80	862.59
2005-06	823.27	1356.03	215.10	2394.40	341.40	599.71	138.77	1079.88
2010-11	766.74	3575.84	550.67	4893.25	496.45	1932.62	406.63	2835.7
2011-12	725.45	4432.98	638.23	5796.66	445.17	2690.30	465.80	3601.27
2012-13	—	5224.78	746.61	—	—	3534.25	552.55	—

Source: Reserve Bank of India—*Handbook of Statistics on Indian Economy*.

The year 2003-04 and 2004-05 can be said as the period of transformation when the Commercial Banks took over from the Cooperative Banks as the lead purveyor of the institutional credit to sector. This transformation coincides with the implementation of 'Doubling of agricultural Credit' which was aimed at doubling the flow of agricultural credit during 2006-07 over 2003-04 base.

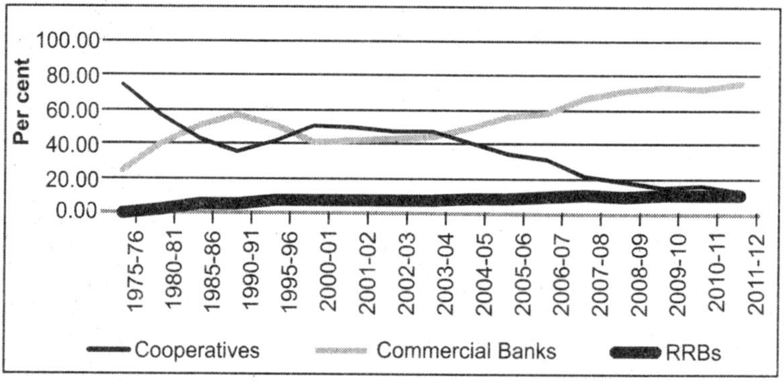

Fig. 11.2: Share of Agencies in Credit to Agriculture (ST+LT)

Coverage of Marginal and Small Farmers by Institutional Finance

As a whole, about 80 million agriculture accounts were found to have been financed during 2013-15 indicating coverage of about 42.6 per cent of the operational holdings by institutional credit in the country (assuming number of accounts as a proxy of number of farmers and dividing it by the number of operational holdings in the country).

Table 11.5: Physical and Financial coverage of Farmers: All India

Year	*No. of agriculture accounts (lakhs)*	*Amount disbursed (Rs. crores)*	*Per account (in Rs.)*
2006-07	423.13	229400	54215
2007-08	439.34	254658	57964
2008-09	456.1	301908	66193
2009-10	482.3	384514	79725
2010-11	549.6	468291	85206
2011-12(P)	646.57	511029	79037
2012-13(P)	703.57	607375	86328
2013-14	799.68	711621	88988

Source: Government of India and IBA (data sourced from Corporate Planning Department, NABARD HO, unpublished).

According to an estimate, currently around 43 per cent of the small and marginal (SF/MF) operational holdings have access to institutional credit. The small and marginal farmers together account for 60 per cent of total accounts and 40 per cent of total credit disbursed by the financial institutions covering cooperative, commercial banks and regional rural bank over the period of 2007-08 to 2013-14. The following chart indicates that coverage of marginal and small farmers by RRBs (67% account and 61% credit goes to small and marginal farmers) is better as compared to commercial banks in which case small and marginal farmers account for 55 per cent to their total agriculture account and just 34 per cent of total credit disbursed by them.

Table 11.6: Share of SF/MF in number of accounts and amount disbursed (Average for the period 2007-08 to 2013-14)

Particulars	*Cooperatives*	*RRBs*	*Comm. Banks*	*Total*
Share of SF/MF in total number of agricultural accounts	64	67	55	60
Share of SF/MF in total agricultural credit disbursed	53	61	34	40

Source: Government of India and IBA (data sourced from Corporate Planning Department, NABARD HO, unpublished).

The foregoing sections have clearly highlighted that the tiny operational holdings are uneconomic and require innovative measures to expand their credit base for agricultural operations. More importantly, there is no way small and marginal farmer households can improve their living standards by depending solely on agricultural incomes. Therefore, there is the need for supporting off-farm and non-farm activities, again with innovative methods of financing. Further, it is also necessary to single out the marginal farmers as distinguished from other farmers, even as distinguished from small farmers. The literature generally brackets small and marginal farmers together, as though their landholding profile is similar, if not identical. This is in fact not so; there are a few distinguishing features which stand out in further analysis. As shown in Table 11.3, the average size of landholding amongst marginal farmers constitutes just a little over one-fourth of the average holdings amongst small farmers. The innovations like SBLP, JLGs and POs have given much hope to these small and marginal farmers who have expanded their livelihood support base beyond agricultural fields either as owner or agricultural labourers.

Some Observations on 'Doubling of Agricultural Credit' Programme (2004-05 to 2006-07)

Observing the steep decline in the growth of agriculture credit to 10.9 per cent per annum in nominal terms and to 2.3 per cent per annum in real terms during the reform period of 1990s, Government of India announced 'Doubling of Agricultural Credit Programme, in 2004. The banks were required to double the credit disbursement during 2006-07 over the 2003-04 base.

The trend in performance of various agencies under 'Doubling of Agricultural Credit Programme' (Table 11.8) indicates that the amount disbursed to agriculture sector has shown an increasing trend from 2003-07 to 2006-07. However, the number of accounts financed during 2005-06 was less than the number of accounts financed during 2004-05 which had increased slightly during 2006-07.

A study to review the implementation of doubling of agricultural credit programme in Uttar Pradesh was conducted by Bankers Institute of Rural Development, Lucknow in 2008 in order to understand the reasons behind the decline in the number of accounts. The performance of the doubling of credit programme in UP is presented in Table 11.9.

Table 11.7: Average Annual Growth Rates in Agricultural Credit of Scheduled Commercial Banks during Certain Distinct Phases

(*In Percentages per Annum*)

Credit Series	Mar-73 to Mar-81	Mar-81 to Mar-91	Mar-91 to Mar-01	Mar-01 to Mar-11	Mar-01 to Mar-07	Mar-07 to Mar-11
Nominal Agricultural Credit	28.4	16.9	10.9	24.7	28.2	19.1
Real Agricultural Credit						
(a) Deflated by Deflators of Overall GDP at Market Prices	17.9	7.6	2.7	18.1	22.8	11.1
(b) Deflated by Agricultural GDP Deflators	19.2	7.7	2.3	16.3	22.5	7.0

Note: Periodization has been done not by any statistical method of discerning structural breaks but by visual observations and by a more potent indicator, namely the peaks and troughs of the number of agricultural borrowers as noted in a subsequent section.

Sources: NABARD Occasional Paper No. 59.

Table 11.8: Achievement under Doubling of Credit Package: All India

Particulars	*Cooperative banks*	*Regional rural banks*	*Commercial banks*	*Total banks*	*Rs. Loan per A/c*
2003-04					
Amount (Rs. Crores)	26,875	7,581	52,441	86,981	—
2004-05					
Numbers (millions)	—	—	—	41.3	
Amount (Rs. Crores)	31,231	12,404	81,481	1,25,309	30341
2005-06					
Numbers (millions)	19.3	4.6	14.6	38.5	
Amount (Rs. Crores)	39,404	15,223	1,25,477	1,80,489*	46880
2006-07					
Numbers (millions)	18.9	6.2	14.6	39.7	
Amount (Rs. Crores)	42,480	20,435	1,40,382	2,03,297	51208
2006-07 as multiple of 2003-04 (amt)	**1.58**	**2.70**	**2.68**	**2.34**	**42602 (Avg 3 yrs)**

Source: Study on 'Doubling of Credit Programme, BIRD, Lucknow (2008), unpublished.

The analysis indicates that, in UP, the number of accounts financed under the programme during 2005-06 declined slightly (by about 10%) but increased again during 2006-07 showing an overall increase of 2.8 per cent over 2004-05 level. The decline in accounts during 2005-06 from the 2004-05 level was mainly contributed by the cooperative banks (declined by 13.6%) and commercial banks (declined by 9.9%). The following factors were also found to have contributed to the decline in accounts during 2005-06 and 2006-07:

(i) The farmers who were holding the KCC from two agencies, viz. Cooperative banks and Commercial banks/Regional Rural banks, had either surrendered the KCC account maintained with cooperative banks or abstained from availing credit from cooperative banks (despite holding cards from cooperative banks) as they were able to get sufficient credit from RRBs/ Commercial banks.

Table 11.9: Achievement under Doubling of Credit Package: Uttar Pradesh

Particulars	*Comm banks*	*RRBs*	*Cooperative banks*	*Total*
2003-04				
Amount (Rs. Crores)	3948.38	1918.91	1980.78	7848.07
2004-05				
Number of A/c	10,18,473	8,65,905	37,38,883	56,23,261
Amount (Rs. Crores)	6208.18	2619.31	2135.05	10962.53
2005-06				
Number of A/c	9,18,127	9,34,266	32,31,950	50,84,343
Amount (Rs. Crores)	8487.82	3205.76	2058.05	13751.63
2006-07				
Number of A/c	13,50,739	10,95,297	33,36,282	57,82,318
Amount (Rs. Crores)	11165.53	4011.61	2309.44	17486.58
2006-07 as multiple of 2003-04 (Amt.)	**2.83**	**2.09**	**1.17**	**2.23**
Total Credit Disbursement under the Programme				
Number of A/c	**32,87,339**	**28,95,468**	**1,03,07,115**	**1,64,89,922**
Amount (Rs. Crores)	**25861.53**	**9836.68**	**6502.54**	**42200.74**
Agency-wise share in total Achievement (%)				
Number of A/c	**19.94**	**17.56**	**62.51**	**100.00**
Amount	**61.28**	**23.31**	**15.41**	**100.00**
Amount of loan disbursed per account (Rs. Per A/c)				
2004-05	60956	30249	5710	19495
2005-06	92447	34313	6368	27047
2006-07	82662	36626	6922	30241
Three years average	**78670**	**33973**	**6309**	**25592**

Source: Study on 'Doubling of Credit Programme', BIRD, Lucknow (2008), unpublished.

(ii) Some farmers who were waiting to get credit from cooperative banks could not get it in time due to shortage of fund with cooperative banks despite their willingness to avail credit from cooperative banks. Some PACS members could not get fertilizer from the society during 2005-06 due to shortage of fertilizer in the season and therefore the accounts of such farmers turned out to be in-operative during 2005-06 and could not be counted into "total number of accounts having availed loan from the bank".

(iii) There was migration of farmers (mostly from cooperative banks to other two agencies) which was mainly on account of less loan disbursement per account both in case of short term as well as investment activity. The unit cost for most of the investment activities was fixed at lower amount by Cooperative (long term) banks as compared to that by RRBs/Commercial banks.

(iv) The higher increase of scale of finance during 2004-05 over 2003-04 period was also one of the reasons for more credit flow particularly during 2004-04 over 2003-04 level (39.7% in UP).

(v) The frequency of transactions in case of KCC accounts had increased significantly during second and third years of implementation of programme (2005-06 and 2006-07). This strategy helped in increasing the GLC during the next two years as every disbursement was considered as a fresh loan/ disbursement and therefore, the loan disbursed was more than the sanctioned amount in a particular loan account without an increase in number of accounts in the same proportion. In nut shell, the increase in credit flow to agriculture sector was supported by increased scale of finance during first year and increased frequency of credit withdrawals from KCC accounts during second and third years.

(vi) The decline in number of loan accounts in case of cooperative banks (Long Term) was also on account of introduction of a very liberal term loan schemes by RRBs/CBs for extending credit up to Rs. 5.00 lakh without much emphasis on appraisal and asset creation of a particular investment activity. Otherwise also, KCC was considered convenient and sufficient to take care of other investment needs of the farmers.

Another interesting feature which came to notice that despite tremendous increase in GLC to Agriculture Sector during the policy package period (2.34 times), the average yield of foodgrains declined continuously during this period. In fact, 51 out of total 70 districts recorded decline in Yield of foodgrains in UP during this period.

A sample study in connection with the above mentioned 'Doubling of Credit Programme' was also undertaken covering some selected farmers (230 each) in the districts of Moradabad and Jhansi representing west UP and Bundelkhand Region of Uttar Pradesh respectively indicated that about 64.5 per cent of total withdrawal in KCC accounts in Moradabad and 35.7 per cent of total withdrawal in Jhansi district were genuinely used for crop production. Majority of the farmers (42%) withdrew the entire amount of the limit sanctioned in one go, a portion of that is utilized for consumption purpose also. Some of the farmers were found to have parked the entire/major portion of their total withdrawal in their saving bank account.

It was observed that the majority of the farmers (42%) withdrew the entire amount of the limit sanctioned in one go, a portion of that is utilized for consumption purpose also. Some of the farmers were found to have parked the entire/major portion of their total withdrawal in their saving bank account. The portion of crop loan was also found to have been used for investment in agriculture in activities like purchase of pump-sets, dairy animals etc. This diversion of production credit to investment credit was obvious on account of price differential between two products (production loan and investment loan). These diversions has also resulted into lower off-take of investment credit although actual amount of investment has been high than what is being reported by the financial institutions. This highlights the need for re-working on the strategy of having large difference in prices of production loan and the investment loan.

The findings of the study suggest that inflow of agricultural credit beyond a limit do not help farmers in raising the agricultural production in a linear manner unless a suitable policy package to raise the absorption capacity of the farmers is designed and implemented. Development of region-specific technology mix for various parts of the state/country appears to be a prerequisite for the growth of the agriculture sector.

Credit Absorption Capacity of Farmers

The various efforts made by financial institutions to upscale the credit disbursement to rural segment have been basically aimed at managing the supply side factor and improving the credit delivery systems. Most of the time, these actions does not take into account the demand side constraints defining the upper ceiling/cap on the amount of total credit which can be absorbed by a rural farmer. An analysis of credit absorption capacity of the farmers (Gyanendra Mani, 2010) was made by assuming that a farmer tries to balance the following equation in a normal situation to support his livelihood on sustainable basis:

Short Term Credit availed for crop production and consumption purpose + income from farm activity + income from other sources

<=>Repayment of principle and interest + cash required to meet the expenses of family.

The result of the study (300 sample in Moradabad, Sultanpur and Jhansi districts of UP in 2007-08) estimated that the credit absorption capacity per farm per annum of marginal, small, medium and large farmers was found to be Rs. 32,563, Rs. 62,362, Rs. 1,31,853, Rs. 1,98,049 respectively. It means the farmers falling under various sub-categories (except the marginal farmers) can productively employ the above mentioned amount for crop production and generate sufficient income even to take care of their consumption needs. The marginal farmers were needed to generate additional income of Rs. 21,303 to satisfy family consumption needs from some other sources including wage labour. The estimate of maximum credit for consumption purpose for various farm categories have been arrived at by taking into account the net surplus left with the farmers after servicing the short term credit loan. The estimates of total credit left for saving for various class categories have been estimated and it was found that small, medium and large farmers were able to generate surplus over and above the amount required for servicing of production loan and consumption requirement (equivalent of consumption needs estimated by NSSO).

SHG Bank Linkage Programme (SBLP)

The Self-Help Group—Bank Linkage Programme beginning with a pilot project of linking of 500 SHGs of rural poor in 1992, is now considered as the world's largest micro-finance initiatives with over 7.4 million SHGs representing 97 million rural households directly becoming part of this great movement.

Although an excellent progress has been made as far as credit linkage is concerned, there has been some concern regarding increased level of NPA under SBLP in recent past in almost all the regions except southern region. The overall NPA percentage of loans to SHGs has come down marginally from 7.08 per cent as on 31.3.2013 to 6.83 per cent during the current year thereby reversing upward swing in NPA observed during the last few years (Table 11.11). However, the level of NPA is still alarmingly high compared to position 5 years back (2.9% as on 31.3.2010).

Region-wise analysis shows that Southern Region with 11679 savings linked SHGs per lakh BPL families is well above the rest, the lowest being the Central Region with 724 SHGs. The corresponding figures of credit linked SHGs are 7018 and 443. In spite of an excellent progress

Table 11.10: Overall Progress under SHG-Bank Linkage for last 3 years

(Amount Rs. in crores/Numbers in lakhs)

Particulars		2011-12		2012-13		2013-14	
		No. of SHGs	*Amount*	*No. of SHGs*	*Amount*	*No. of SHGs*	*Amount*
SHG Savings with Banks as on 31st March	Total SHG Nos.	79.60 (6.7%)	6551.41 (–6.7%)	73.18 (–8.1%)	8217.25 (25.4%)	74.30 (1.53%)	9897.42 (20.45%)
	Of which NRLM/SGSY/Other Govt. spons. programmes	21.23 (5.0%)	1395.25 (–23.2%)	20.47 (–3.6%)	1821.65 (30.6%)	22.62 (10.46%)	2477.58 (36.01%)
	% of NRLM/SGSY/Other Govt. spons. programme Groups to Total	26.7	21.3	28.0	22.2	30.45	25.03
	All women SHGs	62.99 (3.3%)	5104.33 (–3.7%)	59.38 (–5.7%)	6514.86 (27.6%)	62.52 (5.27%)	8012.89 (22.99%)
	Percentage of Women Groups	79.1	77.9	81.1	79.3	84.15	80.96
Loans Outstanding against SHGs as on 31st March	Total No. of SHGs linked	43.54 (–9.0%)	36340.00 (16.4%)	44.51 (2.2%)	39375.30 (8.4%)	41.97 (–5.71%)	42927.52 (9.02%)
	Of which NRLM/SGSY/Other Govt. spons. programmes	12.16 (–5.4%)	8054.83 (2.9%)	11.93 (–1.9%)	8597.09 (6.7%)	13.07 (9.55%)	10177.42 (18.38%)
	% of NRLM/SGSY/Other Govt.spons. programme Groups to Total	27.9	22.2	26.8	21.8	31.1	23.7
	No. of all Women SHGs linked	36.49 (–8.4%)	30465.28 (16.6%)	37.57 (2.9%)	32840.04 (7.8%)	34.06 (–9.34%)	36151.58 (10.08%)
	Percentage Of Women SHGs	83.8	83.8	84.4	83.3	81.2	84.2

(Figures in the parentheses indicate growth/decline over the previous year).
Source: *Status of Microfinance in India* 2013-14.

Table 11.11: Region-wise NPAs of Bank Loans to SHGs

(Rs crore)

Region	*Loans Outstanding against SHGs-Position as on*			*Amount of NPAs reported as on*			*Percentage of NPAs to Loan Outstanding as on*		
	31.3.12	31.3.13	31.3.14	31.3.12	31.3.13	31.3.14	31.3.12	31.3.13	31.3.14
North	1178.28	1160.68	1100.64	81.55	129.87	150.46	6.92	11.19	13.67
North East	993.27	796.76	753.80	51.33	68.23	66.96	5.17	8.56	8.88
Eastern	4629.80	5538.13	4944.63	337.08	570.56	547.42	7.28	10.30	11.07
Central	2780.29	2776.85	2696.66	367.03	479.76	508.98	13.20	17.28	18.87
Western	1363.78	1467.52	1640.46	112.14	126.57	182.26	8.22	8.63	11.11
South	25394.59	27635.36	31791.33	1263.59	1411.93	1476.57	4.98	5.11	4.64
All India	**36340.00**	**39375.30**	**42927.52**	**2212.73**	**2786.92**	**2932.66**	**6.09**	**7.08**	**6.83**

Source: Status of Microfinance in India 2013-14.

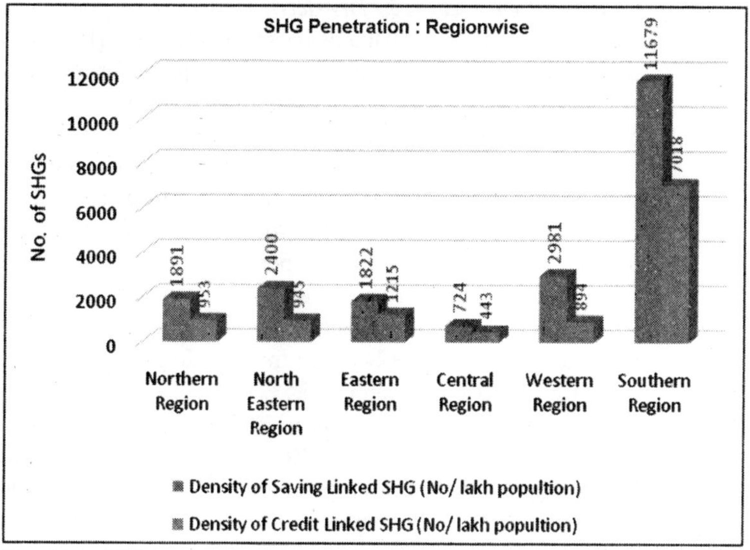

Fig. 11.3: SHG Penetration in different regions

overall of SBLP, the issue of low penetration level coupled with high NPA level in the Central Region followed by Northern region must be viewed very seriously in order to have equitable growth of all the region. It is expected that this particular issue will also find priority under Nation Rural Livelihood Mission (NRLM).

Implementation of National Rural Livelihood Mission (NRLM) and Scheme for Interest Subvention to Women SHGs

The Ministry of Rural Development (MoRD), Government of India launched the National Rural Livelihood Mission (NRLM) by restructuring Swarnajayati Gram SwarojgarYojana (SGSY) with effect from 1 April, 2013 (RBI circular no. RBI/2012-13/559 dated June 27, 2013). NABARD issued guidelines to RRBs and Cooperative Banks on the issue. NABARD is also implementing the Interest Subvention Scheme for Women SHGs under the National Rural Livelihood Mission (NRLM), for Regional Rural Banks (RRBs) and Cooperative Banks.

The two categories of districts covered under the scheme are as follows:

(i) *Category I Districts (150 Districts):* Under the scheme, in the identified 150 districts all the Women SHGs promoted by NRLM or other Central or State Government line departments or NGOs or by NABARD under WSHG programme of DFS and other promotional programmes, which are linked with the banks, will be eligible to avail the benefits of the scheme. Accordingly, Women SHGs, are provided loans up to Rs. 3 lakh at an interest rate of 7 per cent per annum. The Women SHGs are also eligible for an additional subvention of 3 per cent if they repay in time, thereby reducing the effective rate of interest to 4 per cent per annum. The RRBs and Cooperative Banks will be subvented to the extent of difference between their lending rates and 7 per cent (subject to a maximum of 5.5% p.a).

(ii) *Category II Districts (other than 150 Districts):* In Category II (other than 150 districts), all NRLM compliant Women SHGs will be eligible for interest subvention for loans up to Rs. 3 lakh at an interest rate of 7 per cent. In these districts, banks will charge the SHGs as per their respective lending norms and the difference between the lending rates and 7 per cent will be subvented to the loan accounts of the SHGs by the State Rural Livelihood Missions.

Financing of Joint Liability Groups

Joint Liability Groups are informal groups of 4-10 members who are engaged in similar economic activities like crop production and who are willing to jointly undertake to repay the loans taken by the Groups from the Banks. Unlike in the case of SHGs, JLGs are intended basically as credit groups for tenant farmers and small farmers who do not have proper title of their farm land. Regular savings by the Group is purely voluntary and their credit needs are to be met through loans from financial institutions and such loans could be individual loans or group loans. Financing of JLGs was introduced as a pilot project in 2004-05 by NABARD in 8 States with the support of 13 RRBs. Apart from extending 100 per cent refinance support to Banks, NABARD also extends financial support for awareness creation and capacity building of all stakeholders of this Programme. Besides, NABARD extends grant support for formation and nurturing of JLGs to banks and other JLG promoting agencies. As against 196539 JLGs promoted during 2012-13, JLGs promoted during 2013-14 were 207871 making the cumulative number of JLGs promoted by NABARD and financed by banks to 672224 as at the end of March 2014. Acknowledging the good work done by NABARD and the financing banks in financing of JLGs, the Hon. Finance Minister announced (see Fig.11.4) financing of 5 lakh JLGs of *Bhoomiheen Kisan* (Landless Farmers) during 2014-15 while presenting the Union Budget for the year in July 2014. Fresh loans to the extent of Rs. 2220.16 crore were disbursed to these groups during the year (Rs. 1837.64 crore during the previous year—an increase of 21%) taking the cumulative loan disbursed to JLGs to Rs. 6775.72 crore. The growth of JLGs in the country is less skewed than the SHGs. Although the Southern States top the list with over 2 lakh JLGs organized so far, Eastern Region closely follows with 1.80 lakh JLGs. In terms of actual loans disbursed, however, Southern States account for nearly 55 per cent of the total loans disbursed.

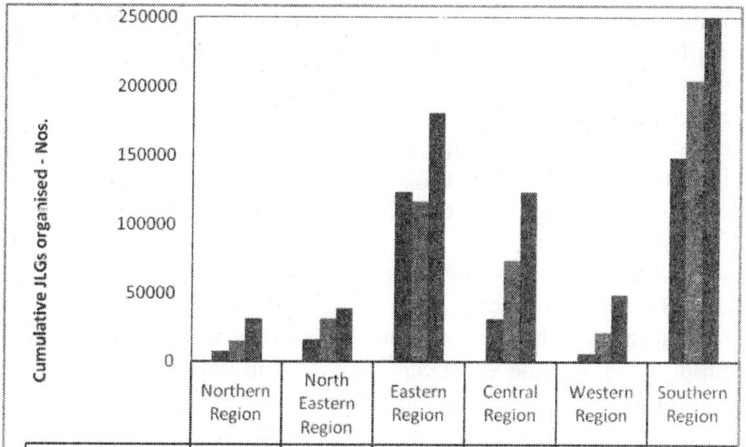

	Northern Region	North Eastern Region	Eastern Region	Central Region	Western Region	Southern Region
As on March 2012	7613	16474	123132	31236	6133	148119
As on March 2013	15422	31461	116887	74160	22045	204378
As on March 2014	31490	38639	180489	123213	48912	249481

Fig. 11.4: JLGs Organised

New Initiative in form of Producers' Organisations

A Producer organization is a registered body and a legal entity which is formed and owned by a group of producers for either farm or non-farm activities. The producers are the shareholders in the organisations. In order to provide Credit Support for financial intervention; support in the form of grant, loans, or a combination of these for capacity building and market interventions and also some contribution to the share capital of the 'Producers Organisations' having low capital base, "Producers Organization Development Fund" (PODF) has been set-up in NABARD w.e.f. 01 April, 2011, with an initial corpus of Rs. 50 crore. Any registered Producers Organization viz., Producers Company (as defined under Sec. 581A in Part IXA of Company's Act, 1956), Producers Cooperatives, registered Farmer Federations, MACS (Mutually aided cooperative society), industrial cooperative societies, other registered federations, PACS, etc. set-up by producers are eligible under the fund. As on 31 March, 2014, the cumulative sanctions to POs amounted to Rs. 210.81 crore and disbursements to Rs. 74.08 crore covering a total of 91 POs in the country has been sanctioned.

In order to encourage the POs to actively participate inthe supply/ value chain management two major initiative were taken during 2013-14. These were: (i) providing short-term loan repayable in 18 months for meeting working capital requirements of the POs so as to upscale their business activities in a systematic manner; (ii) providing financial assistance to POs through implementing agencies like financial institutions, NGOs, voluntary agencies and Trusts so that they can access credit and leverage on the business and technical capabilities of the facilitating agencies.

The major activities of the POs so far assisted by NABARD include dairy (including collection, processing and marketing of milk), fishery (including purchase of fishing equipment and commercial fish production), organic vegetable cultivation in poly houses, marketing infrastructure (including agri-malls), and agro-processing infrastructure (including cotton ginning, spice processing units), horticulture and marketing of lac and forest produce, setting up of poultry and small ruminants units, setting up of pelletised cattle feed plant, promotional interventions in pilot project on value chain management, modernisation of cold storage, etc. Besides, eco-tourism, embroidery, jewellery making and support to potters area few initiatives relating to non-farm sector support under Producer Organisation Development Fund (PODF). The financial support provided to the bamboo societies in Assam for supply of raw materials to Hindusthan Paper Corporation and the loan provided for encouraging scientific orange cultivation in three blocks of Jhalawar district of Rajasthan, through the facilitation of Jhalawar CCB are the special highlights of the initiatives taken during 2013-14.

The Finance Minister in his 2014-15 Budget speech highlighted the issue of profitability of small holding based agriculture in view of increasing proportion of small and marginal farmers in the country and proposed to supplement NABARD's Producers' organization development fund for Producer's development and upliftment called PRODUCE with a sum of Rs. 200 crore which will be utilized for building 2,000 producers organizations across the country over the next two years.

Conclusion

To conclude, a large number of financial innovations viz., SHG-Bank Linkage programme (SBLP), JLG, and Producers Organization have now been developed to help the vulnerable sector of the society to come out of vicious cycle of low standard of living. The basic philosophy behind all these interventions is to encourage farmers and landless women to come together and manage their pooled resources by applying their collective wisdom. At the same time, these innovations help reduce distress and poverty among small and marginal farmers.

References

Datt, Ruddar and K.P.M. Sundharam (1989). *Indian Economy.*

Mani, Gyanendra (2010). "Credit Absorption Capacity of famers in Uttar Pradesh", *Indian Journal of Agricultural Economics.*

NABARD (2014). Occasional Paper No. 59, Agricultural Credit in India: Trends, Regional Spreads and Database Issue by *EPW* Research Foundation (p. 12).

Singh, J.S. (2006). "Changing Agrarian Relationship in Rural India", *Indian Journal of Agricultural Economics*, 61(1), pp. 39-40, 43.

12

Technological and Institutional Innovations for Sustained Increase in Farmers Income

Brajesh Jha

Introduction

The majority of the poor resides in rural India, and more than one-third of the rural poor are farmers.[1] Small and marginal holdings dominate Indian farm conditions, and a significant percentage of these farmers is in distress.[2] The size of agricultural holdings and a dearth of non-farm income in the rural vicinity are two of the many reasons for distress. Allied activities (land saving) therefore make up an important source of farmer's household income but not so good return of farmers from different land saving activities accentuates farmer's distress. The frequency of incidence of distress fluctuates over years. Numerous factors determine farmer distress; few government efforts have counterfactual relations. The higher price of agricultural commodities has a trade-off between the interests of farmers and consumers. Incidentally, a large number of farmers in India are also consumers. The counterfactual relations can also be understood from the fact that the increased expenditure by the government for farmers depletes development expenditures largely for farmers (Government of India 2013).[3] Therefore, agricultural technologies are important; it can simultaneously increase farmers' income and reduce consumer expenditure on that commodities. Technology reduces farmer's distress since it is related to shortfall in household income from agriculture.

Risk is another important determinant of a farmer's decision analysis. The down-side risk in return affects farmers the most (Jha, 1995). Farm decisions have strong implications for the status of natural resources in the regions. But how far technology and technological options can uplift farmers and protect them and the environment is the subject of the present chapter. Many agricultural technologies are scale-biased. Since Indian farmers are small land holders, they require supportive institutions to adopt agricultural technologies. Considering the size of operational holdings, off-farm income is important for small holders.[4] But, for adoption of agricultural technologies, it is important that the major share of a farmer's household income comes from agriculture. Besides introduction and conclusion three are three sections in this chapter. The next section illustrates existing return and distress of farmers from crops. Section three discusses agricultural technologies and technological options for increasing the return of farmers. Section four delineates institutions that may help farmers in increasing their return. Section five presents the conclusion and suggestions.

Farmers Return from Crops and Distress

In agriculture, the problem at the aggregate level is the desired rate of growth; at the micro-level farm level return is important. Though many studies report technology-specific return from a crop, the present study assesses farmer's return with the cost and return of crops as obtained from the Report of Commission for Agricultural Cost and Price (CACP). The CACP report has many advantages over the individual technology-specific studies.[5] The important trends in the productivity and price of agricultural commodities having implications for farm-level return and return related distress of farmers is discussed in the present section. This section also discusses the pattern of cost and return having implications for the cost of adoption of technology in agriculture.

Famers' Return from Principal Crops

Return on farm is investigated with the costs and return of crops in paddy, wheat, *bajra*, gram, rape-mustard, and cotton. These crops represent a wide range of crop groups like fine and coarse cereals, pulses, oilseeds and commercial crops. Table 12.1 A compares yield (production per hectare of land), cost (as represented through C2 in the CACP report) and income (as represented through return over cost) of chosen crops in the selected states of India. The CACP data on costs and return aspect of a crop represent different groups of states. States also present different kinds of situation for crops. Paddy, for example, is highly water-intensive and its cultivation in Northwest India is beyond the carrying capacity of the region, but quite within that of Bihar which is situated in the humid/ sub-humid region of the country. The concern for natural resource encourages cultivation of paddy in the latter kind of states. Similarly, the cost of cultivation data of wheat is available for States like Haryana, Rajasthan and Madhya Pradesh (MP). Most of these states, except MP, are known for growing Mexican wheat. The concern for diversity requires that non-Mexican-wheat may be continued to be cultivated in MP. The coarse cereals, believed to be neglected crops in certain states, are becoming important in some other states. It may be that supporting institutions of specific coarse cereals exist in certain states to make these crops profitable. Markets for certain coarse cereals exist in certain states as this is part of the dietary constituents of the people in that region. Price and profitability of specific coarse cereals also depend on the extent of their alternative use; for example, maize and *bajra* are increasingly used to make poultry feed.

Gram and *urad* are important pulse crops of India. In gram, inter-state variation in income is more on account of difference of gross return rather than cost in states during the reference period. As per the earlier trend in coarse cereals and pulses, return in rape-mustard was increasing across reference states. Interestingly, return in rape-mustard is increasing in Rajasthan. It is reported that the National Research Centre for Rapeseed and Mustard at Bharatpur, Rajasthan has developed the country's first hybrid mustard called '*Shankar Sarson*', which further increases the profitability of rapeseed and mustard by 20 per cent. Cotton is an important commercial crop in many regions in India. It is now a recognized fact that the yield induced risk has started decreasing after adoption of Bt cotton in the early 2000s. The profitability of cotton has increased on the account of the adoption of Bt cotton. Price, another important component of gross return, fluctuates following the prevalence of distortions in the world cotton market.

The above discussions clearly show that the income referred here is return over cost (C2) in a crop varies across states; income of gram in Bihar and rape-mustard in Rajasthan was significant. The return from the same crop (gram, rape-mustard) has however decreased in Haryana. The differential profit of a crop across states provides reasons for growing specialization of crops in selected states of the country. Such profit may not necessarily be due to the natural resource endowment of a region; favourable institutions play an important role in increasing the return to farmers. The natural resource base in a region can be made elastic by

Table 12.1A: Productivity, Cost and Return of Peddy (P) and Wheat (W) in the selected States

	Productivity, Cost and Return of Paddy								*Productivity, Cost and Return of Wheat (P)*							
	Andhra Pradesh (P)		*Bihar*		*Haryana (P)*		*Tamil Nadu (P)*		*Bihar (W)*		*Haryana (W)*		*Rajasthan (W)*		*Madhya Pradesh (W)*	
	2000	2009	2000	2009	2000	2009	2000	2009	2000	2010	2000	2010	2000	2010	2000	2010
Parameteres on the basis of per hectare	02	11	02	11	02	11	02	11	02	12	02	12	02	12	02	12
Productivity (yield in quintals)	48	54	20	19	39	41	47	50	48	54	20	19	39	41	18	31
Cost (C2) Operational Cost+Fixed Cost)	26608	52854	11937	20362	27755	50293	28306	48796	26608	57854	11937	20862	20662	50203	17368	30454
Operational Cost (A2+Family Labour)	17844	35098	8192	14991	14831	28195	20848	36428	17844	36068	8192	14991	14991	28195	7845	16688
Retunrs over Cost C2 (in INR)	380	7229	-1140	91	5674	22550	1195	6210	380	7229	1140	91	5674	22560	820	16033

Source: Ministry of Agriculture, Govt. of India.

Table 12.1B: Productivity, Cost and Return of Maize (M) and Urad (U) in the Selected States

Parameters on the basis of per hectare	*Productivity, Cost and Return of Maize*						*Productivity, Cost and Return of Urad*							
	Andhra Pradesh (M)		*Bihar (M)*		*Rajasthan (M)*		*Andhra Pradesh (U)*		*Maharashtra (U)*		*Odisha (U)*		*Rajasthan (U)*	
	2000-02	*2009-11*	*2000-02*	*2009-11*	*2000-02*	*2009-11*	*2000-02*	*2009-11*	*2000-02*	*2009-11*	*2000-02*	*2009-11*	*2000-02*	*2009-11*
Productivity (yield in quintals)	22	48	23	36	12	17	7	7	4	5	4	4	3	4
Cost (C2) (Operational Cost+ Fixed Cost)	11686	42397	13220	22490	13330	25257	9094	23107	8445	20298	6110	12299	7203	13912
Operational Cost (A2+Family Labour)	8140	28908	9532	16695	10661	19275	4887	13592	6599	14913	3915	7412	5796	9962
Returns over Cost C2 (in INR)	-1430	4677	-1336	26141	-4240	-996	4075	10954	347	1052	-1966	5769	207	4522

Source: Ministry of Agriculture, Govt. of India.

Table 12.1C: Productivity, Cost and Return of Cotton (C) and Rapeseeds and Mustard (R&M) in the selected States

Parameters on the basis of per hectare	*Productivity, Cost and Return of Cotton*								*Productivity, Cost and Return of Rapeseeds and Mustard*					
	Andhra Pradesh (C)		*Gujarat (C)*		*Haryana (C)*		*Rajasthan (C)*		*Haryana (R&W)*		*Madhya Pradesh (R&W)*		*Rajasthan (R&W)*	
	2000-02	2009-11	2000-02	2009-11	2000-02	2009-11	2000-02	2009-11	2000-02	2010-12	2000-02	2010-12	2000-12	2010-12
Productivity [yield in quintals	13	16	4	20	7	18	11	20	17	20	13	14	13	15
cost (C2)] (Operational Cost+Fixed Cost)	22317	46160	13730	48511	17004	47868	13149	40211	16763	35759	11556	24435	13011	25771
Operational Cost (A2+Family Labour)	14745	28033	11038	32056	12347	30620	8602	2418	9638	16280	15470	14307	13355	12820
Returns Over Cost C2 (in INR)	2569	10784	2061	33718	2834	18879	6518	39611	3218	26435	17505	4423	3553	17103

Source: Ministry of Agriculture, Govt. of India.

public investment. Considering the average size of agricultural holding, income from crops is not sufficient for farmers.

Distress as Downside Risk to Farmers

The present sub-section studies farmers' distress as the downside risk in return from crops. The same has been studied by calculating the probability of failure (PF) in yield and return of crops. The down side risk is average negative deviation (AND) from expected return and yield of crops. The PF indicates frequency of shortfall in expected return and yield, whereas AND works out the magnitude of downfall in return and yield. Expected value (return or yield) is based on trend value (return and yield), and the study argues that expected value as observed through historical trend in value is the most important determinant of farmers' expectation. Many farm decisions, including investment, are often based on their expectations. The downside risk in yield is indicative of production induced distress whereas the difference between yield and return indicates the magnitude of market induced stress for farmers. With the commercialization of agriculture, market induced risk has assumed importance. The downside risks have, therefore, been calculated separately for the reference periods; the first period is years between 1997-98 and 2004-05 while the second period is between the years 2004-05 and 2011-12. The downside risk is calculated for important crops separately in states (Bihar, Gujarat and Haryana) and the same is presented in Table 17.2.

A glance at these estimates indicates a general decline in downside risk in important crops; gram is an exception. The decline was in terms of frequency (PF) and also magnitude (AND) of downside risk. Irrigation appears to be an important factor of the amelioration of production-induced risk. Caution needs to be exercised in estimating the risk for paddy, as it is not exactly a commodity, but a commodity group in a state like Haryana. Rice, for instance, consists of *basmati* and non-*basmati* rice in Haryana. The production and price estimates for these commodities vary depending on their proportion to the commodity aggregate (rice). In other words, the price and the yield of rice (commodity aggregate) varies every year. A significant decline in the downside risk of cotton is owed to the arrival of Bt cotton in the study area. Technology (Bt) is thus an important factor of reduction in the risk of crops.

The estimates of downside risk in gross return present a different picture. The down-side risk in the selected coarse cereals (maize), pulses (gram) and oilseeds (rape-mustard seed) has increased over the reference

Table 12.2: Downside Risk (DR) in Yield and Return of the selected Crops, States and Periods

Crops	Bihar				Gujarat				Haryana				Tamil Nadu			
	DR in Yield (in kg)		DR in Return (in INR.)		DR in Yield (in kg)		DR in Return (in INR)		DR in Yield (in kg)		DR in Return (in INR)		DR in Yield (in kg)		DR in Return (in INR)	
	1997-98 to 2004-05	2004-05 to 2011-12	1997-98 to 2004-05	2004-05 to 2011-12	1997-98 to 2004-05	2004-05 to 2011-12	1997-98 to 2004-05	2004-05 to 2011-12	1997-98 to 2004-05	2004-05 to 2011-12	1997-98 to 2004-05	2004-05 to 2011-12	1997-98 to 2004-05	2004-05 to 2011-12	1997-98 to 2004-05	2004-05 to 2011-12
Paddy	153.0	329.5	487.5	2992.4	243.9	89.5	1843.8	3473.5	177.78	153.8	5721.1	10043.1	444.6	329.3	2459.33	7140.0
	(0.13)	(0.25)	(0.13)	(0.13)	(0.25)	(0.13)	(0.25)	(0)	(0.13)	(0)	(0.25)	(0.13)	(0.25)	(0.13)	(0.25)	(0)
Wheat	167.0	161.3	4039.1	3887.6	240.7	219.0	6890.1	6407.8	286.7	194.6	9419.2	8934.9	—	—	—	—
	(0.13)	(0)	(0.25)	(0)	(0.25)	(0.13)	(0.13)	(0.13)	(0.25)	(0)	(0)	(0)				
Maize	132.5	262.5	503.7	3556.9	362.7	269.9	2008.2	3378.7	255.0	215.2	1834.4	3997.4	54.7	1091.3	492.4	1586.6
	(0.13)	(0.13)	-0.13	(0.13)	(0.25)	(0.38)	(0.5)	-0.25	(0.13)	(0.38)	(0.13)	(0.25)	(0.13)	(0.25)	(0.25)	(0.13)
Gram	80.6	128.8	1754.5	5243.2	157.7	99	2029.2	4482.8	88.1	153.0	1598.3	4227.1	19.9	18	1086.7	3021.5
	(0.38)	(0)	(0.13)	(0)	(0.13)	0	(0.25)	(0.13)	(0.25)	(0.38)	(0.25)	(0.38)	0	0	(0.25)	(0)
Rapeseed and Mustard	54.8	57.0	1010.4	4987.3	159.7	134.9	2797.5	7219.9	189.4	237.9	2488.8	9030.8	—	—	—	—
	(0.25)	(0.13)	(0.25)	(0)	(0.13)	(0.13)	(0.38)	(0.13)	(0.25)	(0.25)	(0.25)	(0.13)				
Cotton	—	—	—	—	95.2	64.3	1269.7	6963.3	95.2	64.3	1269.7	6963.2	40.9	108.0	609.7	4813.5
					(0.25)	(0.13)	(0.25)	(0.13)	(0.25)	(0.13)	(0.25)	(0.13)	(0.13)	(0.13)	(0.38)	(0)

Note: Probability of failure (PF) in parentheses is the incidence of downside risk while figure without parentheses is the magnitude of downside risk.

Source: Estimated by the author.

period. The role of market-induced risk in the latter reference period is also evident. A careful analysis of these estimates indicates that production induced risk in the late 1990s nineties was marginally lower than in between 2004 and 2012. The down-side risk in return has increased substantially in the latter period. The data for return is generated from yield and price, and in a multiplicative model a high downside risk in return is possible if the risk in one variable (market, productivity) is very high. Downside risk in return is high in the latter reference period as the difference in prices was very high during the reference periods. Downside risk is particularly high for cotton, followed by oilseed (rape-mustard), pulse (gram), coarse cereal (maize) and fine cereal (wheat).

The estimates of downside risk show that farmers cultivating crops other than fine cereals are liable to greater down-side risk. Besides the effectiveness of minimum support price (MSP) for fine cereals, production-induced risk is also less in fine cereals since these are not cultivated as neglected crops (as gram in Haryana). Wheat is also cultivated significantly in the assured irrigated area. During the reference period, production induced risk has decreased and market induced risk has emerged as important.

Structure of Agricultural Cost and Implications for Adoption of Technology

The cost of cultivation data shows significant variation across states in productivity and also cost of commodity. This variation has various factors; at times, it is because of bad weather and yield. The periodic downfall of yield in some commodities important for maintaining food security is often met with the practice of bonus above MSP in certain states. The bonus price above MSP is primarily to achieve the decentralized procurement of food grains in the referred states. Considering periodic variation in the yield and importance of decentralized procurement, the practice of bonus price above MSP is not unjust. If such bonus above MSP is the cause of price inflation of food grains for consumers, and the government is too concerned about its effect, the government should devise ways of subsidizing the price of food grains. The price of fine cereals is too subdued to that of other agricultural commodities.

The cost data shows that the share of human labour in the total cost of production has increased substantially during the reference period. In human labour, the share of casual worker has increased in the reference period and that of family labour has decreased significantly (*See* Figures in Appendix 12.1A and 12.2A). Many agricultural commodities in India were competitive despite its low productivity in world agriculture; the competitiveness in such commodities was on account of the low cost of family labour engaged in agriculture (Jha, 2001).

The share of machines in the total cost of production of a commodity has increased during the reference period. This is true for all reference states and commodities considered in the study. The factor endowment suggests that India is rich in unskilled human labour. The cost of unskilled human labour, as reflected with the wage of casual workers engaged in agriculture, increased in the 2000s (Jha, 2007); the wages are reported to have further increased after the initiation of the Mahatma Gandhi National Rural Employment Guarantee Scheme (MGNREGS). The increased cost of labour, besides many factors, has caused increase of mechanization in agriculture (Table 12.3).

Again in mechanization, the share of hired component has increased and that of owned cost of machinery has decreased (*See* Fig. in Appendix Figs. 12.1A and 12.2A). This reflects the dependence of farmers on hiring machines. Any assistance in reducing the cost of hiring the machine will be important. The government may assist individuals (person) or institutions (organization) interested in custom hiring, primarily to reduce its hiring cost for farmers. Jha (2011) suggested that individuals such as agricultural adviser or agri-clinics could encourage custom hiring, while Chahal *et al.* (2014) suggested the merits of cooperatives in the institutionalization of custom hiring services. These suggestions assume that encouraging custom hiring services will reduce the cost of hiring machines and increase farmers' income.

Cotton is generally referred to as a cash crop, since farmers used to cultivate this crop for earning the cash requirement of farm holdings. This concept of cash crops appears to have diminished; now, any crops cereals, pulses, oilseeds wherein farmers are assured of a relatively good price are being cultivated to earn cash. Gram (in Bihar), rape-mustard (in Rajasthan) are examples of crops being cultivated as cash crops (Table 12.1).

The age-old hypothesis of marginal land is still true in large parts of the country; productivity of gram in a prosperous state like Haryana is an example. The government must invest in such land to increase its quality. Such investments will improve the chances of increase in farmer's income. In this context, the present prime minister's goal of providing irrigation facility on every piece of land is important.

Table 12.3: Constituents of Total Cost (C2) of Paddy and Cotton

Parameters	*Andhra Pradesh*		*Bihar*		*Haryana*		*Tamil Nadu*		*Andhra Pradesh*		*Gujarat*		*Haryana*		*Rajasthan*	
	2000-02	*2009-11*	*2000-02*	*2009-11*	*2000-02*	*2009-11*	*2000-02*	*2009-11*	*2000-02*	*2009-11*	*2000-02*	*2009-11*	*2000-02*	*2010-12*	*2000-02*	*2010-12*
Human Labour	33.31	34.15	38.82	41.10	26.04	29.79	32.04	30.19	26.04	29.79	32.04	30.19	40.19	34.09	36.16	35.69
Machinery Labour	6.34	10.21	6.89	8.60	2.45	4.82	8.26	3.90	2.45	4.82	8.26	3.90	5.35	5.73	3.26	4.43
Seeds	3.01	4.46	6.19	6.28	6.20	4.84	5.68	5.03	6.20	4.84	5.68	5.03	2.25	7.41	3.79	5.95
Fertilizers and Manure	10.35	6.62	6.68	5.68	9.29	9.29	11.56	9.12	9.29	9.29	11.56	9.12	4.16	4.34	7.79	6.30
Miscellaneous cost	12.78	7.49	7.26	10.75	18.29	9.49	22.05	17.77	18.29	9.49	22.05	17.77	18.17	11.84	12.92	6.86
Fixed Cost	34.22	37.06	34.17	27.59	37.73	41.77	21.14	33.99	37.73	41.77	21.14	33.99	29.88	36.60	36.09	40.76
Operational Cost	65.78	62.94	65.83	72.41	62.27	58.23	78.86	66.01	62.27	58.23	78.86	66.01	70.12	63.40	63.91	59.24

Source: Ministry of Agriculture, Govt. of India.

The constituent of fixed cost in total cost has increased during the reference period.[6] The increase in fixed cost is true for the most of reference states barring Bihar. The increase in fixed costs in Tamil Nadu, Gujarat and Haryana is more than other States (Bihar). The increase in fixed cost indicates that the rental value of land has increased, in other words, opportunity cost of land has increased in the earlier group of states. This also reflects that the cost of agriculture increased in at least Tamil Nadu, Gujarat and Haryana. Considering the size of holding the fixed costs may be subsidized by the Government. This may be important component of income based support in agriculture.

The statistics related to cost and return indicate that return over cost (C2) in any commodity (data for which is published by the CACP) is not sufficient for farmers to concentrate on the above crop activities only.[7] This is true in all reference states. Farmers are increasingly investing in land saving enterprises like dairy to augment their household income. The growth of dairy activities in the past few decades, despite emerging constraints in the sub-sector reveals this. Of late, farmers' income from some land-saving enterprises has decreased (Jha, 2011). Therefore, farmer's dependence on off-farm income increases. Off-farm income accounts for a considerable proportion of farmers' household income (NSS, 2005).

Technologies and Technological Options for Farmers

The above analysis shows that cultivation of field crops with traditional practices may not transform small-holder based agriculture in India. The chapter, therefore, assesses some land-based technological options and also technologies (farm practices) that can possibly increase the household income of small farmers. Bio-chemical technology is one of the most important technologies. This is being replicated in a big way in large part of the country. Externalities associated with bio-chemical technology are also evident (Jha, 2000); there are also efforts to minimize these externalities. In other words, the health of the agricultural environment will improve. The study, therefore, advocates technologies (farm practices) that minimize the deterioration of the environment. The technological options and technologies (farm practices) are discussed in this sub-section.

The assessment of technological options for farmers is based on a field study in North-west India. Some small farmers in the region were practising vegetable-based crop rotation for markets in town. Though vegetable-based crop rotation is profitable across farm size, profitability is particularly high on small farms, due primarily to better utilization of family labour on small farms and the personal involvement of farmers in marketing of those crops. Among vegetables, potato is being cultivated by farmers of different land size. Small farmers generally prefer vegetables those generate regular income and take little time to harvest. Vegetables like pea, tomato, okra, green chilli, spinach, cucumber and gourd provide regular income. Radish and cauliflower harvest quickly too, but provide return at the time of harvesting only.

Since different varieties of a vegetable are grown in different seasons, vegetables may be categorized by season only loosely. Thus, okra and gourd, planted generally in February and harvested by May or June are summer vegetables; and pea, cauliflower, cabbage, and radish planted in September and harvested by February are winter vegetables. Vegetables that resist such categorization are potato (August to November), tomato (January to April), and eddoe (*arbi*: March to August). Figure 12.1 compares relative profitability of vegetables *vis-à-vis* food grains. Figure clearly shows that many of the vegetables are more profitable than cereals. The profitability of vegetables would increase further if we compare a complete rotation of vegetables *vis-à-vis* cereals.

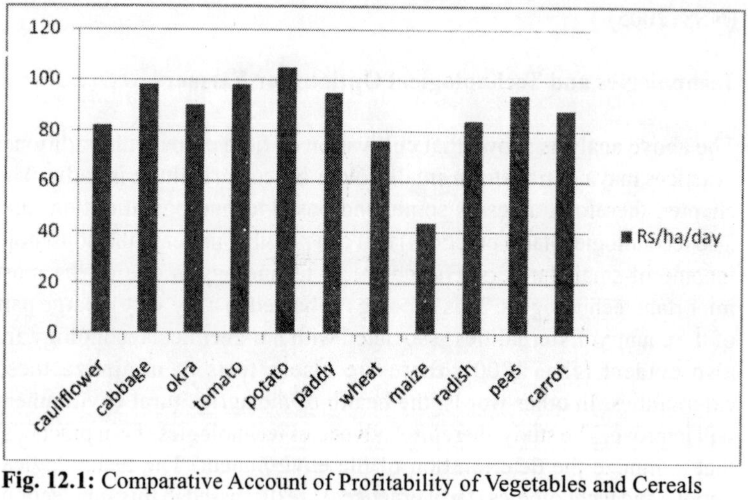

Fig. 12.1: Comparative Account of Profitability of Vegetables and Cereals

Most large farmers grow vegetables for family consumption. Large farmers practice agro-forestry, besides cultivating fine cereals. Some large farmers plant orchards as it consumes less time and family labour. The common plants for agro-forestry in the region are eucalyptus and *poplar*. Many large farmers adopted agro-forestry in the late 1980s and 1990s; whereas, orchards of kinnow and grape are a newer phenomenon. There appears to be sufficient potential to divert resources from fine cereals; however, many farmers have not undertaken these options because of several local level constraints. Some are presented below:

(i) There is no assured price for most crops other than paddy and wheat. The price uncertainty is particularly high for commercial crops like basmati rice and potato. Though price uncertainty is low for pulses, coarse cereals and oilseeds, pulse involves greater yield uncertainty.

(ii) Farmers stressed the need for stabilizing the yield of the existing variety of summer pulse: *moong/urad.* Summer pulse is a short duration crop that can be planted between wheat and paddy crops; however, the yield uncertainty of the existing variety of summer pulse is so high that farmers often have to forgo investments in the form of seeds in cultivating it.

(iii) Hybrid sunflower seeds along with the supportive price policy for oilseeds in the early 1990s, had to some extent replaced late-sown wheat crop in Northwest India. Late-sown wheat is more resource intensive than normal-sown wheat. The growth of acreage under a hybrid crop like sunflower was constrained as spurious seeds were in abundance, whereas quality seeds are very important for hybrid varieties.[8]

(iv) There is limited possibility of diversification towards fruits and agro-forestry on large farms, as these enterprises are less labour intensive. In contrast, small farms, which are labour surplus in general, have potential for vegetables. The selected small farmers have started growing vegetables, but their realization changes with the alteration of marketing channels for vegetables and similar other crops.

(v) The plants generally used for agro-forestry are poplar and eucalyptus. These are required for specialized purposes: poplar in packaging and eucalyptus in paper industry. Farmers, however, complain that the selected industrial units often form a cartel and try to control the market price of logs related to the above. Further intensification of agro-forestry requires specific types of infrastructures in the region.

The above observations are based on field survey of Kurukshetra district in North-West India. It appears from the above investigations that farmers are aware of various technological options for increasing income but in the prevailing price and institutional framework continue with certain technological options because of certainty in income. Similar region-specific constraints need to be addressed and opportunities for other crops may be created for a diversified commodity basket in the region.

Similarly, different regions have suitable land-based technological options; for example, temperate regions for apple, off-season vegetables and other high value horticultural items; the arid regions for growing organic millets; and the coastal hills for growing spices and rubber. Many of these crops are tradable in international markets and are widely referred as high value commodity following exports. Farmers are, therefore, subjected to fluctuations in farm harvest price primarily on account of volatility of international price of the commodity. They need institution to support them in withstanding price volatility in international markets. The farmer's share in the consumer's expenditure on many high value crops is insufficient. This requires analysis of the value chain of commodities. Such analysis often varies across regions.

Agricultural Technologies and Farm Practices

Agricultural technology is any change in farm input and output combinations that will increase farmers' return on a sustained basis. Technologies are of different types, based on the factors of production these are land, labour and water-saving and -intensive crops or enterprises. On the basis of end use, agricultural technologies are biochemical, mechanical and related to natural resource management. Though bio-chemical technology is relatively scale-neutral, mechanical technologies are scale-biased. Adoption of scale-biased mechanical technology requires supportive institutions. Many natural resource related technologies, however, require better price signal for adoption. Adoption of some of the resource conserving technologies, like drip irrigation, requires investment.

Most agricultural technologies are knowledge- and skill-intensive. For farmers to concentrate on agriculture to adopt technology it is important that a major share of their household income comes from agriculture. Against this background, how far technologies can improve the viability of small farms and the agricultural environment of the region is the subject of the present discussion.

Though many bio-chemical technologies are scale neutral, their inimical effects on natural resources are known. Conventional farming methods, which are more responsive to natural cycles and biological interactions, minimize the negative externalities of bio-chemical technologies. Conventional farming is referred to differently in different contexts, for example, as traditional agriculture or conservation agriculture. Organic agriculture, the formalized way of traditional agriculture, is often not encouraged, as productivity in organic farming is perceived to be less than in synthetic farming. Several researchers have, however, found that the transition from inorganic farming to organic farming reduces yield during the initial years, although in the later years the yield is higher in organic farming than in inorganic farming (IFAD 2005, Rai *et al.* 2006).

Box 12.1: Evaluation of Alternative Farm Practices

A comparative study assessed combination of eight organic packages with pure inorganic, pure organic, organic + inorganic and organic bio-fertilizers combinations for high value crops (maize-potato-onion) on the basis of productive efficiency, economic return, B: C ratio and profitability, at C S Azad University of Agriculture and Technology, Kanpur, for three consecutive years. The results showed that sixth treatment consisting of 1/3 recommended N through FYM + 1/3 through vermi composts + 1/3 through neem cake + Azobactor PSB provides highest yield at par to pure inorganic in terms of potato equivalent. But, in economic terms, treatment 5 (50% N through FYM + azobactor + rock phosphate +PSB) provides highest economic returns with the highest BC ratio. In the short run, inorganic treatment has been found to perform well in terms of profitability and return, whereas, in the long run, organic and organic + biofertilizers treatments perform well in terms of productivity, returns and profitability with sustainability of natural resources and safety of environment.

Source: Rai J., M.P. Yadav, U.S. Tiwari, A.K. Singh and B.P. Singh (2006).

The sustainability of an organic production system is often questioned on its ability to fulfil domestic demand for agricultural commodities, while bio-chemical technology has a proven record of achieving it for the country.[9] The present study, therefore, argues for technology that limits the use of synthetic chemicals and water to minimize the negative externalities of biochemical technology and simultaneously increase the natural cycle in production. This limits manipulation of the physical and chemical environment of the soil. At times, this includes strategizing crop-specific farming practices, such as the system of rice intensification (SRI).

The subsequent part of this sub-section discusses some farm practices that can minimize the damage from the use of bio-chemical technology.

Integrated Nutrient and Pest Management (INPM)

The integrated nutrient management system is based on the principle of complementarity between synthetic and organic fertilizers. Though farm yard manure (FYM) has been applied on the field since time immemorial, bio-fertilizers are carrier-based preparations containing effective strains of micro-organisms like bacteria, algae and fungi alone or their combinations, which can provide plant nutrients through microbial activity. On the basis of the type of nutrients provided by bio-fertilizers, they can be categorized as bio-fertilizers for nitrogen fixation, phosphate mobilization and organic matter decomposition.

Besides increasing the cost of production, chemical pesticides also affect human health, as pesticide residue enters the food chain. The chemical pesticides kill some natural enemies of pests. Biological or natural substitutes to pesticides target selected insect pests and benefit plants/crops of economic importance and are referred to as bio-pesticides. For example, *Bacillus thuriengensis* causes disease in butterflies and moth pests; similarly a virus like *nuclear polyhedrosis* causes disease in many insects/pests. However, biological substitutes to pesticides are often not sufficient; therefore bio-pesticides are being used with selected chemical pesticides. Such usage is referred to as integrated pest management (IPM). Integrated pest management relies on natural mortality factors and on weather and crop management, and seeks control tactics that disrupts these factors as little as possible. Besides the advantages of bio-fertilizers and bio-pesticides in integrated nutrient and pest management (INPM), plants are also reported to have benefited as their natural strength is improved. The adoption of such practice, however, requires the patience of farmers. At times, IPM requires the disaggregate production of many bio-pesticides on farm by employing family labour. The adoption of IPM, thus, increases farmers return by employing family labour on small farms. This also has a beneficial impact on the environment (Ramarao *et al.,* 2007). Besides INPM, there are also homoeopathic ways of producing healthy plants by avoiding the use of synthetic chemicals for plants (Details in Box 12.2).

There are many resource-conserving technologies. The System of Rice Intensification (SRI) is one. The SRI attempts to utilize the full genetic potential of a variety by adopting knowledge-intensive agricultural practices that optimize the use of water and other similarly scarce resources (Uphoff, 2002). As compared to conventional farming, the yield under SRI is reported higher with lesser cost of input; therefore the return from crops is significantly higher.

Box 12.2: Homoeopathy for Healthy Farm Produce

In Allahabad, Swami Paramanand, a farm leader, have been practising homoeopathic cultivation since last five years with great success. The farm leader has formulated six mixtures—Homoeo Amrit, Sanjeevani, Rakshak, Sudha, Jeevan and Vardaan; while the first two eliminate the need for DAP and urea, others protect against a variety of plant diseases and pests. Since seeing is believing, over 400 farmers have shifted from chemicals to homoeopathic agriculture in the region. Farmers are finding these mixtures economical, effective and healthy. The homoeopathically grown potato is also reported to have a longer shelf life.

Source: *Hindustan Times*, July 17th, 2007

Technologies that Conserve Water and Other Resources

The benefits of resource-conserving technologies in agriculture can be understood from the fact that agriculture consumes almost 70 per cent of the country's water and a similar proportion of other resources. The adoption of many resource-conserving technologies depends on factor prices and factor price often depends on our priority. Concern for food security in a developing, democratic country like India dominates other problems like food safety and natural resource degradation. Agricultural technology is often evaluated based on productivity per unit of land, as this increases commodity supply at the macro-level. However, it does not necessarily increase farmers' return at the micro-level. The evaluation criterion has in fact replaced much of the wisdom of conventional farming and encouraged resource intensive practices in agriculture. Few activities highlight the need for combining the traditional wisdom of conventional agriculture with the recent research in agriculture.

The IPM method and SRI are some examples of low external input sustainable agriculture (LEISA), which increases farmers' return but is knowledge-intensive and requires trained manpower. Agriculture, therefore, requires farmers' attention. Adoption of a technology requires key investment and incentives. In this regard, a suitable price environment for specific technologies is desired.

Institutions for Adoption of Technology

Institutions that help small farmers to adopt technologies are the ones which increase farmers' assured return on a sustainable basis. Institutions that decrease farmers cost of hiring lumpy goods and increase farmers' share in consumers' expenditure are important. For farmers to adopt agricultural technologies, it is important that a significant part of the household income comes from agriculture. The study also assumes that various allied activities have limited potential to enhance farmers' income. This requires that the production unit or the size of operational holding is not too small. The size of holding that provides farmer's sufficient income to ward off poverty and subsequently invest in their farm varies across regions. The NSS survey of farmers reveals that farms under two hectares do not provide adequate return (NSS, 2005). The subsequent section of this chapter discusses institutions that decrease farmers cost, increases their return and helps them obtain a reasonable size of operational holding of cultivable land.

Institution for Security of Tenure and Size of Land

The agriculture Census shows that all operational holdings barring marginal holdings, have started decreasing after 1991. Like marginal farmers, sub-marginal farmers (with land holdings between 0.01 to 0.4 hectare) also continue to increase in number and proportion.[10] The bulk of these holdings is not viable on their own and requires off-farm income opportunities for sustenance. But a dearth of such opportunities in their vicinity encourages farmers to leave their land under certain sub-optimal arrangements and migrate to distant places for their sustenance. The arrangements are sub-optimal as these deprive farmer of some level of certainty in land tenure. Certainty in tenure is important for farmers to adopt better technology on their farms. Therefore, a flexible land market, which provides certainty to tenure and protects the interests of both lessors and lessees is desirable. The flexible land market may require computerization of land records in states. The above findings show that farm mechanization is important in the dearth of supply of rural labour, and that the consolidation of land holdings is important in the context of mechanization. Productivity on the land of marginal and sub-marginal farmers is important, as they account for a substantial part of the cultivable-land of the country.

Institutions for Organization of Agricultural Production and Technologies

Considering the agrarian structure of the country, cooperatives in one form or another has a definite role in agriculture and the development of the rural parts of the country. Though cooperatives have had limited success in certain parts of the country, this success needs to be replicated country wide. A new form of cooperative in the form of self-help groups (SHGs) has emerged. These consist of small groups of people with a common goal. The goal may be collection and utilization of farmers' resources, as in the micro-credit system to the water users' association formed to encourage efficient use of water in canal irrigated areas. Government institutions like the National Bank for Agriculture and Rural Development (NABARD) encourage the formation of SHGs to expand the micro-credit system, while State government institutions encourage the formation of water users associations (WUAs) to reduce operational and financial inefficiencies of canal irrigation.

Since agricultural technologies are often capital-intensive their adoption requires finance and credit. For asset-poor marginal and small farmers, access to institutional sources of credit is a problem. Much government assistance aims primarily to improve their financial access. An example is the Kisan Credit Card (KCC). In the light of the high cost of credit of non-banking financial institutions (NBFI), different institutional arrangements for credit disbursement are investigated.

Farmers are increasingly adopting mechanical technology as there is a dearth in the supply of reliable labour, however, many of such technologies are difficult to adopt efficiently on a small farm. This increases the cost of adoption of technology on a small farm and also the inefficiency in the use of capital at the macro-level. Researchers have suggested different ways to reduce the cost of technology adoption. Jha (2011) suggested the inclusion of custom hiring of mechanical goods in the services of one of the rural institutions (agri-clinics, village knowledge centres) accessible to farmers. Land holders are aggregating their production units informally to reduce the cost of hiring of lumpy goods. Land collectives represent an example of such activity. Many agricultural technologies are knowledge and skill-intensive; their adoption requires key labour elements at critical junctures (the agronomic stage of agricultural commodities). There are many ways to handle the issue of labour requirement; a labour bank is one (Anoop *et al.*, 2014).

Institution for Marketing Agricultural Production (Technological Options) and Technologies

The marketing of agricultural products is one of the major bottlenecks in small and scattered production units in India. Historically, farmers have grown commodities for which the market is assured. An example is the continuance of paddy-wheat crop rotation in the Northwest India despite the high return from sugarcane and similar other crops. Farmers in the urban vicinity have a higher share in consumers' expenditure on horticultural crops than farmers far away; that is why farmers in far-off rural areas do not grow horticultural crops. Farmers' share also fluctuates with the kind of institutions involved in marketing agricultural products. Prices of horticultural commodities are less assured than fine cereals. To increase farmers share in income and reduce price risk, various options exists in marketing product at a limited scale. These options vary from corporations to cooperatives and farmers' groups (formal and informal). Examples of farmers pooling their land to create agriculture-based enterprises primarily to increase their share in profit of agricultural enterprises are presented in Box 12.3.

Contract farming is one of the important options of assuring farmers a price. There are various models of contract farming. Contract farming based on a mutually beneficial relationship between the producer, consumer and the company has come out with different experiences from various parts of the country. Experiences reveal that small farmers are left out of many contract farming opportunities. A high transaction cost to the company while dealing with small farmer is the most important reason. In this context, a cooperative of small farmers and a subsequent contract with a company would be useful. The contract between a cooperative of farmers and a corporation has been continuing (Box 12.3).

Box 12.3: Examples of Corporate Farming with Cooperative Principles

The famous seed brand "*Pantnager ke beej*" is produced by Tarai Seed Development Corporation (TDC). Farmers of all landholding sizes in the area adjoining to the plant in the Udhamsingh Nagar of Uttarakhand have formed cooperatives. The TDC as per their requirement of seeds in the forthcoming season/year allocate quantity of seed to the head of cooperative societies. The society allocates the individual farmer a quota for supply of seed. Similar method is being practised by sugar processing units in many regions of the country including Uttar Pradesh (UP).

Farmers' return will increase if their share in the corporate profit of an agriculture-based industry increases. The Government is trying to increase farmers' participation in industry or processing units. An example is the amendment of cooperative laws to incorporate producers' associations in the Companies Act, 1956. A new provision has been introduced in the Act that combines the economic advantage of corporate entity with the social benefits of a cooperative. Government has recently mooted the concept of land-based (processing) corporate units, wherein farmers would be share-holders of a company by virtue of the land they own in the unit. Non-farmers would be members (share holders) of a company by virtue of the capital they invest. The community ownership of land, reported from certain parts of country, especially North East India facilitates this process.

Institutions for Farmers to Counter Uncertainty

With the increased importance of trade, the market risk of farmers is increasing. The marketing of many of agricultural commodities, unlike fine cereals, is dependent on private players. Their consideration for profit often makes the farmer's income uncertain. Institutional arrangements like contract farming are perceived to provide farmers a market and reduce market risk in agricultural commodities. At times, these arrangements fail as there is no such institution. Such failure increases market-induced risk further. In agricultural commodities other than those covered under the Price Support Scheme (PSS), the Government undertakes some periodic market intervention programme, but the cost of operating such programmes is significantly higher than the kind of support these schemes provide farmers (Jha, 2013).

The yield risk in agriculture, reported to have been decreasing during the 1980s (Jha, 1995), has not been reported to have declined afterwards. Risk in agriculture has increased recently for many reasons; one is eccentric weather induced by climate change. Farmer's vulnerability to risk increases further with the fragmentation of land holding and the decimation of non-agriculture enterprise in rural areas. All these require strengthening the social security measures for farmers in the unorganized sector (Jha, 2011).

The discussion above highlights the role of institutions in the organization of production and in assuring a better price for agricultural commodities. These are pivotal to increasing returns in agriculture and technology adoption. One form or another of collective (cooperative, farmers group, association etc.) is important to reduce the cost of acquiring technology. Such collectives also help farmers increase their share in the value chain of a commodity. Contract farming helps provide farmers an assured price. The above instances show a ray of hope for a sustained increase in farmers' income, and need to be replicated throughout the country.

Conclusion

Indian agriculture is widely referred to as smallholder agriculture, and the household income of small farmers depends on multiple sources. Despite these multiple sources of income, farmers make up more than one-third of the rural poor in India. The present study, therefore, assesses role of the technology and technological options in increasing farmers' income and improving the health of natural resources. The costs and return data from the CACP shows that the cultivation of traditional crops with the existing method of agricultural practices cannot lift farmers out of poverty, since a considerable percentage of land holdings are less than one hectare in size. This study, therefore, argues for a flexible land market that lets interested farmers increase their land size. Flexibility in the land market also provides the certainty of land tenure to farmers interested in adopting agricultural technologies. Adopting technologies on these farms is important, since small and marginal farmers now account for around half the cultivable land in the country.

The study argues that the cost of adopting lumpy technology will decrease if one form or another of cooperative arrangement of farmers exists. For adoption of land based technological options it is important that the farmer's share in consumer's expenditure for that commodity other than for fine cereals increases. Once market arrangement for these commodities involves small farmers and ensures that their return with such commodities increase, market-induced risk may also decrease. Biochemical technologies have been adopted in specific regions of the country that have certain kinds of infrastructure. Many externalities in agriculture are associated with biochemical technology. These can be reduced by moderating the intensity of agricultural practices. This study argues for an alternative farm practice that incorporates the traditional wisdom of conventional agriculture into existing resource-intensive agricultural practices. Technological practices suggested are largely in the LEISA category. Most agricultural technologies are skill- and knowledge-intensive. Therefore, adopting these requires that bulk of farmers' household income comes from agriculture, and it provides farmers decent living conditions. In the light of increasing vulnerability of farmers, government support to improve safety nets remains important.

Policy Recommendations

Agricultural policies should promote a production system that can help farmers increase their return in the long run. This study shows that many crops can become remunerative at the present regime of relative prices. Suitable incentives for a number of technological options may increase farmers' return and also the agricultural environment of the region. There should be a diversity of crops at the regional level, while individual farms continue to specialize in certain crops, as it can reduce many externalities of resource-intensive agriculture. Government must encourage farmers to develop integrated farming system.

The integrated farming system can increase farmers return and is being practised in certain regions, but the complementary relationship between these enterprises is deteriorating in the light of depleting community assets. Increasingly, farmers depend on off-farm income for their livelihood, but concentration on land-based activities is desired for farmers to adopt agricultural technologies. In addition to a focus on price, government investment in improving, rehabilitating and conserving natural resources like land and water is desired. Limited government resources in the form of a district plan, may address local level needs for agriculture; this may help in the convergence of government plans/scheme related to a sector. Adoption of many resource saving technologies depends on the price of natural resources and also the private cost of adoption of such technology. This also involves subsidies for a technology.

Incentives for farmers can also be in the form of supportive institutions that reduce the cost of adoption of technology and increase the farmer's share in consumer's expenditure for that commodity. Any group of farmers (SHG, association) that helps improve the organization of the production unit and reduce the influence of middleman is desirable. Supportive institutions may also require updation and computerization of land records to practically create land markets in different states of the country. A land market may provide some certainty in the tenurial system and encourage farmers to increase their scale to adopt agricultural technologies.

Appendix

Methods

The frequency of downside risk referred here as distress is measured by the probability of failure (PF). This is the probability of crop return and crop yield falling 10 per cent or more below their respective trend value. In the absence of significant trends (R square less than 0.60) for many crops during the reference year the probability of failure is calculated as 10 per cent or more below their value in the previous year. The magnitude of distress or downside risk is calculated by estimating the average negative deviation from the expected return or yield. The expected return or yield is the trend value of return and yield for individual crop.

Expected annual negative deviation from trend = Average absolute deviation in return/yield of individual crop * probability of shortfall in return/yield from trend

Average absolute deviation = sum $(XT-Xt)/N$

where, Xt is the actual value in t-th year, XT is the trend value in t-th year and N is the number of years in the period.

The Cost Concepts

While fixed and operational costs are relatively uniform throughout the relevant text, the CACP report presents various cost concepts specific to Indian farm conditions. The present study frequently uses costs and cost concepts as used in the CACP Report, and the same is presented below in appendices. The fixed cost in the report is rental value of owned land, rent paid for leased in land, land revenue, cess and taxes, depreciation in implements and farm buildings, interest in fixed capital. The operational costs consist of cost incurred in hiring casual labour, attached labour, family labour, hired bullock labour, hired machine labour; cost incurred in owned bullock labour and machine labour, also cost incurred in procuring seed, fertilizer, manure, insecticide, irrigation charges, working capital, interest on working capital and miscellaneous cost paid out by farmers for the crop.

The cost C2 is often used to present total cost of cultivating the same commodities. The cost C2 consists of all actual expenses in cash and kind incurred in production by owners, interest on value on value of owned capital assets (excluding land), rental value of owned land (net of land revenue), rent paid for leased in land, and imputed value of family labour. The cost A2 often comparable to variable/operational cost consist of including costs incurred in leasing land, land revenue, interest on fixed capital and depreciation and excluding imputed value of family labour over thc operational cost.

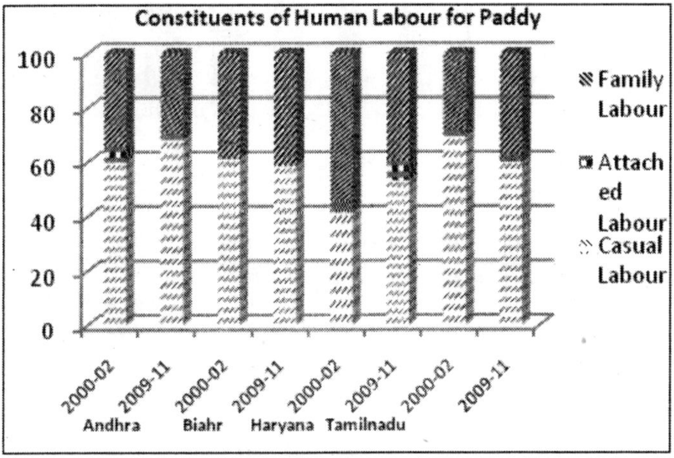

Fig. 12.1A: Constituents of Human Labour for Paddy.

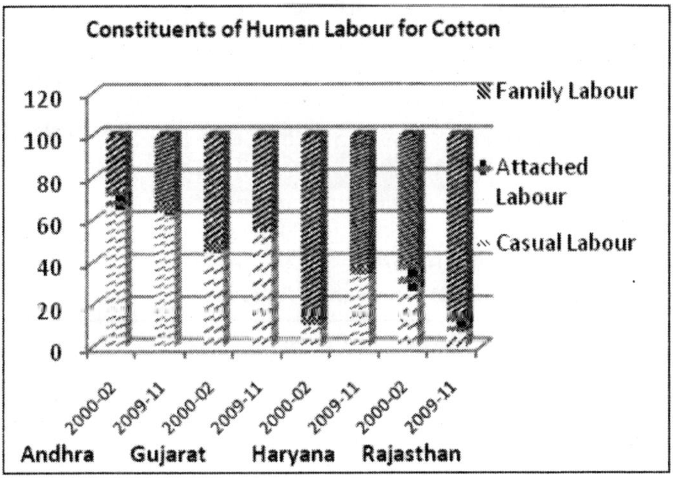

Fig. 12.2A: Constituents of Human Labour for Cotton

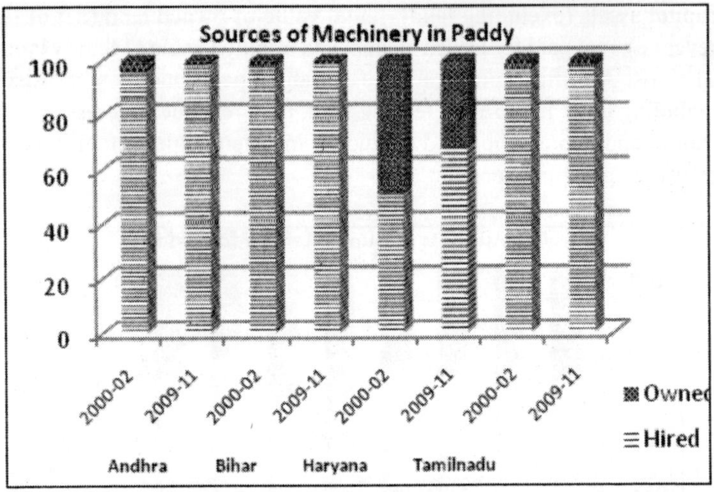

Fig. 12.3A: Sources of Machinery in Paddy

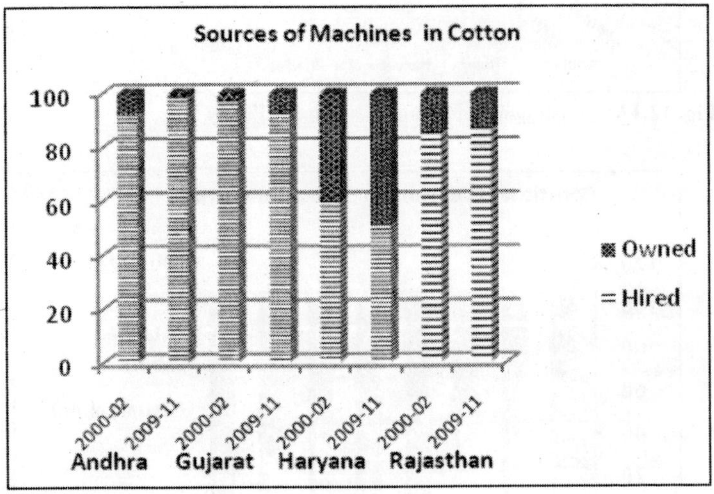

Fig. 12.4A: Sources of Machines in Cotton

Notes

1. *Source*: NSSO 2011-12.
2. Small and Marginal farmers constitute more than 85 per cent of total agricultural holdings in country. The situation assessment of farmers NSS report no. 497 states that average household return of these farmers is negative (*Source*: NSSO 2005).
3. Several issues of Economic Survey show scissor kind of relationship between subsidies and investment in agriculture (Government of India).
4. Several literatures including farmer status report states that significant part of farmer's income is coming from off-farm activities (*Source*: NSS 2005).
5. There are many crop-specific studies, parameters for such individual studies are often different; these studies are not comparable in dearth of suitable meta-analysis. The crop-specific technology based study often relates to primary information and such studies are often impressionistic, unlike the CACP report; it strengthens ideational conflicts.
6. The fixed costs, in addition to operational costs, consist of rental value of land, interest on fixed capital and depreciation on plants and buildings (Appendix).
7. The return over cost (C2) from crops with the national average of agricultural holding of farmer is less than salary of the lowest paid employee in the organized sector. Such farmer's return is less than the minimum wage based salary of unskilled casual workers in organized sector.
8. Farmers in the region reported that the demand for hybrid seeds is significant, since sunflower, being a cross pollinated crop needs to be replaced more frequently. To cater to this demand, many seed companies have emerged. Some of these companies often supply spurious seeds, which has dampened the spirit of farmers. This is evident with the decreasing productivity of sunflower in the region.
9. India was a net importer of agricultural commodities in the 1960s and demand for agricultural commodity in those years was met largely by extensive agriculture in the country. In the subsequent decades following the adoption of bio-chemical technology in agriculture self sufficiency to large extent was achieved in grains.
10. They account for more than two-third of total agriculture holdings and more than one fifth of cultivable land in the country (Government of India, 2014).

References

Anoop M., N. Ajjan and K.R. Ashok (2014). Institutional Interventions in Addressing Labour Scarcity—*A study on Labour Bank Initiatives in Thrissur*

District of Kerala Agricultural Economics Research Review, Vol. 27 (Conference Number), 2014, pp. 11-16.

Arrow, K., B. Bolin, R. Costanza, P. Dasgupta, C. Folke, C.S. Holling, B.O. Jansson, S. Levin, K.G. Maler, C. Perrings and D. Pimentel (1995). Economic Growth, Carrying Capacity and the Environment, *Science*, 268: 520-521.

Chahal S.S., Kataria P., Abbott S. and Gill, B.S. (2014). Role of Cooperatives in Institutionalization of Customs hiring Services in Punjab, *Agricultural Economics Research Review*, Vol. 27 (Conference Number) 2014, pp. 103-110.

Government of India (2014). *Several Issues of Economic Survey*, Ministry of Finance, Government of India, New Delhi.

Government of India (2008). *Agricultural Statistic at a Glance*, Directorate of Economics and Statistics, Ministry of Agriculture, Government of India, New Delhi.

Government of India (2013). *Several issues of the Cost of Cultivation Report*, Ministry of Agriculture, Government of India, New Delhi.

Government of India (2014). All India Report on Number and Area of Operational Holdings, Agriculture Census Division, Ministry of Agriculture, Government of India, New Delhi.

Haque, T. (1995). (ed.) *Small Farm Diversification: Problems and Prospect*, National Centre for Agricultural Economics and Policy, New Delhi.

International Fund for Agricultural Development (IFAD), (2005). *Organic Agriculture and Poverty Reduction in Asia: China and India Focus*, Final Draft, March.

Jaiswal, Pankaj (2007). Homoeopathy for Healthy farm Produce, *Hindustan Times*, July 17, 2007 (http://hpathy.com/homeopathy-papers/agrohomeopathy-tidbits).

Jha, Brajesh (2013). *Evaluation of Price Support Scheme for Pulse, Oilseeds, Cotton and Copra in India*, A seminar of stakeholders including Ministry of Agriculture (Government of India) held on May 25th, 2013 in Institute of Economic Growth, New Delhi.

Jha, Brajesh (2011). *Policies for Increasing Non-farm Employment for Farm Households in India*, IEG Working Paper No. 310.

Jha, Brajesh (2007). Agricultural Wages in India—A State Level Analysis, *Indian Journal of Labour Economics*, 50(2): 293-302.

Jha, Brajesh (2000). Implications of Intensive Agriculture on Soil and Groundwater Resources, *Indian Journal of Agricultural Economics*, Vol. 55 (2): 182-193.

Jha, Brajesh (2000) Towards Measuring Comparative Advantage of Agricultural Commodities in India. *Agricultural Economics Research Review*, 13(20): 159-168.

Jha, Brajesh (1995). The trade off between Return and Risk in farm planning, MOTAD and Target MOTAD approach. *Indian Journal of Agricultural Economics*, 50(2): 193-199.

Jha, Brajesh. (1995) Growth and Instability in Agriculture Associated with New Technology: District Level Evidences, *Agricultural Situation in India*, Vol. 49(7): 517-524.

Joshi, P.K., A. Gulati and Ralph Cummings Jr. (2007) (ed). *Agricultural Diversification and Smallholders in South Asia*, New Delhi: Academic Foundation.

National Sample Survey Organisation (NSSO, 2005) Income Expenditure and Productive Assets of Farm Households 2003, *NSS Report No. 497*, New Delhi.

_______ NSSO 2011. Employment and Unemployment Situation in India, 2009-10, NSS 66th Round (July 2009-June 2010), *NSS Report no. 537* (66/10/1), New Delhi.

Rai, J., M.P. Yadav, U.S. Tiwari, A.K. Singh and B.P. Singh (2006). *An Economic Analysis of Organic Farming Packages for System-based High-value Crops*, paper presented in the 18th Conference of AERA on Sustainable Agriculture and Rural Livelihoods held in Pantnagar (Uttaranchal, India).

Ramarao, C.A., M. Srinivasa Rao, and Y.V.R. Reddy (2007). An Economic Analysis of Adoption of Integrated Pest Management in Cotton, Groundnut, Pigeonpea, an unpublished report submitted to Indian Council of Agricultural Research (ICAR), New Delhi.

Thakur, D.S. and K.D. Sharma (2005). Organic Farming for Sustainable Agriculture and Metting the Challenges of Food Security in the twentyfirst Century: An Economic Analysis, *Indian Journal of Agricultural Economics*, Vol. 60(2): 205-219.

Uphoff, Norman, E.C.M. Fernandes, Yuanlongping, Peng Jiming, S. Rafaralahy and J. Rabenandrasana (2002). (eds.) *Assessment of the System of Rice Intensification: Proceedings of an International Conference*, Sanya, China, April 1-4, 2000. Ithaca, NY: Cornell International Institute for Food, Agriculture and Development (http://ciifad.cornell.edu/sri/).

13

Agrarian Distress and Contract Farming

Implications for Small Farmers of Punjab

Nivedita Sharma

Introduction

Increasing incidences of farmers' suicides from one of the agriculturally developed regions which had strong peasant movements in the past and was a vanguard in the Green Revolution evinces a paradoxical situation (Gill and Singhal, 1984; Suri, 2006). Farmers' suicides represent the extreme manifestation of agrarian crisis (Ghosh, 2005; Suri, 2006). The crisis in the state emerged as an aftermath of the Green Revolution with the dominance of wheat and paddy (Sharma & Singh, 2014). The father of Green Revolution in India, M.S. Swaminathan remarked it as a *Green Revolution* for the intensive application of inputs (Swaminathan, 2007). Intensification of inputs led to deterioration of soil and water resources, farm income and profitability, depeasantization and farmers' suicides (Chand, 1999; Government of Punjab 2013; Sharma and Singh, 2013). The agricultural growth rate of Punjab, which was as high as 5.15 per cent in 1981, began to register a decline from 1990s with 2.16 per cent in 1991 to 1.81 per cent in 2011-12.[1] Further, as per the recent estimates of the 70th round of NSSO, 53.2 per cent of the agricultural households in the state are in debt. Although, there is no consolidated data on the estimates of farmers' suicides, a number of studies (Punjab State Farmers' Commission, 2006; Sidhu *et al.*, 2011; Gill, 2012; Iyer and Arora, 2012) through field reports present the dismal picture of the state. A study carried out jointly by the State Government commissioned survey conducted by three universities of the State – Panjabi University, Patiala; Punjab

Agricultural University, Ludhiana; and Guru Nanak Dev University, Amritsar reported an estimated 5,000 suicides by farmers and farm labourers during 2000 to 2010. Around 60 per cent of the farmers were found to be indebted, majority of them being small and marginal farmers.[2]. It clearly indicates that agrarian distress has worst affected small and marginal farmers. Needless to say, conditions of distress in the farming community are prevalent in the State. A committee headed by Dr. S.S. Johl, popularly known as the *Johl Committee*, in 1986 suggested that 20 per cent of the area under wheat and paddy must be shifted to fruits, vegetables and fodder cultivation (Government of Punjab, 2004; Kumar, 2006). With the persistent crisis, another recommendation was given in 2002 by the Johl Committee, that one million hectares under wheat and rice be shifted to other crops like oilseeds and pulses, which have less water requirement and are ecologically and soil friendly. Farmers who diversify should be compensated by giving Rs. 12,500 per hectare (Kumar, 2006; Witsoe, 2006). Given the above recommendations, the Punjab Government introduced the contract farming programme in 2002-03 as a panacea for the prolonged malaise in the agricultural sector. However, the programme failed to accomplish the objective of diversification and was a big failure. It was finally officially declared "closed" in September 2012 (Sharma and Singh, 2014). However, the practice of contract farming is still prevalent in the state with many agri-business companies contracting directly with the farmers.

The recent corporatization of agriculture with many agribusiness companies making their foray into agriculture has raised the concerns for the small and marginal farmers in developing countries like India. The most exploitative form of this corporatization is corporate farming which is differentiated from contract farming in the second section of the chapter. The concerns for small and marginal farmers are obvious with many land grabbing experiences and marginalization of these farmers from the global value chains. These experiences from the literature with evidences from the field are explicated in the third section of the chapter. The last section concludes the discussion.

Juxtaposing Contract Farming and Corporate Farming

Contract farming is an arrangement for the production and marketing of agricultural produce in which farmers and firms (mainly agro-processing) enter into advance contracts to substitute/replace firms' production by purchasing produce of pre-determined quality and quantity at a pre-determined price and time, with provision of certain services like inputs, technical assistance etc. to the farmers (Minot, 1986; Glover and Kusterer, 1990; Singh, 2002). Corporate farming on the other hand, involves direct ownership or leasing in of farmland by corporate houses in order to produce for their requirements or for the open markets (Singh, 2006). Since, corporate farming is not legal in India, contract farming serves as a best alternative for the business corporations to acquire right over the farm produce and meet its requirement of quality and timely raw material.

Even though corporate farming cannot be practiced at a large scale in India where the agriculture is dominated by small land holdings, several instances about it have been reported. Punjab has experienced corporate farming with Fieldsfresh Foods Pvt. Ltd. (a joint venture between Bharti Enterprises and U.S. based Del Monte Pacific) acquiring 300 acres of land on lease in Ladhowal near Ludhiana in 2004 on a 90 year lease at Rs. 300 per acre[3] (Mahaprashasta, 2013). Governments of Gujarat, Maharashtra, Madhya Pradesh and Andhra Pradesh have also allowed corporate farming for research and development, export oriented production purposes and encourage cultivation on waste lands. Also, the governments of Maharashtra and Gujarat have enacted laws to allow corporate farming on government waste-lands (up to 2,000 acres each) on 20 year lease to the agribusiness companies (Ghosh, 2013). Similarly, Government of Chhattisgarh has doled out 20 lakh ha of land for jatropha cultivation, Gujarat has also offered 2,000 acres of wasteland for horticultural and jatropha cultivation on a 20 year lease at a rate of Rs. 500 per acre to big corporate houses. Concerns have been raised about the denial of rights over wastelands to local people but giving these rights to business corporations. Incidences have also been reported where common land and pastures have been classified as wastelands and handed over to corporation houses (Ghosh, 2013). The corporatization of farming has mooted criticism due to its exploitative nature and profit making intentions creating distress in the farming community. However, corporatization through contract farming and corporate farming are two different stories. The former does not provide any direct land rights to the agribusiness companies, though critics have been vocal about the indirect control over land and production process (Singh, 2002), whereas the latter gives land rights to the companies through purchase or leasing in of land. Thus, corporate farming is more exploitative in approach and is not preferred in India.

However, even the impact of contract farming in the developing countries constitutes a debated ground (Masakure and Henson, 2005;

Winters *et al.*, 2005). Whilst, the proponents of agribusiness promotion highlight a *win-win* situation for both the contracting parties. Participation in contract farming may result in unprecedented opportunities for the farmers like access to markets, inputs, credit, technology, assured prices, increase in income and family employment and better risk management, thus, giving them an edge over the traditional marketing (Goldsmith, 1985; Key and Runsten, 1999; Eaton and Shepherd, 2001; Haque, 2003). Similarly, agribusiness firms also gain an indirect control over the production process with a right over the produce, enabling strategic sourcing of good quality and timely raw material (White, 1997; Eaton and Shepherd, 2001). Also, positive macro-level changes driven by contract farming are increase in food stock, foreign exchange, boost in employment and economic growth (Singh, 2002). Whereas, the critics of agribusiness promotion argue that the exploitative nature of contract farming displaces the decision making powers of the farmers (Wilson, 1986; Reardon and Barrett, 2000). The debate further connects to the future and sustenance of small farmers in the exploitative agro-industrialization. Contract farming may endanger food security by shifting farm production from basic food crops towards high-value crops (Little and Watts, 1994; Singh, 2002). It may also create socio-economic differentiation in the society by favouring large farmers and, thus, may create a dual agricultural development (Singh, 2002). It may encourage gendering of tasks in the fields excluding female labour from better remunerative jobs (Singh, 2003).

Hence, the drivers of this agrarian transformation are extremely varied with the interplay of numerous actors and factors resulting in varied outcomes defying generalizations. The recent concern about the marginalization of farmers from this institutional arrangement is picked as the theme of this chapter as these farmers are the hard hit victims of the agrarian distress. The discussion is carried out in the subsequent section.

Does Contract Farming has a Pro-small Farmers' Approach?: Probing the Type of Farmers Selected by the Agribusiness Companies

The growing internationalization of agriculture has raised a debate about the impact of contract farming on small-farmers. A firm's decision with whom to contract varies across the regions and crops exhibiting a heterogeneous pattern. Such a decision is definitely influenced by the objective of profit maximization that has basically induced these firms to shift to contract farming rather than depend on open markets. Several sceptics indicate the exclusion of the small farmers from the contractual arrangements (Glover and Kusterer, 1990; Runsten, 1992; Little and Watts, 1994; Key and Runsten, 1999; Kirsten and Sartorius, 2002; Singh, 2002; Kumar, 2006). In a scale dualistic farm economy, a company prefers to work with the large farmers. Also, contract farming is considered as an institutional arrangement that reduces transaction costs for companies by reducing dependence on the open markets (Key and Runsten, 1999). Individual contracts are definitely expensive for the companies due to *ex-ante*[4] and *ex-post*[5] transaction costs (Sáenz-Segura *et al.*, 2009). It is so likely that the companies are bound to be opportunistic and will prefer to work with those farmers who maximize their welfare.

However, participation of farmers is not only restricted by their land size, but also by the lack of possession of assets. Ownership of assets is described as a significant indicator of contract participation in the studies of Warning and Key (2002); Winters *et al.* (2005); Boughton *et al.* (2007) and Miyata *et al.* (2009). Despite the recognition in the recent studies that a key indicator of contract participation is possession of assets by the farmers, the process of empirical determination of this issue still remains an unexplored area in the state. The studies on the contract farming in Punjab, in particular, assess the role of small farmers' participation from the aspect of land size classification. Disappointingly, there is no study in the literature that captures the asset position of the contract and non-contract farmers in Punjab. Therefore, there is a need to study the impact of contract farming on the farmers' welfare by taking into account the non-random selection of farmers.

An analysis was carried out with the data based on primary research of three districts of Punjab, namely Ludhiana, Jalandhar and Fatehgarh Sahib carried out during June 2012 to July 2012. The selection of the districts was based on the proportionate area under contract farming to the Gross Cropped Area (GCA) on the information based on Statistical Abstract of Punjab, 2010. The districts were divided into low, medium and high share of contract farming to GCA and further, area under various crops was observed as a selection criterion. Potato and Basmati rice were found to be the two major crops contracted in the state. Based on these crops selection, Jalandhar, Ludhiana and Fatehgarh Sahib were selected from high, medium and low categories respectively. A total of 300 contract and 150 non-contract farmers were interviewed. A list of contracting farmers for each crop was obtained from the contracting companies during the period of survey. From the list, the sampled contract farmers were randomly selected for the interview. The study employs wealth index (WI) (*vide* Appendix A1) to capture economic differentiation between the contract and the non-contract farmers and for determining the factors for participation in contract farming, the study uses Probit model (*vide* Appendix A2).

Results and Discussion: A firm may select a farmer in the wake of reduction of transaction costs or a farmer may self-select into a contractual arrangement for the benefits accruing from it. In either the case, differentiation is likely to occur between the contract and non-contract farmers (Winters *et al.*, 2005). These differences needs to be accounted which will determine if the participation in contract farming is a self-selection of the farmers or it is a non-random selection by the companies. An insight into these issues will also help in understanding the status of small farmers in the contractual arrangements. There may be several observable characteristics, such as, education, age, household size, possession of machinery etc. that may affect the contract farming participation. It is essential to look at these basic demographic and farm characteristics of the contract and the non-contract farmers.

Table 13.1 show that the contract and non-contract farmers are

Table 13.1: Description of characteristics of the farmers

Characteristics	*Test of difference*	*Contract farmers*			*Non-contract farmers*		
		Potato	*Basmati rice*	*Overall*	*Potato*	*Basmati rice*	*Overall*
Average age (yrs.)	0.278^{ns}	43.53	46.19	44.86	42.91	45.82	44.37
Average household Size	-0.937^{ns}	7.00	5.59	6.29	6.71	7.02	6.87
Average number of education years	3.79***	13.87	13.04	13.46	12.37	11.16	11.76
Average own area (acres)	3.47***	14.85	12.34	13.59	10.58	8.79	9.69
Average leased in area (acres)	3.70***	30.69	22.05	26.37	15.67	12.24	13.95
Average operational area (acres)	4.48***	45.48	34.74	40.11	26.13	21.00	23.56
Average number of agricultural machinery	4.241***	12.74	10.87	11.81	9.10	8.78	8.94
Average number of livestock	1.056ns	8.63	7.36	7.99	7.80	6.94	7.37

*** Significant at the 1% level; ns is non-significant, test of difference between the contract and the non-contract farmers using t-test (two-tailed) as appropriate.

Source: Field Survey, 2012.

homogeneous in terms of age, household size and average number of livestock. However, the participants differ significantly from the non-participants in terms of average number of education years. The average nunber of education years for the contract farmers is 13.46 years, while for the non-contract farmers, it is 11.76 years. Average area owned by a contract farmer (13.59 acres) is larger than that of a non-contract farmer (9.69 acres). A contract farmer, on an average, leases in 26.37 acres and a non-contract farmer leases in 13.95 acres of land. Average size of the operational land holding of a contract farmer is 40.11 acres, while that of a non-contract farmer is 23.56 acres. Significant difference between the two categories in terms of agricultural machinery is also observed. Table 13.1 shows that on an average, a contract farmer holds 11.81 number of agricultural machinery, whereas, the corresponding number of machines possessed by a non-contract farmer is 8.94. A perusal of the data given in the table reveals that participation in contract farming is not randomly distributed. It is influenced by education, land owned and operated, and agricultural machinery of a farmer. Further, a look at the land distribution pattern of the sampled farmers reveals (Table 13.2) that only one per cent of the sampled potato contract farmers are small farmers. Whereas, for Basmati rice, 7.7 per cent of the sampled contract farmers are marginal and small. Large farmers constitute a significant proportion of the contract population with a share of 55.80 per cent and 49 per cent for potato and Basmati rice respectively.

Table 13.2: Percentage Distribution of the Sampled Contract Farmers by Different Operational Land size

Crops	*Marginal (less than 2.5 acres)*	*Small (2.5-5 acres)*	*Semi-medium (5-10 acres)*	*Medium (10-25 acres)*	*Large (greater than 25 acres)*
Potato	0.00	1.00	5.10	38.10	55.80
Basmati rice	1.90	5.80	12.50	30.80	49.00

Source: Field Survey, 2012.

The impact of contract farming on rural development largely depends on the type of farmers contracted. If the farmers selected are wealthier, then poor farmers are less likely to draw benefits from these contractual arrangements (Warning and Key, 2002). Exclusion of resource poor farmers from contract farming is thus, likely to foster income and asset inequalities in the society (Goldsmith, 1985; Watts, 1992; Key and Runsten, 1999; Bellemare, 2012). This is likely to occur as the large farmer is able to invest more in the productivity-enhancing capital as a consequence of increased incomes where as, a small farmer is just able to meet his basic needs. This observation about contract farming is however, largely based on anecdotal literature and lacks an empirical examination. Therefore, WI is used to capture the economic differentiation between the contract and the non-contract farmers to confirm the proposition for the region.

The wealth position of a household determines its risk managerial capability, social status in the society and participation in agricultural markets like contract farming (Winters *et al.*, 2005; World Bank, 2007). WI, used to assess the standard of living of masses, has been calculated to find out difference in possession of wealth between the contract and the non-contract farmers. Information on various significant indicators of wealth was obtained during the field research. The factor scores (weights) for the asset variables obtained through PCA were obtained and are presented in Table 13.3.

Table 13.3: Factor scores for different asset variables

Asset variables	*PCA factor scores*
Submersible pump	0.2615
Tractor	0.3925
Trolley	0.3708
Combine	0.1651
Irrigation generator	0.3513
Rotavator	0.2785
Planter	0.316
Digger	0.3304
Cattle	0.123
Air conditioner	0.2562
Car	0.1963
Number of rooms in the house	0.1572
Area owned by the farmer	0.2424

Source: Own estimation based on Field Survey, 2012.

The table above demonstrates that tractor, trolly, planter, digger and irrigation generator have high factor scores and are strong indicator about the possession of other assets by the farmers. Low factor scores indicate equal possession or no possession of an asset by the farmers implying even distribution in the either case. For instance, combine was possessed by very few farmers and car and cattle were possessed by almost all the farmers generating low factor scores for these variables. Further, these factor scores were used to calculate WI as a summation of product of values of different asset variables with their respective factor scores (*see* Table 13.4).

Table 13.4: Mean wealth index of contract and non-contract farmers

Quintiles	*Mean WI*		*Maximum WI score*		*Standard deviation*	
	Contract farmers	*Non-contract farmers*	*Contract Farmers*	*Non-contract farmers*	*Contract farmers*	*Non-contract farmers*
Q1 (Poorest)	0.69	0.52	5.01	4.26	0.87	0.89
Q2 (Second)	0.93	0.84	6.73	5.55	0.55	0.35
Q3 (Third)	1.28	0.83	8.92	6.72	0.63	0.35
Q4 (Fourth)	1.16	1.02	11.40	8.36	0.93	0.54
Q5 (Richest)	2.46	1.57	35.79	33.54	4.84	5.50
Overall	1.31	0.96	35.79	33.54	4.91	3.91

Source: Own estimation based on Field Survey, 2012.

The mean WI is higher for the contract farmers in all the quintiles than their counterpart. The overall mean WI for the contract farmers is 1.31, which is higher than WI of the non-contract farmers (0.96). Also, WI is much higher for the fifth quintile of the contract farmers (2.46) than the non-contract farmers (1.57), reflecting wide disparity in the rich farmers of both the categories. Standard deviations of WI are higher for the contract farmers in all the quintiles. Also, it is highest in the richest category (both contract and non-contract farmers), reflecting wide variations in possession of wealth.

Thus, it can be inferred from the discussion that the contract farmers are better resource endowed than the non-contract farmers as indicated by the WI. Thus, clearly, contract farming participation is not a self-selection of farmers. It is a deliberate non-random selection of the firms that determines the type of participants. The impact of socio-economic factors on contract farming participation is also determined using Probit analysis.

Several studies have examined the issue of contract farming participation using explanatory variables as per the applicability of the study area. Taking reference from the literature and nature of data extracted from the field, Probit analysis is carried out and results are presented in Table 13.5. The dependent variable is dummy for contract participation (yes = 1, no = 0). The effect of the explanatory variables on the dependent variable used in our study is explicated below:

Age: It is hypothesized that age will positively affect contract farming participation, as, it proxies experience of farmer in the agriculture (Randela *et al.*, 2008).

Education: It is likely to affect contract farming participation positively, as, education improves the skills and managerial ability of a farmer (Cai *et al.*, 2008; Sharma, 2008).

WI: It represents the resource endowments of a farmer and, thus, will encourage contract farming participation.

Proportion of adults: It was calculated as a ratio of member in the age group 16-65 years[6] to the total members of a household. It proxies for family labour (Bagamba *et al.*, 2009; Yusuf *et al.*, 2011) and is expected to affect contract participation positively.

Loan Limit per acre: As per the field survey, loan limit per acre represents the credibility of a farmer in the region. The larger the operated land size, the larger is the loan limit of a farmer. Therefore, it is likely to have a positive impact on contract farming participation.

Off-farm income source: Farmers with off-farm income sources are expected to be better endowed with capital (Sharma, 2008) and, thus, will be preferred by the companies. Thus, it may affect contract farming participation positively.

Table 13.5 shows the results of the Probit regression analysis. It is evident from the findings that the WI has a positive and significant impact on contract farming participation. It implies that a better resource endowed farmer is preferred by the companies. Similarly, the proportion of adults

Table 13.5: Binomial Probit estimation Participation in contract farming as a dependent variable (yes-1, no-0)

Predictor (s)	*Coefficient*	*S.E.*
Age (years)	0.3618	0.0379
Age square	–0.0003	0.00040
No. of education years	0.4667	0.05383
Education years square	0.0003	0.00245
WI	0.0455***	0.01638
Prop of adults	1.6640***	0.54170
Loan limit per acre	5.63e-06**	2.53e-06
Off-farm income source	0.1359	0.19952
Constant	–2.7649***	1.05257
Number of observations = 450		
Chi-square 46.41***		

Source: Own estimation based on Field Survey, 2012.

and loan limit per acre are also found to have significant influence on the contract farming participation. Other variables like age, education, off-farm income sources turn out to be insignificant in causing variations in contract farming participation. This indicates that the agribusiness companies in the state are mainly considering economic indicators in the selection of farmers that reflect the ability of the farmer to invest ample resources in contract farming. Nonetheless, the exclusion of small farmers with less asset base from contract farming is based on the practical reasons of transaction cost approach. Such exclusions of small farmers from the market models also reflect policy bias resulting in corporatization of the farm sector.

Although, the findings of this study cannot be generalized for the contract farming model as a whole but these findings are in line with the extant literature on Punjab (Singh, 2002; Asokan and Singh, 2006; Dhillon and Singh, 2006; Kumar, 2006; Sharma, 2008) that indicates an inclination of the agri-business companies towards medium and large farmers. Therefore, at least for the state like Punjab, which is popular for làrge and wealthy farmers, it can be inferred that the companies marginalize the small farmers. But a pro-small farmer approach is also significant for Punjab, where nearly 34 per cent of the land holdings are constituted by small and marginal farmers (Government of Punjab, 2012).

There are also cases, such as, Key and Runsten (1999) for Mexico; Warning and Key (2002) for Senegal; Birthal *et al.* (2005) for India; and Miyata *et al.* (2009) for China that present the cases of the inclusion of small farmers in the contract farming schemes. Suffice to claim that the evidences on the inclusion or exclusion of small farmers from the remunerative value chains are mixed, reflecting a theoretical ambiguity. The ill-experiences of contract farming causing exploitation of farmers through land grabbing in Andhra Pradesh (Ramakrishnan, 2013), reneging on prices and procurement, manipulation of quality standards (Witsoe, 2006; Chandrasekhar, 2013), environmental degradation as in Andhra Pradesh (Rao, 2013), delays in payments, reneging on contracts in situation of glut as done by Pepsi in Punjab and Wineries in Maharashtra (Panwar, 2013); failure of contract farming programme in Punjab due to indifferent attitude of the government (Witsoe, 2006; Sharma and Singh, 2014) point towards the cruel tactics of the corporate world. However, farmers also breach the contract by side selling their produce if the market prices are found to be higher, using the inputs provided by the firms on non-contracted crops and not following the production schedule prescribed by the companies (Barrett *et al.*, 2012).

Therefore, literature on contract farming is flooded with mixed experiences but definitely the benefits accruing to this institutional arrangement cannot be ignored. The success or failure of contract farming largely depends on how the contracts are managed in which government can play an effective role. The experience of Thailand in successful contract farming is attributed to the intrusive approach of the government determining prices, grading standards, extensive research and protecting the growers and companies from the competition (Benziger, 1996; Singh, 2005). The study of Herath and Weersink (2009) points out that the role of the government has been significant in protecting the interests of tea contract farmers in Sri Lanka through fixation of floor price. Similarly, Minten *et al.* (2009) accentuate the role of the government in initiating the growth of contract farming in Madagascar by establishing export processing zones and encouraging foreign direct investment through tax incentives. The role of the government has also been significant in Africa, as apparent in the estimates of Watts (1994), who claims that 70 per cent of the contract farming schemes are jointly owned by the government. These case studies are the reflection of the explicit role of the government in fostering the agrarian transformation process in the economies.

However, role of the government in Punjab in the proliferation of contract farming has been futile. The recent initiative through Punjab Contract Farming Act, 2013 also indicates lack of pragmatic approach by the government. The Act has included plethora of crops with wheat and rice in the list of the crops to be contracted. Wheat-paddy dominance, which is considered to be the major cause of agrarian distress in the farming community of the state, is not given due consideration before framing the Act. Setting up of commission like PCFC with similar powers accorded to Punjab Agro-Foodgrain Corporation (PAFC) in the contract farming programme is a sheer tautology. Settlement of disputes by the officer appointed by the government as done by PAFC also indicates redundant approach of the government. This may not actually win the confidence of the farmers as earlier experience of the contract farming programme did not inflict faith in the farmers towards PAFC for settlement of disputes. Farmers believed that PAFC favoured the companies and this prohibited farmers to approach PAFC for redressal of their grievances (Witsoc, 2006). It is also silent on critical issues of quality and price determination which is a major source of exploitation of farmers. The Act does not encourage the small farmers' participation in any way. Thus, the approach of the government does not appears to be pro-small farmer. In fact, it appears to be an action of haste taken in view of the earlier failed attempts.

For a contract farming to be a win-win situation for the either parties, Eaton and Shepherd (2001) point out some pre-conditions. These are ensured profitability for the either parties; suitable physical environment that ensures good yield, quality and profitability; better rural infrastructure like roads, transport, telecommunication services, power and water supplies; farmers' unrestricted access to land; availability of inputs; government support and efficient legal framework. Hence, it is wrong to claim that contract farming *per se* is faulty, it is the faulty implementation that brings out the worst outcomes.

Concluding Remarks

Despite the contested role of the contract farming, the benefits accruing from it for the farmers can be immense if it is implemented effectively. Contract farming, definitely, cannot be left on its spontaneous mechanisms as its monopsonistic tendency may render the contractual relation exploitative. The major reasons for such exploitative nature of contract farming are lack of legal adjudication and less interventionist approach of the government. Such approach of the Punjab Government is also evident in the present study from the discriminatory approach of the agri-business companies in preferring large and better resource endowed farmers. Exclusion of small farmers from modern marketing channels such as contract farming is a matter of concern as these farmers have been worst affected by the agrarian distress. The exclusion of small farmers from the global value chains implies that they can survive in the competitive market only with the cushion of the government's policies. Therefore, government intervention to limit the role of contract farming in facilitating agrarian transformation and cutting back on its dire consequences is imperative.

The biggest challenge for the sustenance of contract farming in the state is to find an optimal mechanism that conflates farmers' interests with that of firms and still maximize each entity's payoff. Since, contracting with too many small farmers increases transaction costs; 'group farming' by the small and marginal farmers may be encouraged through the policy support. Further, the companies can provide agricultural equipment required for the production of the crop on nominal rental values. This will not only encourage the participation of these asset poor farmers, but will also reduce their cost of production. Hence, the study concludes that contract farming can be helpful for the state's distressed agrarian economy, if the government plays a strategic role in encouraging small farmers' participation in the global value chains.

Appendix

A1: Wealth Index: In the past decade, development economists have increasingly advocated the use of assets to complement income and consumption-based measures of welfare and wealth in the developing countries (Filmer and Pritchett, 2001; Moser and Felton, 2007). Principal Component Analysis (PCA) has been widely used and recommended in number of studies to measure WI (Filmer and Pritchett, 2001; Rutstein and Johnson, 2004; Vyas and Kumaranayake, 2006). PCA is a multivariate statistical technique used for extracting from a set of variables those few orthogonal linear combinations of the variables that capture the common information most successfully (Vyas and Kumaranayake, 2006). From a set of variables-X_1 to X_n,

$$PC_1 = a_{11} X_1 + a_{12} X_2 + \dots\dots\dots\dots + a_{1n} X_n$$
$$PC_m = a_{m1} X_1 + a_{m2} X_2 + \dots\dots\dots + a_{mn} X_n$$

where a_{mn} represents the weight for the *m*th principal component (PC) and nth variable.

The components are ordered so that the first component (PC_1) explains the largest possible amount of variation in the original data, subject to the constraint that the sum of the squared weights equals one. Each component captures an additional dimension in the data, while explaining smaller proportions of the variation of the original variables (Vyas and Kumaranayake, 2006). PCA works best when the asset variables are correlated with a varied distribution of variables across cases (Vyas and Kumaranayake, 2006). Assets which are more unequally distributed between households are given more weight in PCA (McKenzie, 2005). Also, the factor score of anyone variable is related to how much information it provides about the other variables. For instance, if ownership of one type of asset is highly indicative of the ownership of the other assets, then it receives a positive coefficient. If ownership of an asset contains almost no information about what other assets the household owns, then it receives a coefficient near zero. And if ownership of an asset indicates that a household is likely to own few other assets, then it receives a negative coefficient (Moser and Felton, 2007). Correlation matrix was used to ensure that all data have equal weights. The orthogonal rotation was done by varimax option to obtain uncorrelated components. First PC is considered a measure of wealth (Vyas and Kumaranayake, 2006). Factor scores derived from PC_1 were thus, used to generate WI.

Thus, WI can be calculated as

$$W_i = A_{i1} F_{i1} + A_{i2} F_{i2} + \ldots + A_{ij} F_{ij} \quad \ldots(1)$$

where, W_i is the *i*th household's wealth index score ($i = 1, 2, 3\ldots n$), A_{ij} is the asset variable of *i*th household for *j* type of assets and F_{ij} is the factor score or weight for each asset variable derived from PC_1 (Moser and Felton, 2007).

Further, for an ease of interpretation and tabulation, quintiles are used. The quintile classification is based on the distribution of the household population. The cut points in the WI to establish quintiles are calculated by obtaining a weighted frequency distribution of the households. Weight is the product of the number of members of the household and the sampling weight of the household. The individuals are then ordered by the WI score, and the distribution is divided at the points that constitute the five 20-per cent sections. Further, the household score is recoded into the quintile variable so that each household member receives that household's quintile category (Rutstein and Johnson, 2004).

A2: Probit model: The dependent variable used in the model for the study has binary outcomes–a farmer will participate in contract farming or will not participate. There is a vector of explanatory variables X, affecting the outcome Y. To run the model, literature is borrowed from Greene (2003) and Hansen (2012).

The model takes the form,

$$P_r = (y_i = 1|x_i) = \phi\,(x''_i \beta)$$

where, P_r denotes the probability and ϕ is the Cumulative Density Function (CDF) of the standard normal distribution. This model is identical to the latent variable model,

$$y_i^* = x'_1\beta + e_i, \text{ where } e \sim N(0, 1).$$

$$y_i = \begin{cases} 1 & \text{if } y_i^* > 0 \\ 0 & \text{otherwise} \end{cases}$$

For then,

$$\begin{aligned} P_r = (y_1 = 1|x_i) &= P_r(y_i^* > 0|x_i) \\ &= P_r(x'_i\beta + e_i > 0|x_i) \\ &= P_r(e_i > -x'_i\beta\,|x_i) \\ &= 1 - F(-x'_i\beta) \\ &= F(x'_i\beta) \end{aligned}$$

The estimation is by maximum likelihood. To construct the likelihood, conditional distribution of an individual observation is required. In the binary choice model, y_i is conditionally Bernoulli[1] with $P_r(y=1)=p$ and $P_r(y=0)=1-p$, then density of y as,

$$f(y) = p^y (1-p)^{1-y}, y = 0, 1.$$

In the binary choice model, y_i is conditionally Bernoulli with $P_r = (y_i = 1|x_i) = p_i = F(x_i' \beta)$. Thus, conditional density is,

$$\begin{aligned} f(y_i)|x_i) &= p_i^{y_i} (1-p_i)^{1-y_i} \\ &= F(x_i'\beta)^{y_i}[(1-F(x_i'\beta)]^{1-y_i} \end{aligned}$$

Hence, the log-likelihood function is,

$$\begin{aligned} \log L(\beta) &= \sum_{i=1}^{n} \log f(y_i \mid x_i) \\ &= \sum_{i=1}^{n} \log \left\{ F(x_i'\beta)^{y_i} \left[1 - F(x_i'\beta)^{1-y_i} \right] \right\} \\ &= \sum_{i=1}^{n} \left\{ y_i \log F(x_i'\beta) + (1-y_i) \log\left[1 - F(x_i'\beta)\right] \right\} \\ &= \sum_{y_i=1} \log F(x_i'\beta) + \sum_{y_i=0} \log\left[1 - F(x_i'\beta)\right] \end{aligned}$$

The MLE is the value of $\hat{\beta}$ which maximizes log L (β). Standard errors and test statistics are computed by asymptotic approximations.

1. If y_i is Bernoulli, such that $P_r(y=1)=p$ and $P_r(y=0)=1-p$, then density of y can be written as, $f(y)=p^y(1-p)^{1-y}$, $y = 0,1$ (Hansen, 2012).

Notes

1. *Statistical Abstract of Punjab* (several issues).
2. http://www.tribuneindia.com/2012/20120527/main3.htm. Retrieved on October 14, 2014.
3. Market price was Rs. 35,000 per acre.
4. *Ex-ante* transaction costs arise before assigning the contract, such as, costs of searching, collecting information, drafting documents, negotiating, and safeguarding an agreement (Kutlu, 2012).
5. Ex-post transaction costs occur after assigning the contract, which includes costs of monitoring and enforcement (Kutlu, 2012).
6. The cost of cultivation surveys of India define adult as 16-59 years of age. However, farmers above 60 years of age were also found to be actively involved in agricultural activities in the field. Therefore, the age group is increased to 65 years.

References

Asokan, S.R., and Singh, G. (2006). *Contract Farming of Medicinal Herbs and Organic Crops in India*. CMA monograph No. 224, CMA, Ahmedabad: IIM.

Bagamba, F., Burger, K., and Kuyvenhoven, A. (2009). Determinant of smallholder farmer labour allocation decisions in Uganda. International Food Policy Research Institute (IFPRI). Available at www.ifpri.org/sites/default/files/publications/ifpridp00887. pdf Retrieved on April 17, 2014.

Barrett, C.B., Bachke, M.E., Bellemare, M.F., Michelson, H.C., Narayanan, S., and Walker, T.F. (2012). Smallholder participation in contract farming: Comparative evidence from five countries. *World Development*, 40(4), 715-730.

Bellemare, M.F. (2012). As you sow, so shall you reap: The welfare impacts of contract farming. *World Development*, 40(7), 1418-1434.

Benziger, V. (1996). Small fields, big money: Two successful programmes in helping small farmers make the transition to high value-added crops. *World Development*, 24(11), 1681-1693.

Birthal, P.S., Joshi, P.K., and Gulati, A. (2005). *Vertical Coordination in High Value Food Commodities: Implications for Smallholders*. Markets, Trade and Institutions Division Discussion Paper No. 85, International Food Policy Research Institute, Washington DC. Available at: http://www.ifpri.org/divs/mtid/dp/mtidp85.htm. Retrieved on March 25, 2014.

Boughton, D., Mather, D., Barrett, C.B., Benfica, R., Abdula, D., Tschirley, D., and Cunguara, B. (2007). Market participation by rural households in a low-income country: An asset-based approach applied to Mozambique. *Faith and Economics*, 50, 64-101.

Cai, J., Ung, L., Setboonsarng, S., and Leung, P. (2008). *Rice contract farming in Cambodia: Empowering farmers to move beyond the contract toward independence* (No. 109). ADB Institute Discussion Papers. Available at http://www.adbi.org/discussionpaper/2008/06/05/2582.rice.contract.farming.cambodia/ Retrieved on February 3, 2014.

Chand, R. (1999). Emerging crisis in Punjab agriculture: Severity and options for future. *Economic and Political Weekly, 34*(13), A2-A10.

Chandrasekhar, C.P. (2013). Locked into business. *Frontline* (July), 8-9.

Dhillon, S.S., and Singh, N. (2006). Contract Farming in Punjab: An Analysis of Problems, Challenges and Opportunities. *Pakistan Economic and Social Review,* XLIV (1), 19-38.

Eaton, C., and Shepherd, A. (2001). *Contract Farming: Partnerships for Growth* (No. 145). Food and Agriculture Organisation, *Services Bulletin* 145.

Filmer, D., and Pritchett, L.H. (2001). Estimating wealth effects without expenditure data—or tears: An application to educational enrollments in states of India. *Demography, 38*(1), 115-132.

Ghosh, J. (2005). The Political Economy of Farmers' Suicides in India. India International Centre, Lodi Estate, New Delhi. Available at http://www.macroscan. net/pdfs/freedom_hunger.pdf. Retrieved on October 10, 2014.

Ghosh, J. (2013). Recipe for Crisis. *Frontline* (July), 26-28.

Gill, A. (2012). Punjab peasantry: a question of life and debt. In R. S. Deshpande and S. Arora (eds.), *Agrarian Crisis and Farmer Suicides* (pp. 312-328). New Delhi: Sage Publications.

Gill, S.S. and Singhal, K.C. (1984). Farmers' agitation response to development crisis of agriculture. *Economic and Political Weekly*, XIX (40), 1728-1732.

Glover, D.J., and Kusterer, K. (1990). *Small Farmers, Big Business: Contract Farming and Rural Development*. New York: St. Martin's Press.

Goldsmith, A. (1985). The private sector and rural development: Can agribusiness help the small farmer? *World Development, 13*(10), 1125-1138.

Government of Punjab. (2004). *Human Development Report Punjab*. New Delhi: New Concept Information System Private Limited.

Government of Punjab. (2012). *Statistical Abstract of Punjab*. Chandigarh: Economic and Statistical Organisation.

Government of Punjab. (2013). *Agriculture Policy for Punjab*. Committee for Formulation of Agriculture Policy for Punjab State. Available at http://agripb.gov.in/ Retrieved on November 20, 2013.

Greene, W.H. (2003). *Econometric Analysis.* Pearson Education India.

Hansen, B.E. (2012). *Econometrics.* University of Wisconsin.

Haque, T. (2003). Reforms for agricultural growth and rural development. *Economic and Political Weekly*, XXXVIII (48), 5031-5033.

Herath, D. and Weersink, A. (2009). From plantations to smallholder production: The role of policy in the reorganization of the Sri Lankan tea sector. *World Development,* 37(11), 1759-1772.

Iyer, K.G. and Arora, S. (2012). Indebtedness and farmers' suicides. In R.S. Deshpande and S. Arora (eds.), *Agrarian Crisis and Farmer Suicides* (pp. 264-291). New Delhi: Sage Publications.

Key, N., and Runsten, D. (1999). Contract farming, smallholders, and rural development in Latin America: The organization of agro-processing firms and the scale of outgrower production. *World Development*, *27*(2), 381-401.

Kirsten, J., and Sartorius, K. (2002). Linking agri-business and small-scale farmers in developing countries: Is there a new role for contract farming? *Development Southern Africa*, *19*(4), 503-529.

Kumar, P. (2006). Contract farming through agri-business firms and state corporation: A case study in Punjab. *Economic and Political Weekly*, Vol. XLI (52), 5367-5375.

Kutlu, A.C. (2012). Outsourcing contracting strategies from supplier's side: A real options approach based on transaction costs. *Procedia-Social and Behavioural Sciences*, 58, 1601-1610.

Little, P.D., and Watts, M. (1994). *Living under contract: Contract farming and agrarian transformation in sub-Saharan Africa*. London: University of Wisconsin Press.

Mahaprashasta, A. (2013). Growing concern. *Frontline* (July), 18-23.

Masakure, O., and Henson, S. (2005). Why do small-scale producers choose to produce under contract? Lessons from non-traditional vegetable exports from Zimbabwe. *World Development*, 33(10), 1721-1733.

McKenzie, D.J. (2005). Measuring inequality with asset indicators. *Journal of Population Economics*, 18(2), 229-260.

Minot, N. (1986). *Contract Farming and Its Effect on Small Farmers in Less Developed Countries* (No. 54740). Michigan State University, Department of Agricultural, Food, and Resource Economics.

Miyata, S., Minot, N., and Hu, D. (2009). Impact of contract farming on income: Linking small farmers, packers, and supermarkets in China. *World Development*, 37(11), 1781-1790.

Moser, C.O., and Felton, A. (2007). *The Construction of an Asset Index Measuring Asset Accumulation in Ecuador*. Working Paper 87, Washington DC, USA: Chronic Poverty Research Centre.

Panwar, S. (2013). Towards Corporatocracy. *Frontline* (July), 13-15.

PSFC (2006). Suicides in rural Punjab. Punjab State Farmers' Commission and Institute for Development and Communication (IDC).

Ramakrishnan, V. (2013). Calling in the corporate. *Frontline* (July), 4-7.

Randela, R., Alemu, Z.G., and Groenewald, J.A. (2008). Factors enhancing market participation by small-scale cotton farmers. *Agrekon*, 47(4), 451-469.

Rao, A.P. (2013). Exploitative Models. *Frontline* (July), 10-12.

Reardon, T., and Barrett, C.B. (2000). Agroindustrialization, Globalization, and International Development: An overview of issues, patterns, and determinants. *Agricultural Economics*, 23(3), 195-205.

Runsten, D. (1992). *Transaction costs in Mexican fruit and vegetable contracting: Implications for association and participation.* Paper Presented at the XVIII International Congress of the Latin American Studies Association, Atlanta.

Rutstein, S.O., and Johnson, K. (2004). *DHS Comparative Reports: The DHS Wealth Index.* ORC Macro, Calverton, Maryland USA.

Sáenz-Segura, F., D'Haese M., and Schipper, R.A., (2009). A seasonal model of contracts between a monopsonistic processor and smallholder pepper producers in Costa Rica. *Agricultural Systems,* 103, 10-20.

Sharma, N. and Singh, S. (2013). Agricultural diversification and contract farming in Punjab. *Journal of Economic* and *Social Development, IX* (1), 38-51.

Sharma, N. and Singh, S. (2014). Agricultural diversification in Indian Punjab: An assessment of government intervention through contract farming. *Journal of Agricultural* and *Food Information,* 15(3), 191-213.

Sharma, V.P. (2008). India's agrarian crisis and corporate-led contract farming: Socio-economic implications for smallholder producers. *International Food and Agribusiness Management Review*, 11(4), 25-48.

Sidhu, R.S., Singh, S., and Bhullar, A.S. (2011). Farmers' suicides in Punjab: a census survey of the two most affected districts. *Economic and Political Weekly*, XLVI (26 and 27), 131-37.

Singh, S. (2002). Contracting out solutions: Political economy of contract farming in the Indian Punjab. *World Development*, 30(9), 1621-1638.

Singh, S. (2003). *Contract Farming in India: Impacts on Women and Child Workers.* International Institute for Environment and Development. Gatekeepers Series 111. Available at http://www.iied.org/pubs/pdfs/9281IIED.pdf Retrieved on November 21, 2013.

Singh, S. (2005b). Contract farming system in Thailand. *Economic and Political Weekly*, XL (53), 5578-5586.

Singh, S. (2006). *Corporate Farming in India: Is it must for agricultural Development?* Working Paper No. 2006-11-06. Ahmedabad: Indian Institute of Management.

Suri, K. (2006). Political economy of agrarian distress. *Economic and Political Weekly,* XLI (16), 1523-1529.

Swaminathan, M.S. (2007). *Agriculture cannot Wait: New Horizons in Indian Agriculture.* New Delhi: Academic Foundation, National Academy of Agricultural Sciences.

Vyas, S., and Kumaranayake, L. (2006). Constructing socio-economic status indices: How to use principal components analysis. *Health Policy and Planning*, 21(6), 459-468.

Warning, M., and Key, N. (2002). The social performance and distributional consequences of contract farming: An equilibrium analysis of the Arachide de Bouche Programme in Senegal. *World Development*, 30(2), 255-263.

Watts, M.J. (1992). Peasants and flexible accumulation in the Third World: Producing under contract. *Economic and Political Weekly*, XXVII (20), PE90-PE97.

White, B. (1997). Agroindustry and contract farmers in upland West Java. *The Journal of Peasant Studies*, 24(3), 100-136.

Wilson, J. (1986). The political economy of contract farming. *Review of Radical Political Economics*, 18(4), 47-70.

Winters, P., Simmons, P., and Patrick, I. (2005). Evaluation of a hybrid seed contract between smallholders and a multinational company in East Java, Indonesia. *The Journal of Development Studies*, 41(1), 62-89.

Witsoe, J. (2006). India's second Green Revolution? The socio-political implications of corporate-led agricultural growth. In D. Kapoor (Ed.), *India in transition: Economics* and *Politics of Change*. Philadelphia, PA: Centre for the Advanced Study in India.

World Bank. (2007). *Development and the Next Generation*. Washington DC: The International bank for Reconstruction and Development.

Yusuf, H.A., Omokore, D.F., Akinola, M.O., and Omolehin, R.A. (2011). Socio-economic characteristics influencing farm household participation in gandu system in Charanchi Local Government Area of Katsina State, Nigeria. *Journal of Agricultural Extension and Rural Development*, 3, 220-223.

14

Seeds, Soil and Society

Strengthening the Law of Return for Sustainable Livelihoods

Vaibhav Singh and *Kunwar Jalees*

I

Introduction

Green Revolution penetrated the agriculture of India in 1965-66. The State of Punjab became the epicentre of Green Revolution. However, 45 years after the beginning of Green Revolution in the State, the misery of Punjab is beyond comprehension and beyond the limited scope of human understanding. Later in our discussion, we will try to portray a clearer picture of the state of affairs in Punjab as diligently as possible. Green Revolution was declared successful according to deliberately chosen criteria that were not only ill-founded but also misleading. The propaganda of Green Revolution was marketed craftily, and Green Revolution was sold to or imposed upon the Indian farmers, the majority of whom have been small and marginal. Green Revolution was implemented with the aim of making India food independent and food sufficient, eliminating hunger and malnutrition from the society, and conferring prosperity upon the farmers and the nation. The effects of Green Revolution were further supplemented by the policies of economic liberalization that India adapted in 1991. Subsequent to liberalization, India witnessed the rise of nationally owned global multinational corporations, the "GDP" crossed the magical $1 trillion mark, and the middle class in India expanded. The world recognized India as an emerging global superpower. The magical recipe of combining Green Revolution with economic liberalization seems to have worked. However, four and a half decades after Green Revolution was started, India is dependent upon foreign countries for pulses and oil seeds. India has also the maximum absolute number of hungry and malnourished humans in the world and in the last 19 years India has witnessed 2,84,000 farmers' suicides the majority of which could be traced back to the chronic debt trap attributed to costly external inputs essentially required in the Green Revolution model of agriculture. Is this food sufficiency and is this economic growth?

In order to manipulate the success of Green Revolution, yields of a few selected crops were chosen out of thousands of different varieties of crops grown in India–Green Revolution increased the production of hybrid rice and wheat in India but completely destroyed the production of pulses, oil seeds, and other native food crops. Increase in productivity of rice and wheat, which was mainly related to the increase in acreage diverted to growing these crops, was wrongly projected as the supernatural phenomenon underlying the science of Green Revolution. On the contrary, Green Revolution is more a pseudoscience than an authentic science because it catastrophically assumes that agricultural ecosystems will eventually return to its equilibrium state, which is considered static, after application of external inputs. The proponents of Green Revolution intentionally ignore the fact that, like any other living ecosystems, agricultural ecosystems have an equilibrium state that is dynamic and agricultural ecosystems shift to a new equilibrium after application of external inputs. Yield per acre of a monoculture is measured as a benchmark criterion to reveal the efficiency of the agriculture systems. We, at Navdanya, conducted a study, Health Per Acre–Organic Solutions to Hunger and Malnutrition, to weed out the confusion that yield per acre data presents. Logically, it is unfair to compare the output of 1 acre of farmland used for conventional mono cropping of 5 quintals of maize with that of 1 acre of farmland used for organic mixed cropping of 4 quintals of maize, 2 quintals of radish, 100 bundles of mustard leaves, and 2 quintals of peas. If only maize is taken into consideration, conventional mono cropping seems to be more productive. However, the productivity picture changes if radish, mustard leaves, and peas are taken into consideration. To remove ambiguity from yield per acre data, we propose a more uniform comparison benchmark, one which measures nutrition produced per acre of farmland. For example, 5 quintals of maize produced per acre in conventional mono cropping and translated into production of 55.5 kg of protein, 331 kg of carbohydrate, and 18 kg of fat can be compared to 4 quintals of maize, 2 quintals of radish, 100 bundles of mustard leaves, and 2 quintals of peas produced per acre in organic mixed cropping and translated into production of 64.2 kg of protein, 304 kg of carbohydrate, and 17.2 kg of fat. Similar comparison can be made across all 24 nutrients including vitamins, minerals and trace elements.

Methodology

An initial study—Biodiversity Based Organic Farming: A New Paradigm for Food Security and Food Safety—was conducted at Navdanya (Shiva V. and P. Pandey, 2006) They adopted standard scientific procedures in the research which included preparing a questionnaire, pretesting, visiting site and interviewing farmers. Farmers from four different regions of the country were chosen and compared. Data was collected from 1 organic farmer and 1 conventional farmer from Sikkim, Northeast Himalayas, 3 organic farmers and 3 conventional farmers from Rajasthan, Northwest arid region, 5 organic farmers and 3 conventional farmers from Uttarakhand, Northern Himalayas, and 3 Navdanya farms, each utilized for such traditional mixed cropping as Baranaja, Navdanya and Septrashi, and 3 local conventional farms. Data was collected on the basis of social, cultural economic, and environmental aspects with special emphasis on farming inputs, seed availability, choice of fertilizer regime, duration of involvement in mixed cropping or homestead style of farming, on farm crop biodiversity, yield or output per acre and the trade practices prevalent in the region. Subsequently, the another study–Health Per Acre: Organic Solutions to Hunger and Malnutrition–was conducted at Navdanya (Shiva and Singh, 2011) in which the above data was segregated into twelve case studies and comparison was made between organic farms and conventional farms in these 12 case studies with each case study comparing nutrition produced per acre farmland and revenue generated per acre farmland in organic mixed cropping and in conventional mono cropping. Yield or output data collected in the initial study was consolidated, analysed and translated into nutrition per acre data. The analysis was further extended to estimate the revenue per acre. While quantitatively transforming yield per acre data into nutrition per acre data, we assumed that there is no qualitative difference between a unit output of convention food and a unit output of organic food realizing with clarity that our assumption is debatable. We used nutritive value of food items as given in a report published by Indian Council of Medical Research (ICMR), National Institute of Nutrition, Hyderabad—Nutritive Value of Indian Foods. In each of the twelve case studies, the yields of different crops produced in one acre of farmland in conventional mono cropping and in organic mixed cropping were broken down into the quantities of 24 nutrients produced per acre. The 24 nutrients compared are protein, carbohydrate, fat, carotene, thiamine, riboflavin, niacin, vit B6, folic acid, vit C, choline, calcium, iron, phosphorous, magnesium, sodium, potassium, chlorine, copper, manganese, molybdenum, zinc, chromium and sulphur. Yield per acre data was also converted to revenue per acre data assuming that farmers sold all their produce at prices that prevailed in February 2013. Nutrition produced per acre of farmland and revenue generated per acre of farmland across all the conventional and organic farms were systematically tabulated and compared.

Results

Yield per Acre data collected in the initial study, Biodiversity based organic farming: A New paradigm for food security and food safety, was first converted in the current study to estimate the nutrition produced per acre of farmland. Yield per acre data from 1 organic farmer and 1 conventional farmer in Sikkim converted into and compared for nutrition produced per acre farmland is presented as an example below:

Sample Case Study, Sikkim

Table 14.1: Yield per acre

Organic Mixed Cropping	*Conventional Mono Cropping*
Maize = 4 QT Radish = 2 QT Mustard leaves (*Saag*) = 100 bundles Peas = 2 QT	Maize = 5 QT
Total = 9 QT	Total = 5 QT

Table 14.2: Nutrition per acre

Macronutrients	*Organic Mixed Cropping*	*Conventional mono cropping*
Protein	64.2 kg	55.5 kg
Carbohydrate	304.0 kg	331 kg
Fat	17.2 kg	18 kg
Energy	16,22,000 kcal	17,10,000 kcal

(Contd.)

Table 14.2 (*Contd.*)

Vitamins	*Organic Mixed Cropping*	*Conventionalmono Cropping*
Carotene	3154 mg	450 mg
Thiamine	2330 mg	2100 mg
Riboflavin	460 mg	500 mg
Niacin	9800 mg	9000 mg
VIT B6	—	—
Folic Acid	80 mg	100 mg
Vit C	81,000 mg	0 mg
Choline	1,66,000 mg	0 mg

Minerals	*Organic Mixed Cropping*	*Conventional Mono Cropping*
Calcium	305 g	50 g
Iron	29.3 g	11.5 g
Phosphorous	1740 g	1740 g
Magnesium	626 g	695 g
Sodium	145.2 g	79.5 g
Potassium	1878 g	1430 g
Chlorine	172 g	165 g

Trace Elements	*Organic Mixed Cropping*	*Conventional Mono Cropping*
Copper	6420 mg	2050 mg
Manganese	3030 mg	2400 mg
Molybdenum	790 mg	190 mg
Zinc	14240 mg	14000 mg
Chromium	48 mg	20 mg
Sulphur	645000 mg	570000 mg

Average Nutrition Produced per Acre Across 12 Case Studies

The comparison shown above is made across 12 case studies each comparing nutrition produced per acre farmland in organic mixed cropping and in conventional mono cropping in different regions of the country. The next step in the study involved calculating average nutrition produced per acre farmland in organic mixed cropping and in conventional mono cropping; we simply took the average amount of nutrient produced across 12 case studies in organic mixed cropping and in conventional mono cropping. The Average Nutrition produced per acre of farmland across 12 case studies in organic mixed cropping and in conventional mono cropping is presented in the table below.

Table 14.3: Average Nutrition Produced per Acre

Macronutrients	*Organic Mixed Cropping*	*Conventional Mono Cropping*
Protein	240 kg	116 kg
Carbohydrate	833 kg	785 kg
Fat	66 kg	23 kg
Energy	4914270 kcal	3711475 kcal

Vitamins	*Organic Mixed Cropping*	*Conventionalmono Cropping*
Carotene	2919 mg	745 mg
Thiamine	6550 mg	3911 mg
Riboflavin	3179 mg	1685 mg
Niacin	31443 mg	28381 mg
Vit B6	821 mg	475 mg
Folic Acid	878 mg	328 mg
Vit C	24145 mg	36833 mg
Choline	680675 mg	537527 mg

Minerals	*Organic Mixed Cropping*	*Conventional Mono Cropping*
Calcium	2166 mg	731 mg
Iron	82 g	43 g
Phosphorous	5158 g	3117 g
Magnesium	1866 g	1496 g
Sodium	197 g.	158 g
Potassium	6076 g	3465 g
Chlorine	323 g	320 g

Trace elements	*Organic Mixed Cropping*	*Conventional Mono Cropping*
Copper	12591 mg	6101 mg
Manganese	25124 mg	15629 mg
Molybdenum	3694 mg	1077 mg
Zinc	43977 mg	26769 mg
Chromium	345 mg	157 mg
Sulphur	1.64 kg	1.303 kg

Finally, the yield per acre data from the initial study, Biodiversity Based Organic Farming: A New Paradigm for Food Security and Food Safety, was converted into Revenue Generated per acre farmland by assuming that farmers sold all their produce in the market at prices that prevailed in February 2013. The comparison of Revenue Generated per acre farmland between organic mixed cropping and conventional mono cropping is presented in the table below:

Case Study Number	*Revenue Generated in Organic Mixed Cropping (INR)*	*Revenue Generated in Conventional Mono Cropping (INR)*
Case Study 1	16000	11000
Case Study 2	54100	21600
Case Study 3	32220	18900
Case Study 4	42950	30800
Case Study 5	72350	42600
Case Study 6	36950	42600
Case Study 7	100100	19500
Case Study 8	46500	19500
Case Study 9	37700	32500
Case Study 10	121423	48092
Case Study 11	133385	36450
Case Study 12	100690	72900
Average Revenue Generated Across 12 Case studies	**66197**	**33037**

Underneath are some graphs (Fig. 14.1-Fig 14.8) that compare important nutrients of public health significance produced on an average per acre farmland in organic mixed cropping and in conventional mono cropping. The last three graphs (Fig. 14.9-Fig 14.11) show revenue generated per acre farmland in each case study and on an average.

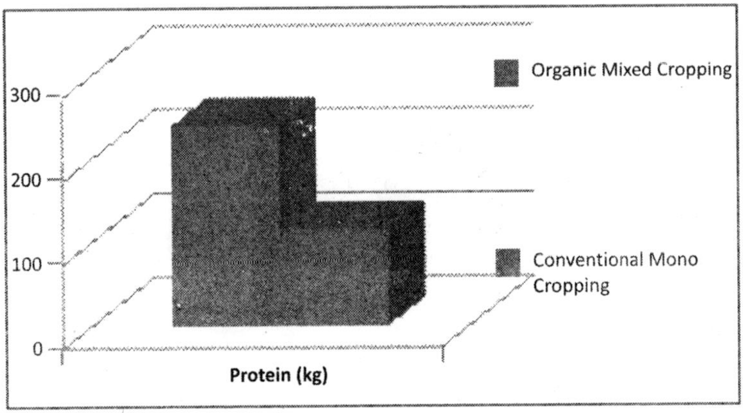

Fig. 14.1: Average Protein per Acre

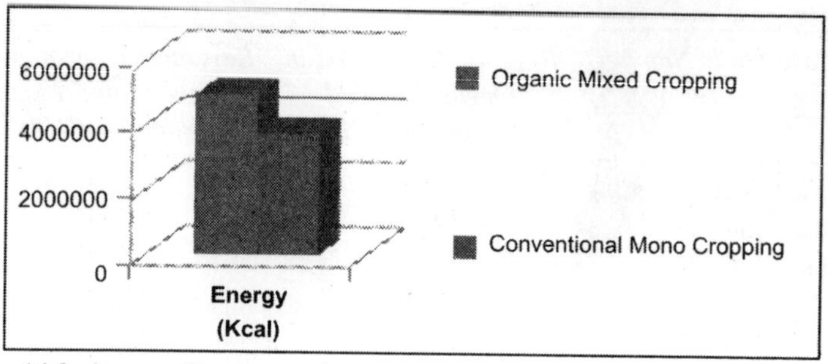

Fig. 14.2: Average Energy per Acre

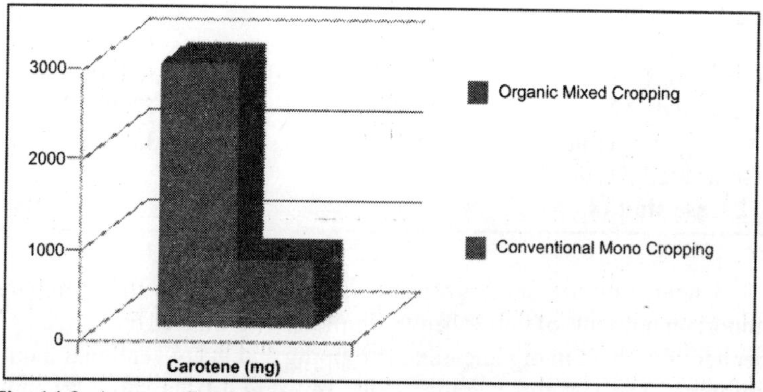

Fig. 14.3: Average Carotene per Acre

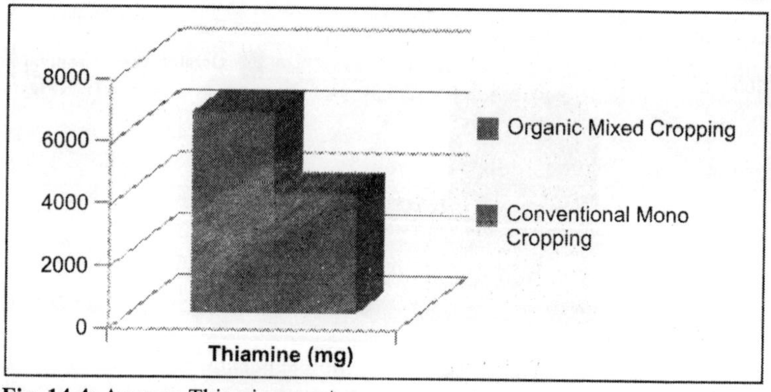

Fig. 14.4: Average Thiamine per Acre

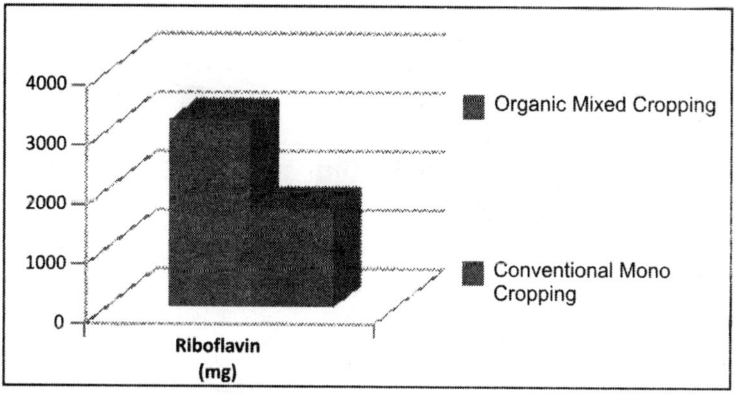

Fig. 14.5: Average Riboflavin per Acre

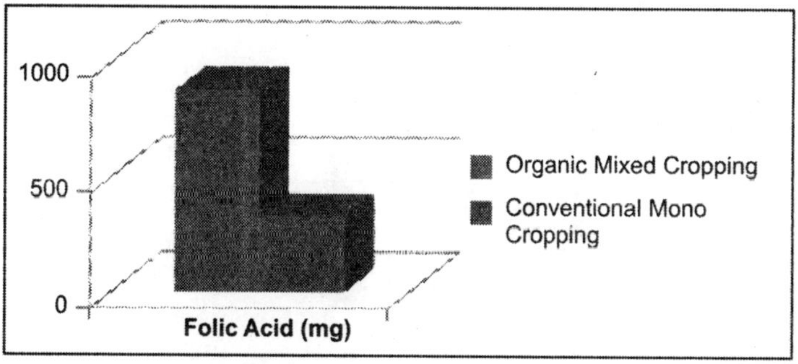

Fig. 14.6: Average Folic Acid per Acre

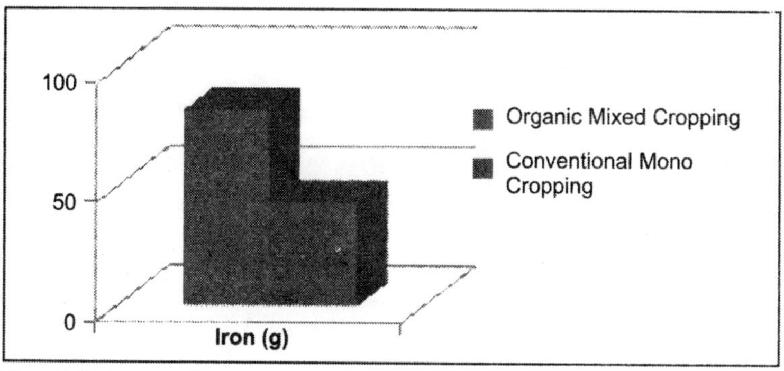

Fig. 14.7: Average Iron per Acre

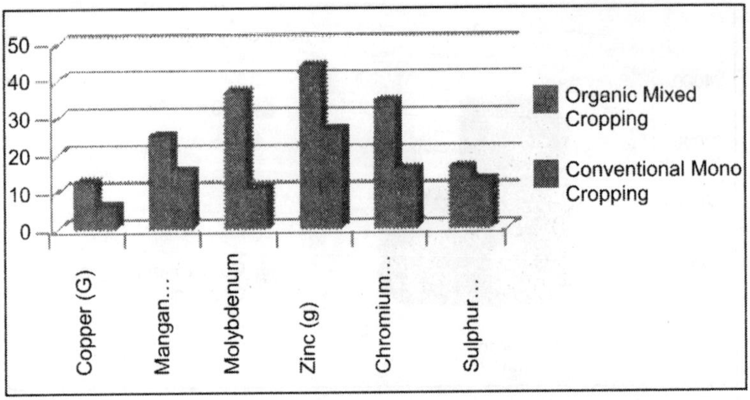

Fig. 14.8: Average Trace Elements per Acre

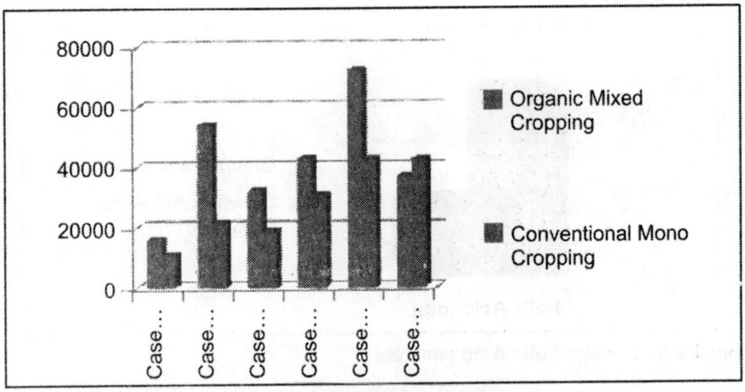

Fig. 14.9: Revenue per Acre in First Six Case Studies

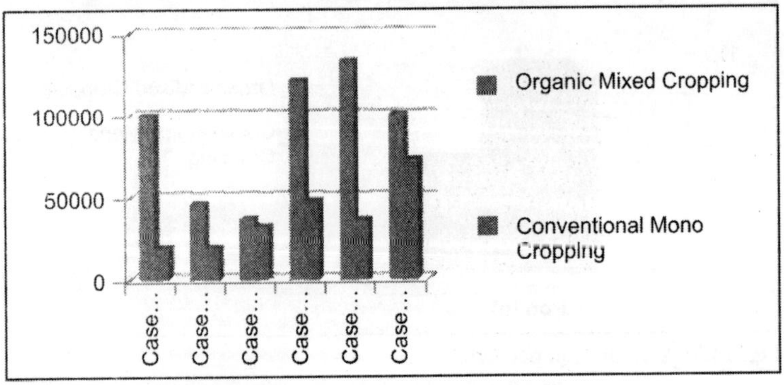

Fig. 14.10: Revenue per Acre in Second Six Case Studies

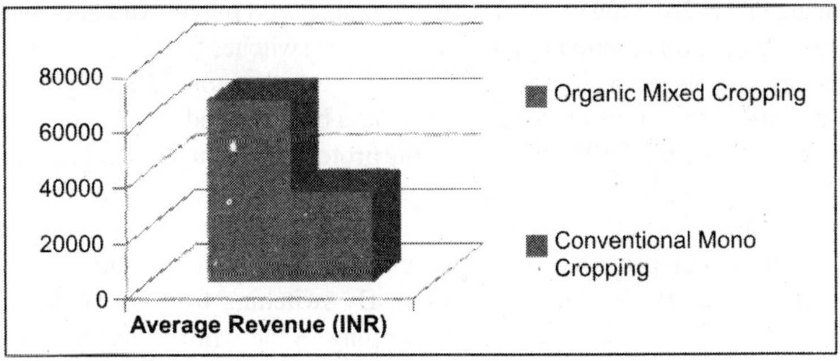

Fig. 14.11: Average Revenue across 12 Case Studies

Discussion

Protein Energy Malnutrition has become a crisis in India. According to NFHS-3, 51.1 per cent children are undernourished in India. Risk factors of protein energy malnutrition include:

Risk Factors of PEM

1. Young age of mother
2. Female gender
3. Rural area
4. Birth order >3
5. Birth spacing < 47 months
6. Low birth weight
7. Illiterate mother
8. Schedule caste/schedule tribe
9. Underweight mothers

The total cultivable land in India, according to Central Water Commission 2003-04, is 183 million hectares. If we assume that the results presented in the 12 case studies above can be replicated across the entire nation and that distribution of food is fully efficient, the impact of farming practices on the nutritional health of population can be fairly estimated. By switching from conventional mono cropping to organic mixed cropping, 1 acre of farmland produces 124 kg of extra proteins and 240 kg of total protein, 12,02,795 kcal of extra energy, 2174 mg of extra carotene, 2639 mg of extra thiamine, 1494 mg extra riboflavin, 550 mg extra folic acid, 39 g of extra iron, 6490 mg extra copper, 9495 mg extra manganese, 2617 mg of extra molybdenum, 17208 mg of extra zinc, and 188 mg of extra chromium. If the entire nation switches from conventional mono cropping to organic mixed cropping, 56 million tons of extra protein and 108 million tons of total protein would be produced. 56 million tons of extra protein is sufficient to fulfil the protein requirement of additional 2.5 billion adults for 1 year. A total of 108 million tons of protein is sufficient to fulfil the protein requirement of 5 billion adults for 1 year. This is enough protein to eradicate protein energy malnutrition from the entire planet, if distributed efficiently. By switching from conventional mono cropping to organic mixed cropping in the whole country, extra energy produced would be sufficient to supply 2500 kcal to 600 million additional adults for 1 year; the nation would produce enough food calories to feed 2.4 billion adults for 1 year. The prevalence of clinical and subclinical vitamin A deficiency in India is among the highest in the world. With respect to vitamin A deficiency, the most vulnerable groups are young children, adolescent girls, and pregnant woman. India is home to more than a quarter of world's preschool children with subclinical vitamin A deficiency and a third of preschool children with Xerophthalmia. By nationally switching from conventional mono cropping to organic mixed cropping, 982670 tons of extra carotene would be produced, a quantity of carotene enough to fulfil needs of an additional 750 million adults; enough carotene would be produced in the country to fulfil the needs of 1 billion adults for a year. Similarly, transition of the entire country from conventional mono cropping to organic mixed cropping would produce extra thiamine sufficient to feed 2.6 billion adults for 1 year and a total thiamine enough for 5 billion adults for 1 year, extra riboflavin sufficient for 1.2 billion adults for a year and a total riboflavin sufficient for 2.5 billion adults a year, and extra folic acid sufficient for 1.7 billion pregnant women for 1 year and a total folic acid enough for 2.7 billion pregnant women a year who require 4 times as much folic acid as normal adults do. The prevalence of iron deficiency anaemia is very high in India. 58 per cent of pregnant women, 50 per cent of lactating mothers, 56 per cent of adolescent girls, 30 per cent of adolescent boys, and 80 per cent of children under 3 years of age are anaemic, anaemia being the underlying cause of 20-40 per cent maternal deaths in India that amounts to 80 per cent of maternal deaths due to anaemia in South Asia. Consuming insufficient quantity of iron rich foods is one of the primary reasons for iron deficiency. Transition to ecological agriculture nationally would produce extra amount of iron enough for 20 billion hypothetical lactating mothers and a total amount of iron sufficient for 42 billion hypothetical lactating mothers. Organic mixed cropping compared to conventional mono cropping produces on an average 106 per cent more copper, 61 per cent more manganese, 243 per cent more molybdenum, 64 per cent more zinc, 120 per cent more chromium, and 72 per cent more trace elements when taken together. Transition to ecological agriculture would generate extra revenue of Rs. 33,160 per acre farmland which, on a national scale, would generate extra revenue of Rs. 15 trillion or $ 276 billion–15 per cent of Indian GDP in 2011-12.

There are two aspects of profit, benefit, or net gain both of which each need individual attention. One aspect of benefit is the net gain or the profit to the farmer. The other aspect of benefit is net gain or net profit to the society. Many economic activities, including agriculture, have costs that are internal to the system of production or consumption and costs that are external to the system of production or consumption. Similarly, many such activities have benefits that are internal to the system of production or consumption and benefits that are external to the system of production or consumption. Hence, we can classify costs and benefits as following:

Costs associated with production	*Benefits associated with production*
Internalized production costs	Internalized production benefit
Externalized production costs	Externalized production benefit
Costs associated with consumption	*Benefits associated with consumption*
Internalized consumption cost	Internalized consumption benefit
Externalized consumption cost	Externalized consumption benefit

Externalized costs are also called negative externalities and externalized benefits are also called positive externalities. In order to arrive at the net benefit or net profit to the farmer, we have to subtract internalized production cost from internalized production benefit–we have to compare the cost and benefit internal to the system of production. Externalized cost is borne neither by the producer nor by the consumer, and externalized benefits are enjoyed neither by the producer nor by the consumer. Since, it is the society that bears or enjoys the externalized costs or externalized benefits respectively, we ignore them for the time being to understand the net profit, the net benefit, or the net gain to the farmer. Internalized production cost is the sum of costs of all external inputs that are needed to grow a particular crop–cost of seeds, chemical fertilizers, pesticides, irrigation, human labour, energy, mechanization, etc. Internalized production benefit is the market value of all the produce whether consumed or sold in the market. By subtracting internalized production cost from internalized production benefit, we arrive at net profit or gain to the farmer. Conventional mono cropping is a high external input farming with huge costs associated with the system of production. Huge production cost in conventional mono cropping can be attributed to:

- Hybrid or gmo seeds
- Chemical fertilizers
- Pesticides
- Heavy irrigation
- Use of fossil fuel
- Excessive mechanization
- Occupational diseases due to exposure to chemicals

As a result of high costs incurred due to purchase of costly external inputs and reduced levels of revenue generation, net profits in conventional farming are small compared to organic farming. Organic farming not only generates higher revenue but also drastically reduces the cost of production to farmers because it relies on inputs intrinsic to the agro-ecosystem instead of dependence on purchase of external inputs. For example, organic farming relies on farm manure, cow urine, and organic biomass produced and composted within the farm for fertilization instead of application of purchased chemical fertilizers. Conventional farming operates according to reductionist principle in which symptom management is the priority–if there are pests, spray pesticides, if there is a fungal disease outbreak, spray fungicide, if the soil is unfertile, spray chemical fertilizers, and if there are weeds, spray herbicides. As a result of the reductionist approach to manage and treat symptoms, the principles adapted by conventional farming not only fail to go to the root of the problem but also increase the cost of production tremendously. On the contrary, organic farming adapts a holistic approach to farming based on Agro-ecological principles in which system management is the priority. Organic farming is primarily based on the mixed cropping choosing the combination of crops and plants that are resilient to diseases and pests, that interact with each other synergistically and symbiotically, and that minimize the risk associated with farming through diversification of crops and animals on the farm. If paddy grown as a monoculture succumbs to pest attack, the farmer suffers a catastrophic loss. However, if paddy grown with a variety of other crops and plants succumbs to pest attack, the other crops on the farm still generate some revenue for the farmer thereby avoiding a catastrophic loss-risk is minimized through diversification of crops and plants at the farm. Since pest control, disease control, weed control, fertilization, etc. are built within the agro-ecosystem in organic farming, the purchase of external inputs is almost always avoided. Since most farmers in India are small and marginal, they use family labour at the farm; the opportunity cost of family labour is very low and money is saved by the household by supplying labour at the farm. The table underneath reveals the quantities of different fertilizers

Table 14.4: Cost of the Total Amount of Fertilizer Used in India (Chanda, Sati, Robertson, and Arora, 2012)

Fertilizer type	*Quantity consumed in 2011-12 (in tonnes)*	*Retail cost of consumption in 2011-12 in rupees*
Urea	29565320	158470115200
Ammonium sulphate	509390	5526881500
Ammonium chloride	58010	870150000
Calcium ammonium nitrate	109200	1298934000
Single super phosphate	4746010	20702095620
Tsp	78130	978890770
Muriate of potash	3028930	27381527200
Sulphate of potash	30790	369480000
Di-ammonium phosphate	10191190	158208033560
Mono ammonium phosphate	281430	5375313000
Rock	46690	479903165
NPK- 10-26-26	1581670	22494510740
NPK- 12-32-16	1205940	16773419460
NPK- 14-35-14	315800	4569152300
NPK- 15-15-15	500270	5574258475
NPK- 16-20-0	523580	5738436800
NPK- 17-17-17	5370	91844900
NPK- 19-19-19	12130	163322363
NPK- 20-20-0	5416950	63622077750
NPK- 23-23-0; 24-24-0	173160	1908742680
NPK- 28-28-0	289180	4370088160
NPK- 16-16-16	47440	721088000
NPK- 13-33-0-6	112400	1955760000
NPK- 14-28-14	210810	3226025430
Total =	**10394700**	**Rs. 510870051073= $9.29 billion approx.**

used in India in 2011-12, the expenditure by farmers on each type of fertilizer, and the total expenditure by farmers on fertilizers of all types.

Conventional farmers incur this huge cost of $ 9.29 billion because of their reliance on such chemicals. The cost of fertilizers and other agro-chemicals increase the production cost internal to the system. Because of greater production cost, reduced level of revenue, higher risk associated with conventional farming, it has become a debt trap creating mechanism for the farmers of the developing world. 2,84,000 Indian farmers have committed suicide since 1995, most of these suicides can be traced to debt leading to loss of land, and most of these debts have originated from purchase of costly external inputs required in industrial model of farming. The cost internal to the system of production in conventional farming is further exacerbated by occupational diseases associated with handling of agricultural chemicals. Indian farmers poorly follow the safety guidelines due to a low literacy rate. Pesticides and other agricultural chemicals are applied without safety gear because of two reasons–first that India is a tropical country with a hot and humid weather and second that majority of Indian farmers are too marginal to afford expensive protective gears. Farmers indulge into such practices as over application of pesticides, application of dangerous, unscientific, and spurious mixtures of pesticides, and storage of food in pesticide containers. The agro-chemical corporations aggressively market their products and, often, farmers gather agro-chemical application guidelines from the dealers who have a vested interest in prescribing more of these chemicals to the farmers. As a result of these factors, there are a number of occupational diseases associated with conventional farming some of which are listed below (Shiva, Shiva, and Singh, 2012):

- Skin rash and chronic dermatitis
- Burning eyes, blindness
- Shortness of breath
- Epistaxis
- Pain in stomach, muscles, joints, or bones
- Memory loss, mental illness
- Headache
- Pterygium
- Polyneuropathy
- Kidney diseases
- Abortions, still births, neonatal deaths
- Congenital defects, and mental retardation
- Non-hodgkin's lymphoma
- Leukemia
- Multiple myeloma
- Sarcomas
- Prostate cancer
- Pancreatic cancer
- Lung cancers
- Breast cancers
- Testicular cancers
- Hodgkin's lymphoma
- Liver and kidney cancers
- Rectal cancers
- Stomach cancers
- Endometrial cancers

Production cost and production benefit internal to the system of agriculture is just one aspect of the "Big Picture" the other aspect being production cost and production benefit external to the system. One major production cost external to the system is the government subsidies to the fertilizer corporations. Fertilizers actually cost more than their market price in India. Government offers huge subsidies to the Fertilizer Corporations in order to artificially reduce fertilizer prices in the market. In the year 2011-12, the Government of India offered a fertilizer subsidy to Corporations amounting to Rs. 74,170,00,00,000 or $13.49 billion approximately. This cost of fertilizer in the form of subsidies is borne not by the farmer but by the tax payers of India and the Indian society. These subsidies lead to inefficient overproduction and overconsumption creating a dead weight loss. As a result, subsidies constitute a part of cost external to the system of production, cost that is borne not by the farmer but by the entire society. There are a number of other costs that are external to the system of production and that are tricky to be evaluated in monetary terms. Some such external production costs associated with conventional farming result from environmental degradation due to agro-chemicals leading to loss of species, loss of biodiversity, bee colony collapse syndrome, depletion of soil micro-organism, desertification of land, salinization of land, destruction of river and ocean ecosystem, creation of dead zones in the ocean, creation of super pests and super weeds, global warming, deterioration of public health due to residues of agro-chemicals in the environment, etc. Because farmers do not have to bear these costs external to the system of production, inefficient level of excessive conventional farming is undertaken. Finally, while comparing ecological agriculture with conventional agriculture, we must take into consideration the benefits external to the system of production, positive externalities associated with ecological agriculture. Some such positive externalities are:

- Increased carbon sequestration
- Biodiversity conservation and promotion
- Decreased health hazards due to elimination of harmful chemicals
- Conservation of natural ecosystems
- Conservation of scarce resources
- Diversified nutrition to the society
- Promotion of public health through increased production of good quality nutrition

Because of these positive externalities, less than efficient levels of ecological agriculture is undertaken. Organic farming maximizes overall benefit to the society and to the farmers and promotes equity and efficiency in the society because of high internalized benefit, high externalized benefit, low internalized cost, and low externalized cost.

II

The strengthening of law of return mandates that when we grow crops and plants which derive their nutrition from the soil, we must conduct farming in ways that rejuvenate and regenerate the soil, that when we grow crops for their seeds for consumption, we must save some seeds to be sown in future, and that when the society extracts their food and fibre requirement from agriculture, it must divert some of its resources back into agriculture. Organic farming complies with the law of return as shown in the diagram below:

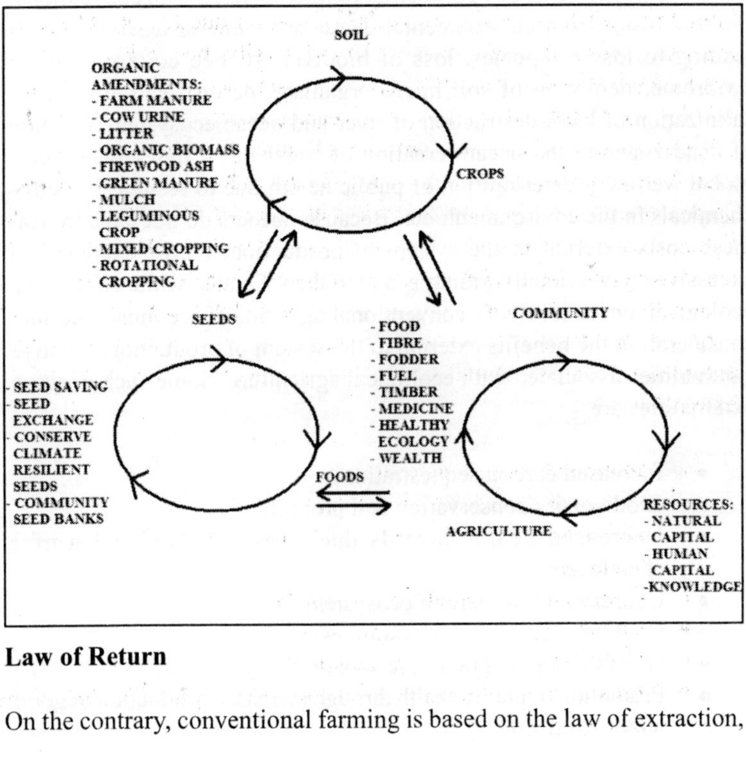

Law of Return

On the contrary, conventional farming is based on the law of extraction, also called the law of exploitation. Excessive extraction from the soil leads to desertification and micronutrient deficiencies in the soil, excessive extraction from the ecosystem leads ecosystem degradation, and excessive extraction from the farming community leads to a debt trap creating system of which the victims are the resource poor farmers of the developing world. Diagrammatic representation of the Law of Exploitation is presented below:

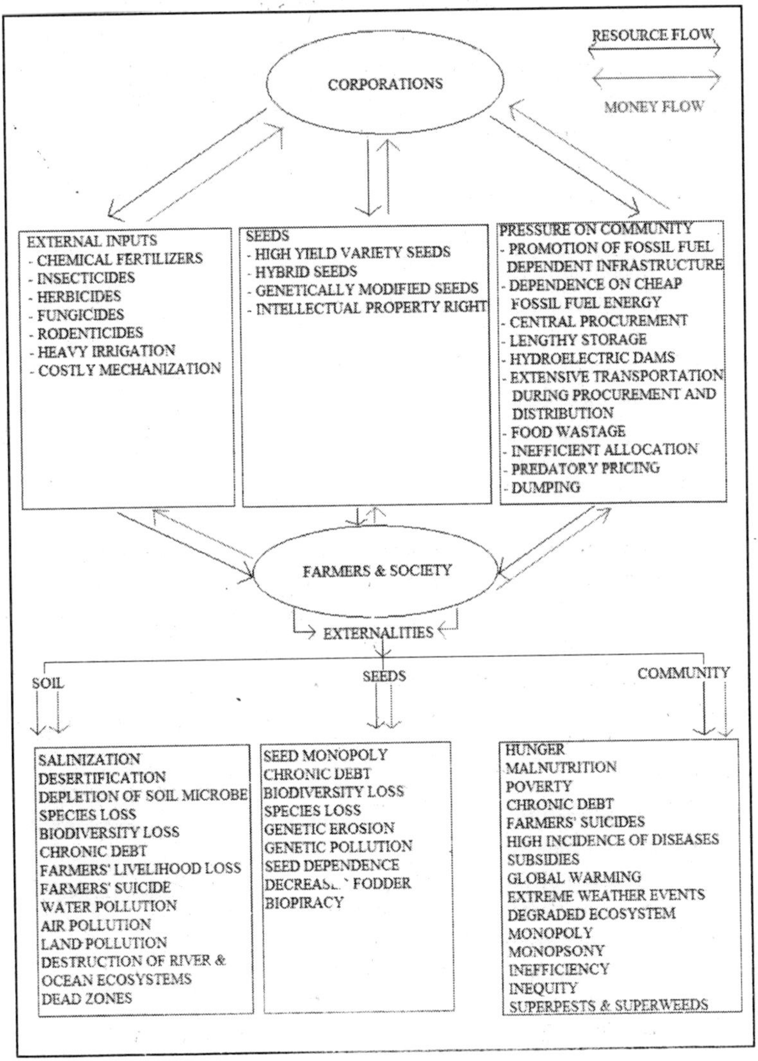

Law of Exploitation

The difference between ecological agriculture and conventional agriculture is primarily philosophical. Ecological agriculture strengthens the law of return, operates towards community empowerment, and recognizes the rights of soil and Mother Earth, rights of animals, and rights of citizens. The focus in ecological agriculture is ecosystem rejuvenation and regeneration and community welfare. Conventional agriculture strengthens the law of exploitation, operates to increase corporate profits giving it more importance than community welfare, and recognizes the right of corporations more than the rights of communities. The traits of a sovereign society are:

- Seed sovereignty
- Food sovereignty
- Sustainable livelihoods
- Ecosystem protection and promotion
- Community health
- Community welfare and prosperity
- Fair trade practices
- Efficiency
- Equity
- Earth democracy

Ecological agriculture works in harmony with nature and community to achieve these goals. Conventional agriculture fails miserably in achieving these goals. Conventional agriculture relies on seed patents and corporate monopolization of the seed sector and the food sector. This system of farming is neither ecologically sustainable nor economically viable in addition to creating severe ecological disturbances, conventional farming has created debt traps for Indian farmers leading to 2,84,000 farmers' suicides since 1995. Conventional agriculture deteriorates community health as is evident in the cotton belt of Punjab, Bathinda and Fajilka districts. So many people are developing cancers and other diseases in Bhatinda and Fajilka that Jodhpur-Abohar passenger train is referred to as "CANCER TRAIN" because hundreds of cancer patients from these regions board the train on most working days of a week to visit a charitable hospital in Bikaner, Rajasthan. Conventional agriculture aggressively endorses monopoly or oligopoly (one or few sellers) and monopsony (single buyer). For example, corporations are trying hard to take full control of the seed sector leading to monopoly, and corporations also enter into dicey contracts with the farmers forcing them to grow monocultures in the entire region and, during harvest, the same corporation is the sole buyer of the produce leading to monopsony. Such trade practices lead to predatory pricing, rent seeking, and unequal playing field between the corporations and the communities. Conventional agriculture breeds inefficiency and inequality and recognizes the right of the corporations totally neglecting the rights of Mother Earth, the rights of communities, and the rights of all species that inhabit this planet. Conventional agriculture is promoted for the sake of economic growth, prosperity, food security, welfare, and wealth generation. In order to negate these assumptions, we present another case study in which we compare the costs and benefits of dominant monocultures grown in conventional farming, as given in published literature and adjusted for inflation, with costs and benefits of ecological alternatives derived from our own experiences with the organic farmers. The four dominant monocultures grown in conventional agriculture are hybrid rice, hybrid soya, hybrid corn, and green peas. Costs and benefits of these monocultures are derived from published literature and adjusted for inflation. Finally, the costs and benefits of these monocultures are compared with ecological alternatives such as native *Dehraduni Basmati* rice, *Rajma* (Kidney Beans), *Ragi* (Finger Millets), Amaranth (*Ramdana*), and *Sukhda* (sun dried vegetables).

Hybrid Rice *versus* Native Organic Basmati Rice

Chengappa, Janaiah, and Gowda (2003) conducted a study in which they assessed the profitability of cultivating hybrid rice compared to inbred

Hybrid Rice

Inputs per hectare	*Nominal cost (2000-01) in INR*	*Real cost (adjusted for inflation till 2013) in INR*
Seeds	1715.98	8892.00
Farm yard manure	1205.20	3083.58
Chemical fertilizers	2396.98	6132.82
Irrigation	1949.91	4988.96
Plant protection	630.68	1613.63
Human labour	7711.53	19730.41
Machine labour	1623.42	4153.62
Miscellaneous	656.10	1678.67
Total	**17889.80**	**50273.69**

varieties. For the purpose of our case study, we will extract costs and benefits of growing hybrid rice from the above mentioned study, adjust these costs and benefits for inflation, and compare it with our own experience with organic farmers who grow native Dehraduni Basmati.

Output per hectare	*Nominal value (2000-01) in INR*	*Real value (adjusted for inflation till 2013) in INR*
Value of 8.41 ton of paddy produced per ha	45598.18	116665.68
Value of straw	2138.08	5470.41
Gross revenue	47736.26	122136.09
Net profit in 2013 inr	29846.46	71862.40

Organic Dehraduni Basmati

Mr. Mukundi Lal an organic farmer member of Navdanya from village Bhoodpur village situated on old Shimla road is one of the happy organic farmers. He joined Navdanya in the year 1998 when he became part of Navdanya campaign against the Basmati Biopiracy by Ricetech Company of Taxas, USA. Since then he became an organic farmer.

Farmer: Mukundi Lal village Bhuddi Year: 2013
Name of the Crop: Dehraduni Basmati Organic (Marigold on bunds)
Land area: 1 Bigha (900 sq. yards)

Item	*Quantity*	*Total (Rs.)*
Costs for field preparation*		
Ploughing Cattle	Four times	—
Ploughing by tractor		700*
Diesel	5 litre	300
Seeds	2.50 kg	125*
Nursery Sowing	1 man day	200*
Transplantation including nursery uprooting	3 man days	450*
Irrigation	4 hours	400*
Chemical Fertilizers	Nil	
Other Fertilizers (FYM etc.)	1 trolley or 10-12 qtl	1000*
Pesticides and weedicides		Nil
Weeding	1 man day	150*
Harvesting	2 man days	300*
Threshing, winnowing etc.	3 man days	450*
Total Production Cost Per Bigha		**4375**
Transportation and labour Cost		200*
Other costs (if any?)	Gunny bags (04)	100
Total Expenditure Per Bigha	—	**Rs. 4675**
Total Expenditure Per Hectare	—	**Rs. 62130.75**

(Contd.)

(*Contd.*)

Item	Quantity	Total (Rs.)
Total Production		
Grains	230 kg @ Rs 46	10580
Straw	300 kg @ Rs 4	1200
Other (weeds)	200 kg x 2	400
Marigold	36 plants	1000
Gross Income Per Bigha	—	**Rs. 13180.00**
Net Income Per Bigha (Gross income-Total expenditure)	—	**Rs. 8505.00**
Gross Income Per Hectare	—	**Rs. 175162.20**
Net Income Per Hectare	—	**Rs. 113031.45**

*Own Source

Mr. Mukundi Lal did not hire any labour or buy anything from outside except for the diesel and gunny bags for the paddy. His income can be calculated as 13180 – 4675 = 8505 × 13.29 = Rs. 113031.45/hectare. Other than this the price quoted is average price in the market for the year 2013. During the past few years his production was always in between 2 - 2.7 qt/bigha which is 33.73 qt/hectare. Even when his neighbours lost the crops due to various reasons including less rains or heavy rains, production of his native *Dehraduni Basmati* was always more or less the same with a difference of 10 to 15 per cent to the maximum.

The year 2013 received lots of rains but production in his field was above average and he also fetched a good price for his *Dehraduni Basmati*.

Economic Analysis of Growing Soybean versus to Native Organic Kidney Beans (Rajma)

Mandal, Saha, Ghosh, Hati, and Bandyopadhyay (2002) conducted a study to examine energy requirement and energy input output relationship of soybean based crop production. For the purpose of our case study, we will extract input requirement, total cost of production, gross revenue, and net profit associated with growing soybean from the above mentioned study, adjust these values for inflation, and compare it with our own experience with organic farmers who grow native kidney beans (*rajma*).

Soybean

Type of input	Amount of input
Land preparation	One summer ploughing with two ploughing with 35 hp tractor drawn cultivator

(*Contd.*)

(Contd.)

Type of input	*Amount of input*
Fertilizer/manure	5 ton/ha of farmyard manure, 30 kg/ha N, 60 kg/ha P_2O_5, 30 kg/ha K_2O
Sowing	By tractor drawn seed drill with seed rate of 80 kg/ha
Weeding and thinning	Manual labour
Irrigation	Rainfed
Plant protection	Chemical control by spraying phosphamidon 85% EC at 0.02%
Harvesting	Manual
Threshing	By 35 hp tractor driven thresher

Soybean	*Nominal value (2000-01) in INR*	*Real value (adjusted for inflation till 2013) in INR*
Gross cost of production per ha	10924.00	27949.70
Gross return from economic product per ha	7905.00	20225.42
Gross return from by products per ha	4138.00	10587.32
•Total return per ha	12043.00	30812.74
Net profit in 2013 inr	1119.00	2863.03

Kidney Beans (*Rajma*)

Navdanya started encouraging farmers to conserve local seeds and to promote organic farming in the region. Before becoming member of Navdanya, they were giving kidney beans to the local trader at the rate decided by him. Now, the group decides its rate collectively. The rate of kidney beans in the village has gone up to Rs. 100 from Rs. 35 in 12 years time. Farmers are more empowered and happy with organic farming and production is also going up year after year, whereas diseases and pests are now quite uncommon. According to Mr. Mohan Singh "*Now, we fix rates for our Rajma, which was earlier done by the traders*". He also adds that "*Navdanya directly deposits money in our group's bank account. Navdanya not only gives us 10 per cent more than the local price but also gives us 5 per cent more when procured through the Farmers group.*"

Farmer: Mohan Singh
Name of the Crop: *Rajma* (Kidney beans)
Village Pokhri, Tehsil Chakrata, District Dehradun
Land area: 15 Nali (3000 sq. metre) 1 Nali = 200sq. Metre

Item	*Quantity*	*Total (Rs.)*
Costs for field preparation*		
Ploughing Cattle	2	2000
Seeds	50 kg	4500
Irrigation	Nil	
Chemical Fertilizers	Nil	
Other Fertilizers (FYM etc.)	100qtl	7000*
Pesticides and weedicides		Nil
Weeding	20 man days	4000*
Harvesting	10 man day	2000*
Threshing, winnowing etc.	5 man days	10000*
Total cost in Production		**29500**
Transportation and labour Cost		1000*
Other costs (if any?)	Gunny bags (20)	400
Total expenditure Per Field	—	**30900**
Total Expenditure Per Hectare	—	**102897**
Total Production		
Grains	1000 kg @ Rs. 95	**95000**
Straw	2000 kg @ Rs. 2	4000
Other (weeds)	1000 kg x 2	2000
Mandwa	100 x 15	1500
Chaulai	110 x 50	5500
Maize	80 x 20	1600
Other Straw	800 x 2	1600
Gross income Per Field	—	**Rs. 111200**
Net income Per Field (Gross income-Total expenditure)	—	**Rs. 80300**
Gross Income Per Hectare	—	**Rs. 370296**
Net Income Per Hectare	—	**Rs. 267399**

*Own Source

Economic Analysis of Growing Hybrid Corn Compared to Finger Millet (*Ragi*) and Amaranth (*Ramdana*)

Pimentel and Patzek (2005) conducted a study to reveal the energy inefficiency of producing ethanol from corn, switchgrass, and wood biomass and the energy inefficiency of producing bio-diesel from soybean and sunflower. For the purpose of our case study, we will extract cost and benefit of growing hybrid corn, adjust the values for inflation, and compare these values with our experience with organic farmers growing native finger millet (*ragi*) and amaranth (*ramdana*).

Hybrid Corn

Type of input	*Amount of input per hectare*	*Nominal cost (2003-04) in INR*	*Real cost (adjusted for inflation till 2013) in INR*
Labour	11.4 hour	200	455.66
Machinery	55 kg	4815.78	10971.74
Diesel	88 l	1621.90	3695.16
Gasoline	40 l	970.53	2211.15
Nitrogen	153 kg	4426.17	10084.10
Phosphorous	65 kg	1880.40	4284.10
Potassium	77 kg	1113.77	2537.49
Lime	1120 kg	513.26	1169.36
Seeds	21 kg	3490.64	7952.69
Irrigation	8.1 cm	5739.18	13075.52
Herbicides	6.2 kg	5785.84	13181.82
Insecticides	2.8 kg	2612.96	5953.08
Electricity	13.2 kwh	42.93	97.81
Total	**—**	**33213.36**	**75669.67**

Yield of corn	*8655 kg/ha*
Market price of yield in 2013 inr	Rs. 106329.27/ha
Net profit in 2013 inr	Rs. 30659.60/ha

Finger Millet (*Ragi/Mandwa*)

Rajeshwari Devi an organic farmer and member of Navdanya of village Maniguha in district Rudraprayag joined the movement in 1998. She stopped using chemicals and converted her farm to organic. With Finger millet she grows Bhat, Kali dal, Chaulai and Bhangjeera. She along with her family members work together in their fields. Some time they help other families in weeding, manuring and harvesting to cut the labour costs. Only input for during the cultivation is gunny bags, which they use occasionally. According to Rajeshwari "*People of our generation never suffered from backache and joint pains as we ate finger millet right from our childhood. It is food as well as medicine, both for growing children to the elderly people.*" She also adds that "*Finger millet grows equally good if there is no rain or too much rain, cattle also eat straw with interest as it is much sweeter in taste than any other straw.*"

Farmer: Rajeshwari - 2013
Village: Maniguha, District Rudraprayag
Name of the Crop: Mandwa (Finger Millet)
Land area: 1 Nali (200 sq. metre)

Item	*Quantity*	*Total (Rs.)*
Costs for field preparation*		
Ploughing Cattle	2	200*
Seeds	1kg	25*
Irrigation	Nil	—
Chemical Fertilizers	Nil	—
Other Fertilizers (FYM etc.)	Nil	—
Pesticides and weedicides	Nil	—
Weeding	1 or 2 man days)	400*
Harvesting of Heads and Straw	2 man day	400*
Threshing, winnowing etc.	3 man days	600*
Total cost of Production		**1625**
Transportation and labour Cost		100*
Other costs (if any?)	Gunny bags (4)	100
Total Expenditure Per Nali	—	**Rs. 1825**
Total Expenditure Per Hectare	—	**Rs. 91250**
Total Production		
Grains	48kg @ Rs. 16	768
Straw	400kg @ Rs. 2	800
Other (weeds)	200 kg x 2	400
Bhat	6 x 60	360
Kali Dal	6 x 85	510
Chaulai	10 x 75	750
Bhangjeera (Perilla)	2 x 100	200
Other Straw	300 x 2	600
Gross Income Per Nali	—	**Rs. 4388**
Net income Per Nali (Gross income–Total expenditure)	—	**Rs. 2563**
Gross Income Per Hectare	—	**Rs. 219400**
Net Income Per Hectare	—	**Rs. 128150**

*Own Source

Rajeshwari gets income of about Rs. 4388 from her 200 sq yard field. The labour as an input cost it is calculated to be Rs. 1625. Another Rs. 200 can be added for transportation and gunny bags to make it Rs. 1825.

Approximate income per hectare Rs. 128150 when labour is calculated as an input. However, Mrs. Rajeshwari did not hire any labour or used any input for cultivation from outside, avoided cost that increases net income.

Amaranth (*Ramdana*)

Susheela Devi a farmer from village Kharsi became Navdanya member in 2004. She grows Amaranth with black *rajma*. Amaranth and black *rajma* grow very well together, as *rajma* is a creeper and it takes support of Amaranth. According to her "*Not only productivity of crops is improving year after year but also pests and diseases are rarely seen in our village, which were quite common before we were using chemicals.*"

Name of Farmer: Susheela Devi
Name of the Crop: Amaranth (*Ramdana*)
Village: Kharsi, Tehsil Chakrata, District Dehradun
Land area: 4 Nali (800 sq. metre) 1 Nali = 200 sq. Metre

Item	*Quantity*	*Total (Rs.)*
Costs for field preparation		
Ploughing Cattle	2	800*
Seeds	2 kg	100*
Irrigation	Nil	
Chemical Fertilizers	Nil	
Other Fertilizers (FYM etc.)	40 qtl	4000*
Pesticides and weedicides	Nil	Nil
Weeding	4 man days	800*
Harvesting	2 man day	400*
Threshing, winnowing etc.	4 man days	800*
Total cost in Production		**6900**
Transportation and Labour Cost		400*
Other Costs (if any?)	Gunny bags (10)	250
Total Expenditure Per Field	—	**Rs. 7550**
Total Expenditure Per Hectare	—	**Rs. 94375**
Total Production		
Grains	400 kg @ Rs 75	30000
Straw	500 kg @ Rs 2	1000
Other (weeds)	600 kg x 2	400
Kali Rajma	70 x 75	5250
Other Straw	130 x 2	260
Gross income Per Field	—	**Rs. 36910**
Net income Per Field	—	**Rs. 29360**
(Gross income–Total expenditure)		
Gross Income Per Hectare	—	**Rs. 461375**
Net Income Per Hectare	—	**Rs. 367000**

*Own source

Net income of Susheela Devi from her 4 nali field totalled Rs. 29,360 (or Rs. 3,67,000/hectare) when labour is calculated as an input. Like other organic farmers, Susheela also worked herself with her husband in her field and did not hire any labour or any external input for cultivation, avoided costs that raise household's net income.

Economic Analysis of Growing Green Peas Compared to Native Organic Dried Vegetables (*Sukhda/Sukhsa*)

Singla, Chahal, and Kataria (2006) conducted a study analysing economics of growing green peas. For the purpose of this study, we extract cost and benefit associated with green peas production from the above mentioned study, adjust these values for inflation, and compare them with our experience with organic farmers growing ***Dried Vegetables*** (*Sukhda/ Sukhsa*).

Green Peas

Type of input per hectare	*Nominal cost (2002-03) in INR*	*Real cost (adjusted for inflation till 2013) in INR*
Seeds	3538.90	8320.65
Urea	833.99	1960.88
Ssp	37.02	87.04
Dap	1227.18	2885.35
Zinc	47.76	112.29
Other micronutrients	68.66	161.43
Farmyard manure	170.74	401.44
Plant protection chemicals	453.45	1066.15
Electricity/diesel	630.99	1483.58
Human labour	8284.93	19479.53
Animal labour	190.80	448.61
Machinery/tractor	1902.80	4473.86
Total	**17387.22**	**40880.83**

Yield per ha of green peas	7369 kg
Nominal gross revenue (2002-03) per ha in inr	57671
Real gross revenue (adjusted for inflation till 2013) in inr	135596.06
Net profit in 2013 inr	94715.23

Sun Dried Vegetables (*Sukhda/Sukhsa*)

For thousands of years people have sundried vegetables to preserve for off season use. Dried vegetables help improve family nutrition because they contain high quantities of vitamins, minerals and fibre. Farmers keep the dried vegetables not only cultivated but also wild vegetables to be used during the off season.

Some of the commonly dried cultivated and wild vegetables are as follows:

- Pumpkin (about 8-10% of fresh weight)
- Lauki (about 8-10%)
- Bitter gourd (10-12%)
- Sweet Gourd (10-12%)
- Greens: (*Rai*, Mustard, Spinach, *Methi* about 7-10%)
- Ferns: Linguda (25%) and Kuthda (50%).
- Mushrooms (Guttchi, *Lactarius, Agaricus*, 10%)
- Stinging Nettle (10-15%)

It takes about a week to dry most of the vegetables. Greens dry faster than the fleshy vegetables. Fleshy vegetables are cut into small pieces or slices like chips for better results. It helps in reducing the wastage of the vegetables as well as nutrition drain. Dried vegetables can be used in place of frozen vegetables which requires more energy to keep them fresh. When soaked in warm water dried vegetables quickly assume the flavour and texture of their frozen or even fresh counterparts. These dried vegetables can also be eaten as a low-calorie, healthful replacement for other crispy snacks without any preparation or rehydration. It is recommended to soak the dried vegetables for about 30 to 90 minutes in cold or warm water before cooking.

Nutritional Value of Dried Foods

Drying, like all methods of preservation, can result in loss of some nutrients. For best retention of nutrients in dried foods, store them in a cool, dark, dry place and use within a year. Nutritional changes that occur during drying include:

- *Calorie content*: does not change, but is concentrated into a smaller mass as moisture is removed.
- *Fiber*: no change.
- *Vitamin A*: fairly well retained under controlled heat methods.
- *Vitamin C*: mostly destroyed during blanching and drying of vegetables.

- *Thiamin, riboflavin, niacin*: some loss during blanching but fairly good retention if the water used to rehydrate also is consumed.
- *Minerals*: some may be lost during rehydration if soaking water is not used.
- *Iron* is not destroyed by drying.

Dried and fresh vegetables have identical calorie counts, but dehydrated vegetables take up about half the volume of fresh. This gives one cup of dried about the same calories as two cups of fresh.

The drying of vegetables not only can help in waste management but can be one of the potential sources of employment generation. Through cultivated and wild vegetable drying nutrition availability can be assured to the masses especially for the small and marginal farmers who cannot afford to buy the fresh vegetables in the off season.

Conclusion

Yield per acre as a benchmark of comparison is misleading. Actual comparison entails comparing the quantities of different nutrients produced in the two system of farming. When the nutrition produced per acre is compared, organic farming proves to be far superior to conventional farming. Except for Vitamin C, all nutrients in organic mixed cropping were produced in quantities greater than 1.5 to 2 times the quantities of nutrients produced in conventional mono cropping. Assuming that similar results could be replicated across the entire nation, the effect of transition to ecological agriculture can be immense on the public health issues of the country such as protein energy malnutrition, night blindness, anaemia, maternal mortality, infant mortality, etc. Moreover, the extra nutrition produced ecologically translates into $ 276 billion worth of agriculture commodities produced over and above the regular conventional production for consumption and for sale. Net incomes of farmers practicing organic farming further increase because costly external inputs such as seeds, fertilizers, pesticides, and irrigation are eliminated and avoided. If we consider the net benefit to the society over and above the farmers' income, organic farming again proves to be far superior to conventional farming because of several positive externalities associated with organic farming and several negative externalities associated with conventional farming. The second part of the chapter questions the assumption that conventional farming generates greater incomes. The average profit per hectare in growing hybrid rice, hybrid soybean, hybrid corn, and green peas is meagre compared to average profit per hectare in growing such ecological alternatives as *Dehraduni Basmati*, Native Kidney Beans, Native Amaranth, Native Finger Millets, and Native Sun Dried Vegetables. The second part of the chapter calls for a need to undergo a detailed and comprehensive study in various regions of the country to establish the true profitability of the two systems of agriculture.

References

Chanda, T.K., Sati, K., Robertson, C., and Arora, C. (2012). *Fertiliser Statistic 2011-12.* New Delhi, India: FAI—The Fertiliser Association of India.

Chengappa, P.G., Janaiah A., and Gowda, M.V.S. (2003). Profitability of Hybrid Rice Cultivation: Evidence From Karnataka. *Economic and Political Weekly*, 38(25).

Gopalan, C., Ramasastri, B.V., and Balasubramanian, S.C. (1989). *Nutritive Value of Indian Foods.* Hyderabad, India: National Institute of Nutrition, Indian Council of Medical Research.

Mandal, K.G., Saha, K.P., Ghosh, P.K., Hati, K.M., and Bandyopadhyay, K.K. (2002). Bioenergy and Economic Analysis of Soybean Based Crop Production Systems in Central India. *Biomass and Bioenergy*, 23(5), 337-345.

Pimentel, D. and Patzek, T.W. (2005). Ethanol Production Using Corn, Switchgrass, and Wood; Biodiesel Production Using Soybean and Sunflower. Natural Resources Research, 14(1), 65-76.

Shiva, V. and Pandey, P. (2006). *Biodiversity Based Organic Farming: A New Paradigm for Food Security and Food Safety.* New Delhi, India: Navdanya.

Shiva, V. and Singh, V. (2011). *Health Per Acre: Organic Solutions to Hunger and Malnutrition.* New Delhi, India: Navdanya/Research Foundation for Science, Technology, and Ecology.

Shiva, V., Shiva, M., and Singh, V. (2012). *Poisons in Our Food-Links Between Pesticides and Diseases.* New Delhi, India: Natraj/Research Foundation for Science, Technology, and Ecology.

Singla, R., Chahal, S.S., and Kataria, P. (2006). Economics of Production of Green Peas (Pisum sativum L.) in Punjab. *Agricultural Economic Research Review*, 19, 237-250.

15

Effects of Climate Change on Small Farms

Can Climate Smart Agriculture Help?

Susmita Mitra

Introduction

Climate change, one of the greatest challenges to global community, has a negative impact on the agriculture of tropical developing countries. India being an agriculture dependent tropical country, this problem has serious implications where agriculture even today constitutes the prime occupation for the majority of the masses. Thus, a detailed study of the nature and climatic change on the agricultural sector and its consequent effect on people's livelihood constitutes a space of great importance. This chapter emphasises on poor and marginalised farmers because on one hand they are often mostly affected by climate change/variability and on the other hand poor economic condition of these farmers is a critical factor that limits the adaptive capacity. Moreover, more than 80 per cent farmers are small and marginal farmers in India. This main emphasis is on how climate smart agriculture can help the small and marginal farmers in India.

It attempts to answer the following questions:

1. What is the theoretical underpinning between climate change and agriculture?
2. What are the impacts of climate change witnessed in India over the last few years?
3. What are the effects of climate change on Indian agriculture, particularly for the two staple food-crops Rice and Wheat?
4. What are the effects of climate change on the productivity and

income of the small and marginal farmers? How the existing problems of small farmers get aggravated due to climate change?

5. What are the government policies at present to help our small farmers?
6. How can climate smart agriculture help small and marginal farmers in future through adaptive measures and mitigation policies?

Theoretical Underpinning between Climate Change and Agriculture

Agriculture plays a two-fold role in climate change: on the one hand, it is severely affected by climate change and on the other hand it is a significant contributor to greenhouse gas (GHG) emissions, which is the primary reason behind climate change.

Effect of Climate Change on Agriculture

The impacts of climate change are being felt all over the world. The temperature is becoming warmer, rainfall is becoming more erratic, the sea level is slowly rising and extreme weather events are becoming more frequent and intense. Variations in average temperature and imbalance in the distributional pattern of rainfall are the two prime consequences of the climatic change, which, as per Fourth Assessment Report (FAR) of Intergovernmental Panel on Climate Change (IPCC), will continue to have a profound impact on human lives and the functional pattern of eco-system. With or without adaptation, negative impacts on average yields become likely from the 2030s and after 2050 the risk of more severe impacts increases. These impacts will be more prominent in the context of rising crop demand, which is projected to increase by about 14 per cent per decade until 2050 (IPCC, 2014).

Agriculture productivity is sensitive to two broad classes of climate-induced effects—(i) direct effects from changes in temperature, precipitation, and carbon dioxide concentrations; and (ii) indirect effects through changes in soil moisture and the distribution and frequency of infestation by pests and diseases. There are many complex ways through which global warming related events impact agriculture. As the climate warms due to greenhouse gases, sea water evaporates faster and increases air moisture that condenses and falls as intense rains. Lack of uniformity in temperature changes in India affects microclimates and makes weather predictions more and more difficult. Increased intense rainfall with less or moderate rainfall could decrease groundwater recharge and soil moisture, affecting agriculture. Decline in soil fertility, increase in pest population and weeds are all inevitable with the increasing temperatures. In certain crops geographical shift to cooler areas will occur. More carbon in the atmosphere might be useful for the crops but this effect will be nullified duc to increase in temperatures. Coastal regions are likely to suffer more (Ninan and Bedamatta, 2012).

Effect of Agriculture on Climate Change

Agriculture sector also contributes in climate change. This is mainly due to methane emission from rice, enteric fermentation in ruminant animals, and nitrous oxides from application of manures and fertilizers to the soils (Aggarwal, 2008). In India, total greenhouse gases emissions 28 per cent is contributed by agriculture sector, which is much higher than the global average of 13.5 per cent.

So, while talking about climate change and agriculture we have to talk about both adaptations (how to reduce the effect of climate change) as well as mitigation (how to reduce GHG emission from agriculture).

Impacts of Climate Change Witnessed in India over the Last Few Years

In the past, the mean annual temperature (both maximum, i.e. day-time temperature and minimum, i.e. night-time temperature) of the country has increased in the period from 1901 through 2007 and more importantly it has increased more in the latter half (Tables 15.1 and 15.2).

In future the increase is temperature is only to aggravate. According to fourth assessment report of IPCC, parts of the Indo-Gangetic basin may receive less rain than in the past and most Indian landmass below the Ganges plain is likely to experience a 0.5 to 1 degrees Celsius rise in average temperatures during 2020-2029 (Acharyya, 2013).

In case of rainfall, we find large coefficient of variation[1] for more or less all the Indian states in all the three agricultural cycles, namely summer, *kharif* and *rabi* from 2004 through 2010. Interestingly, if we take all the 12 months together then annual variation is much less compared to the cycles. This implies that rainfall in the overall year has not varied much, but the rainfall pattern or cycle has changed/shifted or became erratic (Table 15.3).

If we see the rainfall variations over last four decades, then we find

Table 15.1: Change in Maximum Temperature (1901-2007)

All India	*1901*	*1950*	*2007*	*% change 1901-2007*	*% change 1950-2007*
January	22.6	24.5	25.4	12.4	3.7
Feberuary	24.6	25.3	26.8	8.9	5.9
March	30.6	29.5	30.9	1.0	4.7
April	34.6	34.4	35.7	3.2	3.8
May	36.2	36.6	36.5	0.8	-0.3
June	35.7	34.3	34.3	-3.9	0.0
July	32.1	30.6	31.5	-1.9	2.9
August	30.2	30.3	31.1	3.0	2.6
September	31.4	30.2	31.2	-0.6	3.3
October	31.6	30.4	31.5	-0.3	3.6
Novober	28.0	27.4	29.2	4.3	6.6
December	24.8	24.4	25.4	2.4	4.1
Average change				**2.4**	**3.4**

Source: Author's calculation with data from Indian Institute of Tropical Meteorology
(Available at: http://www.tropmet.res.in/static_page.php?page_id=54)

Table 15.2: Change in Minimum Temperature (1901-2007)

All India	*1901*	*1950*	*2007*	*% change 1901-2007*	*% change 1950-2007*
January	10.5	10.6	10.8	2.9	1.9
February	12.3	11.3	13.7	11.4	21.2
March	16.3	15.9	16.6	1.8	4.4
April	20.5	20.0	21.7	5.9	8.5
May	23.4	23.8	24.0	2.6	0.8
June	24.7	24.5	24.6	–0.4	0.4
July	23.9	23.6	24.0	0.4	1.7
August	23.3	23.0	23.6	1.3	2.6
September	22.0	22.1	22.8	3.6	3.2
October	19.7	18.3	18.8	–4.6	2.7
November	14.4	12.9	15.6	8.3	20.9
December	10.7	9.8	11.6	8.4	18.4
Average change				**3.5**	**7.2**

Source: Author's calculation with data from Indian Institute of Tropical Meteorology.
(Available at: http://www.tropmet.res.in/static_page.php?page_id=54).

Table 15.3: Coefficient of Variation (CV%) for rainfall in different States of India (2004-2010)

	Summer				Kharif				Rabi				Annual
	Jan	*Feb*	*Mar*	*Apr*	*May*	*Jun*	*Jul*	*Aug*	*Sep*	*Oct*	*Nov*	*Dec*	
Andaman and Nicobar Islands	67.3	126.0	74.5	77.5	19.9	10.1	34.9	35.7	26.0	26.8	47.4	78.4	**11.7**
Andhra Pradesh	108.1	157.0	146.3	64.2	48.0	45.8	43.3	32.4	27.8	50.0	60.1	132.3	**21.5**
Arunachal Pradesh	87.9	75.6	61.1	29.8	27.6	15.9	19.9	33.6	28.7	38.6	95.4	61.8	**10.8**
Assam	110.7	83.1	78.2	40.8	28.1	15.6	23.2	27.9	30.6	41.6	70.7	66.9	**14.7**
Bihar	130.6	122.2	115.5	71.6	72.3	50.6	30.0	33.9	51.7	43.4	82.6	80.7	**18.6**
Chhattisgarh	163.8	111.6	105.7	99.2	77.8	55.7	18.1	19.2	39.0	63.2	133.0	257.2	**11.1**
Gujarat	128.8	225.0	253.1	225.7	255.1	91.9	30.0	39.6	61.5	84.5	204.2	237.8	**19.4**
Haryana	130.9	160.8	119.9	106.2	88.6	69.8	55.0	66.1	59.6	225.2	77.1	155.3	**21.5**
Himachal Pradesh	72.4	76.7	105.3	39.3	32.9	59.7	25.5	27.2	44.9	128.0	144.8	88.8	**13.4**
Jammu & Kashmir	60.2	62.2	99.5	48.6	60.7	50.7	18.6	35.9	73.9	83.0	82.3	78.9	**15.3**
Jharkhand	112.6	173.8	64.8	95.8	50.0	54.6	37.9	21.5	36.7	75.6	84.6	175.6	**22.3**
Karnataka	149.6	172.5	164.3	59.3	49.9	23.4	27.1	16.6	28.2	31.7	81.8	136.3	**9.9**
Kerala	109.8	132.1	108.4	37.6	73.1	17.6	31.7	22.3	32.1	24.4	52.6	90.6	**10.8**
Madhya Pradesh	125.4	165.3	154.0	72.5	84.4	58.5	29.3	35.0	25.1	105.8	144.0	184.7	**10.8**
Maharashtra	139.4	130.7	108.2	66.1	109.0	29.2	24.0	25.6	24.6	52.9	137.7	154.8	**14.6**
Manipur	157.0	155.5	80.3	93.5	56.2	31.4	30.1	39.1	22.7	61.6	102.3	135.8	**18.3**
Meghalaya	167.2	103.4	75.1	74.9	41.7	26.8	55.2	45.6	54.8	71.7	159.7	87.5	**30.2**
Mizoram	223.9	138.4	119.9	86.8	51.4	50.7	53.3	34.2	47.9	20.0	85.1	231.9	**20.2**
Nagaland	166.7	80.9	123.0	56.6	17.7	28.9	31.3	39.2	23.1	30.0	138.1	90.7	**17.4**
Odisha	127.9	129.2	80.1	77.2	40.3	41.9	31.2	38.2	36.3	66.5	77.8	219.2	**11.6**
Punjab	104.6	108.2	113.9	102.3	84.3	92.1	40.8	41.7	53.0	150.7	125.3	110.2	**18.7**
Rajasthan	96.0	170.6	106.8	114.6	72.7	53.4	28.7	49.5	51.7	172.6	221.2	90.4	**16.8**
Sikkim	90.9	110.9	34.8	32.8	28.7	17.9	18.2	13.1	25.6	34.7	55.1	82.6	**8.4**
Tamil Nadu	50.1	155.4	148.1	76.9	76.2	33.4	32.0	38.7	39.2	37.2	39.8	71.1	**13.9**
Tripura	248.3	120.4	131.8	60.8	32.7	37.2	35.4	19.9	37.6	40.2	190.6	250.3	**16.3**
Uttarakhand	90.7	114.9	133.8	61.5	78.1	43.7	21.1	27.3	56.4	106.0	102.9	113.9	**20.1**
Uttar Pradesh	148.2	155.4	129.1	92.1	38.4	80.7	27.1	21.7	39.4	103.6	120.3	95.2	**16.2**
West Bengal	133.8	143.7	97.1	23.9	32.0	31.1	21.8	12.3	28.2	68.4	108.9	194.8	**9.1**

Source: Author's calculation with data from India Meteorological Department.

variation to be increasing. Table 15.4 reveals the case of only one state, namely Karnataka.

Table 15.4: Coefficient of Variation (CV%) for rainfall in Karnataka (1960s, 1970s, 1980s and 1990s)

	Summer	*Kharif*	*Rabi*	*Annual*
1960-1969	67.0	19.1	35.7	19.1
1970-1979	82.9	13.3	25.7	13.1
1980-1989	123.2	16.6	41.4	13.3
1990-1999	189.3	35.8	36.9	35.7

Source: Author's calculation with data from India Meteorological Department

Apart from the secondary governmental data presented in the above tables, following are some recent evidences of climate change from news articles, which also reveal the changing pattern of monsoon and intensifications of cyclones/rainfalls:

Changing Pattern of Monsoon

Late arrival of monsoon

For the recent few years, places such as Chhattisgarh, Madhya Pradesh, Rajasthan, Gujarat and Kerala which are mostly dependent on monsoon for the agricultural activities are suffering greatly due to its late arrivals. (http://www.hindustantimes.com/business-news/monsoon-may-arrive-in-24-hours-weather-official/article1-1226225.aspx)

Early arrival of monsoon

Early arrival of monsoon is again an issue of concern. The cool showers sometimes spoil the crop-pattern by washing away the seeds not yet ready. However, early arrival of monsoon is sometimes beneficial for *Kharif* crops.

(http://businesstoday.intoday.in/story/early-monsoon-record-agriculture-economy/1/195899.html)

Intensification of Cyclones/Rainfall

Irregular patterns of rainfall are often witnesses to be coupled with the unpredictable patterns of wind. A very recent incident of cyclone being weathering away everything in Andhra Pradesh has created an alarming situation. This cyclone, Hudhud, travelled through the coastal plains of Andhra Pradesh towards the Odisha, Chhattisgarh, Jharkhand, Bihar, West Bengal, and Uttar Pradesh region leaving strong imprints of damage in Vishakhapatnam on the way. Also, recent Rainfall in Uttarakhand and Jammu & Kashmir, Ayla in Odisha created massive agricultural losses.

(http://indiatoday.intoday.in/story/live-cyclone-hudhud-visakhapatnam-andhra-pradesh-odisha-evacuation/1/395343.html)

India is diversified in the agricultural context with 15 agro-climatic zones along with diverse seasons, crops and farming systems. With its close intimacy with nature there is no way in which Indian agriculture can escape the impact of climate change.

Existing Studies on the Effects of Climate Change on Rice and Wheat

There have been a number of works on the effect of climate change on the yield (harvest per hectare) of rice and wheat, specifically for India. The existing literature suggest that due to increase in day-time (maximum) and night-time (minimum) temperature and reduction in total rainfall (June to September) rice yield has already declined. Since the effect of climate change is going to increase much more in coming years, therefore, in future the fall in yield will also be higher. Several studies also point out the adverse effect of climate change on wheat output. This section presents crop specific review of existing literature for rice and wheat.

Rice

At all India level Lahiri and Roy (1985) studied the supply response of rice yields to climate change. Studying the effect of rainfall on yield they suggested that Indian agriculture has become more rainfall sensitive with the spread of High Yield Varieties (HYV) seeds (post-green revolution, i.e. mid-sixties onwards), since steady water requirement has gone up and the spread of irrigation has not kept pace with it. Kanwar (2006) extended Lahiri and Roy (1985) framework to several food grains. He looked at the supply response using a state level panel dataset and reconfirmed that rainfall matters considerably for supply response.

Auffhammer *et al.* (2012) performed a statistical analysis on state-level Indian data and found that both drought and extreme rainfall negatively affected rice yield in nine predominantly rain-fed states of India during 1966-2002. Their results indicated a non-linear impact of monsoon, i.e. the negative impact of reduced rainfall was amplified when rainfall was very low (drought) and the positive impact of higher rainfall reversed sign and became negative when the increase occurred as extreme rainfall. However, drought revealed to have a much greater impact compared to extreme rainfall. Also, night-time (minimum temperature) warming at the end of the growing season revealed to have even greater impact on yield than changes in monsoon characteristics during the time period. Using Monte Carlo simulation, they also found that harvest would have been

5.67 per cent higher in the absence of climate change. Moreover, authors pointed out the concern that the changes on rice yield in the country would likely to be larger than the historical time period covered in their study.

Using district level panel data for India from 1966 through 1999, Gupta *et al.* (2012) found significant impacts of rainfall and temperature on yields of two key food crops, namely, rice and millets. They estimated an agricultural production function with exogenous climate variables, namely precipitation and temperature. Extrapolating forward, they found climate change is likely to reduce yields of these crops significantly. Regional variations, i.e. the uneven impact of climate change across India has also been reflected from their results.

Another disaggregated (district level) study by Pattanayak and Kumar (2013) found the negative impact of higher daytime temperature on rice yield to be much larger than the impact due to higher night time temperature. The study further estimated that average rice yield would have been 8.4 per cent higher had the pre-1960 climatic conditions prevailed during the study period, i.e. from 1969 through 2007. Authors also argued that such significant loss in rice production under climate change conditions in future will have strong implications for India's food security and poverty, given that a large number of producers and consumers depend on rice for their livelihood and sustenance.

Bhattacharya and Panda (2013) investigated the impact of climate change on yield response of rice for the period 1977-2007 at Kharagpur in West Bengal. They found that the grain yield of rice was influenced by a combination of fixed increments in carbon dioxide (CO_2) concentration, temperature and rainfall variability. Their simulations results suggested that maximum and minimum temperatures have significant negative effect on rice yield, and this effect could become more pronounced if temperatures rise above 40 degree Celsius. Sensitivity analysis indicated that CO_2 levels have positive impact on rice yield. Although higher CO_2 levels in the future would balance the detrimental effects of increased temperatures to some extent but it would not be able to offset them. Their results also suggested that changes in rainfall pattern may also adversely affect on rice yield.

Wheat

Aggarwal and Sinha (1993) conducted an all-India level study and suggested that since wheat production takes place in varying latitudes, the magnitude of changing temperature is what is important to note while judging the impacts of climate change on wheat production. They found that small increase in temperature like one degree throughout the crop season will not have significant effect in the growth or quality of the grains. In northern India, a slightly more increase in temperature might reduce the potential yields but will not affect much as the productivity is quite high in such regions. However, the extent of impacts can be witnessed in the central India, where an increase of only 2 degree Celsius will have serious concern since the yield is already low in those regions.

However, when Kalra *et al.* (2008) extended Aggarwal and Sinha (1993) they found that in North-Western regions (where the climate is usually dry) the wheat production is already showing signs of stagnation. This is essentially due to the increase in humidity in the atmosphere which is adversely affecting the production in such regions. As per Kalra *et al.* (2008) the productivity of these regions is likely to suffer even more in near future with the increasing unpredictable features of climate changes. They suggested a shift in sowing-time as an adaptation policy which they feel might help in improving the situation.

Another study by Attri and Rathore (2003) indicated that a constant growth in CO_2 is adversely affecting the wheat production in India. In North-East regions where there is ample amount of rainfall taking place every year, the late/early arrival coupled with short stay and intensified monsoon is likely to effect the wheat production in such regions. Moreover, with the increase in temperature and CO_2 the grain quality is likely to decrease due to increase in the attacks by pests/insects.

According to Saini and Nanda (1995), with every one degree Celsius increase in mean temperature above 17 to 17.7 degree Celsius will result in a yield reduction for wheat. As per Rao and Sinha (1993) wheat yield will decrease due to the adverse effect of temperature during filling and maturity stages of the growth. However, Aggarwal and Kalra (1994) found that with one degree Celsius rise in mean temperature will have no effect on potential yields. Rather irrigated yields will increase in the places where current yields were greater than 3.5 tonnes/hectare. But a 2 degree Celsius rise is likely to reduce potential yield in most of the places affecting majorly the areas with low potential productivity. This decrease in sub-tropical regions would be somewhere around 1.5-5.8 per cent whereas in tropical locations the decrease will be as high as 17-18 per cent (Ninan and Bedamatta, 2012).

Methodology and Results

This chapter analyses the effect of weather input (rainfall and temperature) and other inputs (fertilizer, pesticide and irrigation) on the yield of rice and wheat in India. The yield data was collected from *Handbook of Statistics on Indian Economy* (2012), temperature data from Indian Institute of Tropical Meteorology, rainfall, fertilizer, pesticide and irrigation data from Agricultural Statistics at a Glance (2012). Rice and wheat are important crops not only in terms of higher area coverage, but also for food security of small and marginal farmers. Therefore, any effect of climate change on rice and wheat productivity would also affect the food security situation of small and marginal farmers.

A regression analysis technique was used to check the effect of rainfall, temperature (both average minimum and average maximum temperature) and fertilizers on the yield of rice and wheat. Pesticides and irrigation was left out to avoid multi-collinearity problem as both the variables came out to be highly correlated with fertiliser consumption.

Table 15.5: Regression results: Rice (1991-92 to 2011-12)

	Coefficients	*Standard Error*	*t Stat*
Intercept	1.69	1.79	0.94
% Departure from normal rainfall	0.00***	0.00	3.69
Lnavgmintemp	–1.43*	1.12	–1.28
Lnavgmaxtemp	1.80	1.52	1.18
Lnferti	0.39***	0.06	5.99
Regression Statistics			
Multiple R	0.93		
R Square	0.87		
Adjusted R Square	0.83		
Standard Error	0.02		
Observations	21.00		
ANOVA			
	F	Signifioance F	
Regression	25.88	0.00	

***implies significant at 1%,
** implies significant at 10%,
* implies significant at 15%,

Table 15.5 shows that rainfall and fertilisers have a significant positive impact (significant at 1%) while the rise in minimum temperature has a negative impact (significant at 15%). However, maximum temperature does not reveal to be affecting rice yield significantly. While checking the effect on wheat yield, considering both minimum and maximum temperature the effects were found to be insignificant, thus only maximum temperature was considered in case of wheat.

Table 15.6: Regression results: Wheat (1991-92 to 2011-12)

	Coefficients	*Standard Error*	*t Stat*
Intercept	–0.68	1.79	–0.38
lnrainfall	0.16*	0.10	1.62
lnavgmaxtemp	2.14**	1.16	1.83
lnferti	0.23***	0.04	6.18
Regression Statistics			
Multiple R	0.90		
R Square	0.82		
Adjusted R Square	0.78		
Standard Error	0.01		
Observations	21.00		
ANOVA			
	F	Significance F	
Regression	25.20	0.00	

*** implies significant at 1%,
** implies significant at 10%,
* implies significant at 15%.

The result for wheat (Table 15.6) shows that rainfall (significant at 15%) and fertilisers (significant at 1%) have a significant positive impact. Interestingly, for wheat we find maximum temperature to be affecting the yield significantly (significant at 1%).

The Problems of Small Farmers Due to Climate Change

Farmers with operational holding of less than one hectare are called marginal farmers and farmers with operational holding of more than of one hectare but less than of two hectares are called small farmers. In general they face problems of imperfect market for inputs, low or no access to the credit markets, smaller access to technological know-how, poor human resource base, poorer access to public goods like public irrigation, electricity grid and others (NCEUS, 2008). Moreover, they are generally geographically located in more exposed or marginal areas, such as flood plains or on nutrient-poor soils and mostly monsoon dependent areas. Their biggest problem is income volatility. With poor human, institutional and financial capacity they have very limited ability to cope with climate impacts, and to adapt to a changing hazard burden.

The farmers, particularly the small and marginal farmers feel helpless to deal with the climate change, particularly the irregularity of rainfall in the absence of proper irrigational facilities. They simply try to compensate the loss in yield due to climate change by using more and more fertiliser. Increase in the use of fertiliser in turn makes them more dependent on irrigational facilities. If the irrigation facilities are groundwater dependent then the use of more chemical fertiliser depletes the groundwater level. Thus, although the farmers compensate for the short-run loss by using more fertilizer, the possibility of long-run losses increases due to depletion of soil quality and water level.

Moreover, in case of droughts/floods/shortage in crop production small and marginal farmers are left with limited options like mortgage/selling of lands and assets. Due to climate change, when these natural disasters continue to occur more frequently they are pushed into the debt trap. In the years of losses they are bound to take loans at very high interest rate, and in the years of full production the money lenders often take the entire crop as a compensation of their previous indebtedness. This debt-trap can eventually lead to other complex problems like migration to the cities. Farmers migrating to cities are not new phenomenon. But when the farmers succumb to the debt trap and seek a way out by migrating to the cities after selling their land and assets, it becomes a major concern. Unlike the big land owned farmers the small/marginal scale farmers completely rely on the crop production. If government aid is insufficient for coming out of the debt-trap migration to nearest city after selling off their land might take place. This essentially indicates a complete uprootment of the village life/set-up of that farmer family. They add to the unskilled labour mostly to the construction sector. The condition of the poor farmer becomes worse with family staying on roads and children adding to the out of school beggars or child labours.

At the same time the summary of IPCC latest (fifth assessment) report mentions about the followings risks for the first time, which is more probable to hit the small and marginal farmers the most. The report says that these risks constitute particular challenges for the least developed countries and vulnerable communities, given their limited ability to cope.

Risks More Probable to Hit the Small and Marginal Farmers the Most

(i) Risk of death, injury, ill-health, or disrupted livelihoods in low-lying coastal zones and small island developing states and other small islands, due to storm surges, coastal flooding, and sea-level rise.

(ii) Risk of severe ill-health and disrupted livelihoods for large urban populations due to inland flooding in some regions.

(iii) Systemic risks due to extreme weather events leading to breakdown of infrastructure networks and critical services such as electricity, water supply, and health and emergency services.

(iv) Risk of mortality and morbidity during periods of extreme heat, particularly for vulnerable urban populations and those working outdoors in urban or rural areas.

(v) Risk of food insecurity and the breakdown of food systems linked to warming, drought, flooding, and precipitation variability and extremes, particularly for poorer populations in urban and rural settings.

(vi) Risk of loss of rural livelihoods and income due to insufficient access to drinking and irrigation water and reduced agricultural productivity, particularly for farmers and pastoralists with minimal capital in semi-arid regions.

(vii) Risk of loss of marine and coastal ecosystems, biodiversity, and the ecosystem goods, functions, and services they provide for coastal livelihoods, especially for fishing communities in the tropics and the Arctic.

(viii) Risk of loss of terrestrial and inland water ecosystems, biodiversity, and the ecosystem goods, functions, and services they provide for livelihoods.

Source: IPCC Assessment Report 5: Climate Change 2014: Impacts, Adaptation, and Vulnerability, *Summary for Policymakers*.

Government Policies at the Present to Help Out Farmers

India's National Action Plan for Climate Change (NAPCC)

In June 2008, Prime Minister Manmohan Singh released the first National Action Plan on Climate Change (NAPCC) for India. Currently, all Indian states are preparing State Action Plans on Climate Change (SAPCC) to operationalise the NAPCC. NAPCC outlines how India plans to address national climate change concerns over the coming years without compromising the country's development. Under NAPCC, the National Mission for Sustainable Agriculture supports climate adaptation in agriculture through the development of climate-resilient crops, the expansion of weather insurance mechanisms and innovative agricultural practices. Also the National Mission on Strategic Knowledge for Climate Change seeks to establish a better understanding of climate science, impacts and challenges. This mission envisions a new Climate Science Research Fund, improved climate modelling, and increased international collaboration. It also encourages private sector initiatives to develop adaptation and mitigation technologies through venture capital funds. Apart from this broad umbrella policy, some other policies which are directly linked with the farmers seem to perform better. Some of those are mentioned below:

Kisan Call Centres

Kisan Call Centres, which are considered to be revolutionary step in agriculture and farming sector, have been established around 13 cities of India. These centres are scattered all over the country but can be reached out *via* a common toll-free number. Guidance regarding farming techniques, crop-cultivation, seeds to be used etc. is being provided through these centres by the representatives (who are generally agriculture graduates). The call is formatted in a way that it automatically gets connected to an attended of that particular or surrounding region having the knowledge of farming and agricultural know-hows of that area also knowing the local language. For any major issues/queries call is being transferred to the second level comprising of experts in the field.

Kisan Credit Card

Another revolutionary initiative taken up by the Indian government is the Kisan Credit Card that allows farmers with easy accessibility of capital for any kind of agricultural investment purposes. Repayment is rescheduled if there is a bad crop season, and extensions for repayment are offered up to three to four years. This facility is provided for all the agricultural and allied activities such as livestock and others.

Kisan Television

In the interim budget presented by the current government, Finance Minister stated the plan of launching a television channel exclusively for farmers by 2014. The channel would aim in providing all sorts of precise and even technical information regarding farming and agriculture. Issues such as water conservation and organic training would also be approached and discussed in detail through this channel.

How Can Climate Smart Agriculture Help Small and Marginal Farmers through Adaptation and Mitigation Policies?

Climate Smart Agriculture

Climate-smart agriculture (CSA), as defined and presented by FAO at the Hague Conference on Agriculture, Food Security and Climate Change in 2010, is an approach to develop the technical, policy and investment conditions to achieve sustainable agricultural development for food security under climate change (FAO, 2013). Green Revolution during the 1960s was characterised by the planting of high-yielding crop varieties, with the associated chemical package and irrigation. As a result, cereal food production increased many-folds and a large number of people were saved from famine. However, decades of intensive cropping have degraded fertile land, depleted groundwater, triggered an upsurge in pests, eroded biodiversity and polluted the air, water and soil. Intensive crop production is no longer sustainable, and a new paradigm should emerge—and this is what Save and Grow, i.e. sustainable crop production intensification is about (FAO, 2011).

CSA integrates the three dimensions of sustainable development (economic, social and environmental) by jointly addressing food security and climate challenges. It is composed of three main pillars: (a) Sustainability i.e. increasing agricultural productivity and incomes, (b) Adaptation and (c) Mitigation (FAO, 2013). These three targets are fulfiled through climate-smart crop production system, and the below mentioned are few requirements for that:

Soil Management

A healthy soil is fundamental for sustained agricultural productivity. Diversified production systems and land uses will conserve the fertility of soil and at the same time reduce farmers' risk and vulnerability if one or more crops fail or if other farming enterprises collapse. Management practices that do not deplete soil organic carbon content, but rather increase it from year to year through organic matter management and crop rotation are of utmost importance. They will create productive soils that are rich in carbon, require fewer chemical inputs and maintain vital ecosystem functions, such as the hydrological and nutrient cycles. There is also a need to reduce the use of inorganic fertilizer. It requires building on practical farming skills, observation, personal experience, knowledge sharing and developing local capacity for adapting complex agro-ecosystems to change (FAO, 2013).

Water Management

Increasing the reliability and flexibility of access to water for farmers is of prime importance. Water storage, and the combined use of groundwater and canal irrigation water, can go a long way towards improving the productivity of water used for irrigation. Economic incentives, in particular the use of subsidies for pumping, must be designed in a way that promotes the efficient use of water and avoids wastage of both energy and water resources. Crop selection and changes in crop calendars will help farmers adapt to new temperatures and rainfall patterns. The use of crops or varieties with better resilience to dry spells will be preferred. Making water more accessible contributes to a great extent especially to the regions of Gujarat, Madhya Pradesh, Rajasthan etc. where scarcity of water and rainfall are major concerns. The other way which needs immediate attention is the proper rain-water management system. In case of excessive rainfall we should be able to harvest it to extract the best utility out of it instead of giving in to the natural forces of floods. With intensified rainfall followed by shorter stay of monsoon a vivacious way of water-management system is indeed the need of the hour (FAO, 2013).

Bringing agricultural research, awareness and know-how to the farmers

Promotion of agricultural research on better heat and drought-resistant seeds are required. Maintaining seed banks at the community level are also necessary. Promotion of crop diversification and agro-biodiversity is of utmost necessity to avoid monoculture to reduce the risk of crop failure. At the same time, integrated approaches such as crop-livestock systems, rice-fish systems and agro-forestry should be encouraged. This will not only diversify food sources and consequently strengthen the resilience of farmers' livelihoods but also provide opportunities for mitigating climate change.

However, since poor and mostly illiterate farmers are the people who will be adopting these changes, so providing precise knowledge regarding agriculture and conservation management is definitely going to help increase the agricultural production. Getting the farmers aware of what they should cultivate, the manner of cultivation, the precision of proportions of chemicals, and bio-fertilizers to be used and when—all these little but significant information would empower farmers in their own right which will definitely escalate the growth of agricultural production. With severe climatic changes ahead farmers have to be well-equipped with knowledge of using advanced techniques in order to reduce their vulnerability on natural forces. At the same time, farmers are the primary custodians of knowledge about their environment, agro-ecosystems, crops and cropping patterns, and local climatic patterns. Thus, adapting cropping practices and approaches should be related to local farmers' knowledge, requirements and priorities. There is a need to build on farmers' knowledge and innovations and develop local capacity of land users to manage their soil systems so that they can build resilience and continue to innovate and adapt to a warming climate and changes in production systems.

Mitigating GHG Emission

Application of nitrate fertilizers in aerobic conditions and ammonium fertilizers in wetland crops helps reducing the GHG emission. Curtailing the nitrification process by the use of nitrification inhibitor, particularly the low-cost locally available plant-derived materials, such as neem cake is a good option to mitigate nitrous oxide emission from soil. Improved management of livestock diet through substitution of low digestibility feeds with high digestibility ones, substituting fibrous concentrate with starchy concentrate, supplementation with molasses and changing micro-flora of rumen could lead to reduced methane emission. Approaches to increase soil carbon, such as organic manures, minimal tillage and residue management should be encouraged. Resource-conserving technologies restrict release of soil carbon thus mitigating increase of CO_2 in the atmosphere. Improvement in the efficiency of energy-use in agriculture by using better designs of machinery, increasing fuel efficiency in agricultural machinery anu commercialisation of wind and solar power potential, and use of laser levellers could also lead to mitigation (Aggarwal, 2008).

How to Contextualise CSA for Small and Marginal Farmers in India?

There has been a rapid uptake of the term CSA by the international community, national entities and local institutions. However, implementing this approach is challenging, partly due to a lack of tools and experience, particularly when it comes to resource poor small and marginal farmers of the developing countries say for example India. Therefore, government support, in the form of crop insurance, health insurance, awareness spreading campaigns, technical know-how building institutions are of enormous importance. There is the need for strong government commitment to develop and/or adapt agricultural policies to take into consideration climate change, its potential impact and ways to overcome or minimise its effects. At the same time, adaptation strategies must address the issue of the commons, e.g. climate resilience must be in the public domain instead of only a few who might have purchased some rights.

Another point of concern is that climate change is not a linear phenomenon of warming everywhere, or more rain or less rain, it is a non-linear phenomenon, so different adaptation policies should be taken on a case to case basis analysis. Climate-smart interventions are highly location-specific and knowledge-intensive. In this regard, local expertise offers an immense repository of knowledge—not only about biophysical aspects of agricultural production, but also of the needs of communities and farmers. Thus, locally practiced innovative adaptation policies should be adopted.

Conclusion

Agriculture plays a two-fold role in climate change: on the one hand, it is severely affected by climate change and on the other hand it is a significant contributor to greenhouse gas emissions. Variations in average temperature and imbalance in the distributional pattern of rainfall are the two prime consequences of the climatic change. India being an agriculture dependent tropical country, and with more than 80 per cent small and marginal farmers this problem has serious implications. Due to increase in temperature and reduction in rainfall (with high density of rain) there is a reduction in yield of most of the crops in India. The results of regression analysis further suggest that there is a positive significant effect of rainfall and fertiliser, and negative but significant result of increase in minimum temperature on rice yield. However, for wheat there is a positive significant effect of the increase in maximum temperature. The poor farmers are trying to compensate the loss of yield due to climate change by increasing the use of fertiliser. Increase in the use of fertiliser in turn makes them more dependent on irrigational facilities. If the irrigation facilities are groundwater dependent then the use of more chemical fertiliser depletes the groundwater level. Thus, although the farmers compensate for the short-run loss by using more fertiliser, the possibility of long-run losses increases due to depletion of soil quality and water level. As a result of climate change the existing problems of small and marginal farmers are getting aggravated. Climate Smart Agriculture (CSA) can help to a great extent. However, implementation of CSA is a challenge for the country at present due to lack of adequate tools and experience. Therefore, government support, in the form of crop insurance, health insurance, awareness spreading campaigns, technical know-how, building institutions are of enormous importance.

Note

1. Coefficient of Variation (CV) is the ratio of Standard Deviation (i.e. deviation from the average) and Average in percentage form. CV% measures the variability of the data series. It is also a desired measure in comparison to Standard Deviation because it makes the series comparable by normalizing with Average value.

References

Acharyya, A. (2013), "Groundwater Irrigation in West Bengal: Opportunities for daptation and mitigation to climate change", Working Paper Series: No. 3, 2012-2013, UGC SAP (DRS-I) Department of Economics and Politics, *Department of Economics and Politics*, Visva-Bharati.

Aggarwal, P.K. (2008), "Global climate change and Indian agriculture: impacts, adaptation and mitigation", *Indian Journal of Agricultural Sciences*, Vol. 78, No. 10.

Aggarwal, P.K. and N. Kalra (1994), *Simulating the Effect of Climatic Factors, Genotype and Management on Productivity of Wheat in India*, Indian Agricultural Research Institute Publication, New Delhi (NCEUS, 2008).

Aggarwal, P.K. and S.K. Sinha (1993), "Effect of Probable Increase in Carbon Dioxide and Temperature on Wheat Yields in India", *Journal of Agricultural Meteorology*, Vol. 48, No. 5.

Attri, S.D., and L.S. Rathore (2003), "Simulation of Impact of Projected Climate Change on Wheat in India", *International Journal of Climatology*, Vol. 23.

Auffhammer, M., Ramanathan, V. and J.R. Vincent, (2012), "Climate change, the monsoon, and rice yields in India", *Climatic Change*, Vol. 111, pp. 411-424.

Bhattacharya, T. and R.K. Panda (2013), "Effect of climate change on rice yield at Kharagpur, West Bengal", *Journal of Agriculture and Veterinary Science*, Vol. 4, Issue 2, pp. 6-12.

Kalra, N., Chakraborty, D., Ramesh, P.R., Jolly, M., and P.K. Sharma (2007), "Impacts of Climate Change in India: Agricultural Impacts", *Final Report*, Joint Indo-UK Programme of Ministry of Environment and Forests, India, and Department for Environment, Food and Rural Affairs (DEFRA), United Kingdom. New Delhi: Indian Agricultural Research Institute, Unit of Simulation and Informatics.

Kanwar, S. (2006), "Relative profitability, supply shifters and dynamic output response, in a developing economy." *Journal of Policy Modeling*, Vol. 28, pp. 67-88.

Lahiri, A. and P.L. Roy, (1985), "Rainfall and supply response: a study of rice in India", *Journal of Development Economics*, Vol. 18, pp. 315-334.

Ninan, K.N. and S. Bedamatta (2012), "Climate Change, Agriculture, Poverty and Livelihoods: A Status Report" *Working Paper 277*, Centre for Ecological Economics and Natural Resources, Institute for Social and Economic Change, Bangalore.

Pattanayak, A. and K.S.K. Kumar, (2013), "Weather Sensitivity of Rice Yield: Evidence from India", *Working Paper 81/2013*, Madras School of Economics, Chennai, India.

Saini, A.D. and R. Nanda (1986), "Relationship between Incident Radiation, Leaf Area and Dry Matter Yield in Wheat", *Indian Journal of Agricultural Sciences*, Vol. 56.

Reports and Data Sources

Food and Agricultural Organization (FAO) Report (2011), "Climate-smart agriculture policies, practices and financing for food security, adaptation and mitigation".

Food and Agricultural Organization (FAO) Report (2013), "Climate-smart agriculture sourcebook".

http://businesstoday.intoday.in/story/early-monsoon-record-agriculture-economy/1/195899.html

http://indiatoday.intoday.in/story/live-cyclone-hudhud-visakhapatnam-andhra-pradesh-odisha-evacuation/1/395343.html

http://www.hindustantimes.com/business-news/monsoon-may-arrive-in-24-hours-weather-official/article1-1226225.aspx

India Meteorological Department

Indian Institute of Tropical Meteorology (Available at: http://www.tropmet.res.in/static_page.php?page_id=54)

IPCC Assessment Report 5: Climate Change 2014: Impacts, Adaptation, and Vulnerability, Summary for Policymakers.

IPCC report (2014), "Climate Change 2014: Impacts, Adaptation, and Vulnerability", Inter Governmental Panel on Climate Change (IPCC).

Handbook of Statistics on Indian Economy (2012).

Indian Institute of Tropical Meteorology.

Agricultural Statistics at a Glance (2012).

Index